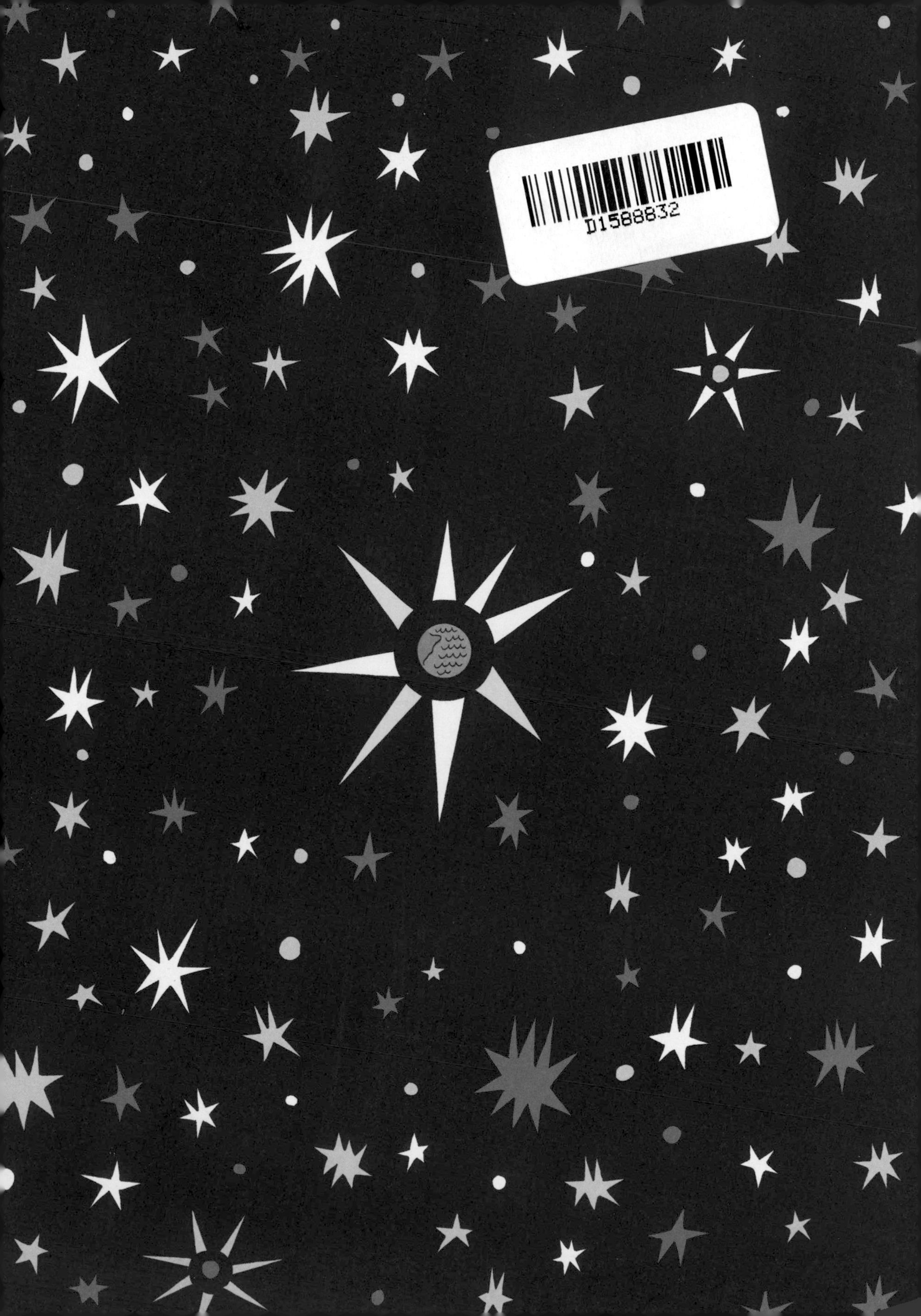

This is a first edition published in 2017 by
Nobrow Ltd. 27 Westgate Street, London E8 3RL.

Published in the US by Nobrow (US) Inc.
Printed in Poland on FSC® certified paper.

MIX
Paper from responsible sources
FSC® C002795

ISBN: 978-1-910620-28-1
Order from www.nobrow.net

Supported using public funding by
ARTS COUNCIL ENGLAND

Supported by
wellcome trust

Daniel Locke with David Blandy

OUT OF NOTHING

Nobrow

LONDON I NEW YORK

TO MY WIFE HANNAH AND CHILDREN POLLY AND FELIX,
WITHOUT YOU I'D BE LOST IN THE DARK.
AND TO DAVID, MY CO-CONSPIRATOR AND GREAT FRIEND.
— D. L.

FOR CLAIRE, PHOEBE AND SAMSON, THE STARDUST I LOVE THE MOST.
AND DAN, FOR ALL THE HOURS, DAYS, WEEKS AND YEARS.
— D. B.

DANIEL AND DAVID WOULD LIKE TO THANK...

EVERYONE AT LIGHTHOUSE, ANDREW SLEIGH, HONOR HARGER, JAMIE WYLD,
NATALIE KANE, MIRIAM RANDALL, PAUL GRAVETT, HANNAH EATON,
HANNAH BERRY, JOE DECIE, IAN WILLIAMS, RICHARD COWDRY, ALEX FITCH,
ILYA, KARRIE FRANSMAN, CORINNE PEARLMAN, DAN HANSEN, LAWRENCE ELWICK,
ALAN TAYLOR, TOM GRIFFITHS, NORMAN CHALLIS, AND JAMES KYDD.

INTRODUCTION

As a geneticist, one can take the view that life as we know it is a remix. You are a remix of your parents' DNA, such that your genome is not only unique today, but it is unique for the history and future of humankind, no matter how long our species endures. Since the 1970s we've been remixing biology in entirely unnatural ways. DNA is the universal code of life. It's the same in bacteria and blue whales and mushrooms and sunflowers and goats and us. Once we had a solid understanding of how DNA works, and how biological information is stored within it, the era of genetic engineering began. We could take DNA from one species and insert it into another. This is remixing evolution. It began with taking single genes from one virus and inserting them into another. Not long after, we could insert genes into bacteria, and then animals. One creature that captures people's imagination to this day – though the technology itself is 20 years old – is the spider-goat. Spider silk remains a material unmatched by human ingenuity, but spiders are tricky creatures to farm as they are mostly solitary and tend to eat each other. So, in the 10,000-year-old tradition of agriculture, scientists in Utah inserted the spider gene for making dragline silk into goats, under the control of their lactation genes. The goats then produce milk replete with spider silk, which we can simply extract by the spool-full.

Daniel Locke, David Blandy and I were put together by the Wellcome Trust a few years ago. We sat for many happy hours in the Lighthouse Studio in Brighton eating noodles, sharing stories and chewing the fat. This is how to collaborate: put people together with different ideas in quiet spaces and feed them noodles. I told them about the earliest figurative art known, a 40,000-year-old Ice Age statue from a cave in Germany. It is a mesmerizing object – beautiful in its own right, and even more so when you consider that

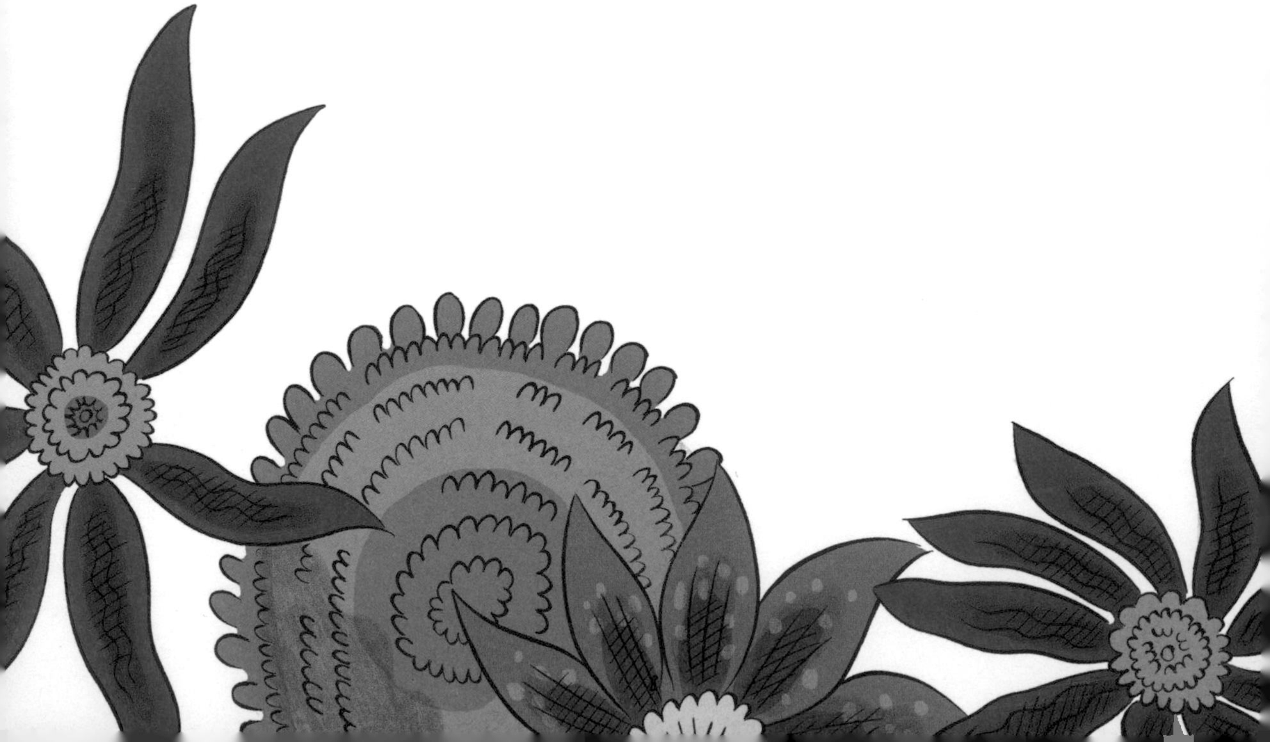

it was carved from mammoth tusk by someone highly skilled and creative, who had imagined a thing that does not exist. It's called the Lowenmensch of Hohlenstein Stadel, and it is a remix: the figure is a lion-man.

We became slightly obsessed with chimera. We couldn't find a culture that doesn't have imagined beasts with characteristics from one impossibly transposed onto another. Sphinx, mermaid, centaur, shedu, Ganesha, Ammit, and my personal favourite: the Wolpertinger – a Bavarian vampire-toothed rabbit-squirrel -deer-duck-footed hybrid. Remixing is as old as culture. The so-called cognitive revolution, some 50,000 years ago, incurred not only a physical change in us, but a suite of behavioural changes; changes in how we create, and how we interact with each other in communities. With these changes came modern human behaviour and abstract art, an ability to imagine things that cannot be.

The idea grew and grew, and couldn't be contained, and mutated into a tangled web, and the Wellcome Trust were happy to fuel us. It expanded to fill the space we had available. Daniel and David incorporated the history of technology, for we are a technological species: a history of writing, of printing, of painting, of general relativity; stories of DNA, genetic engineering, agriculture, Sputnik, the space age, CERN, the Internet, music, but also of gods and monsters. It evolved into a celebration – a history of ideas.

And so that is what this book is. A story created ex nihilo, from the big bang to an imagined future, remixed from its constituent parts, and realized so beautifully by Daniel; science and storytelling and art encompassing 13.84 billion years so that we might know who we are. It is always a privilege to tell stories so that we might know where we have been, and where we are going. This one starts with a bang.

Dr Adam Rutherford, Geneticist
May 2017

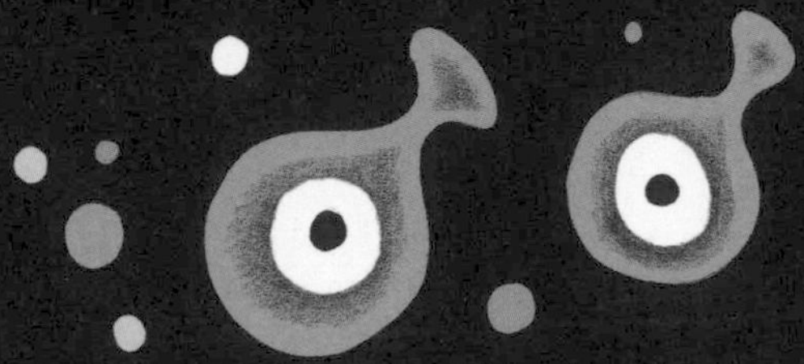

the past,
the deep past...

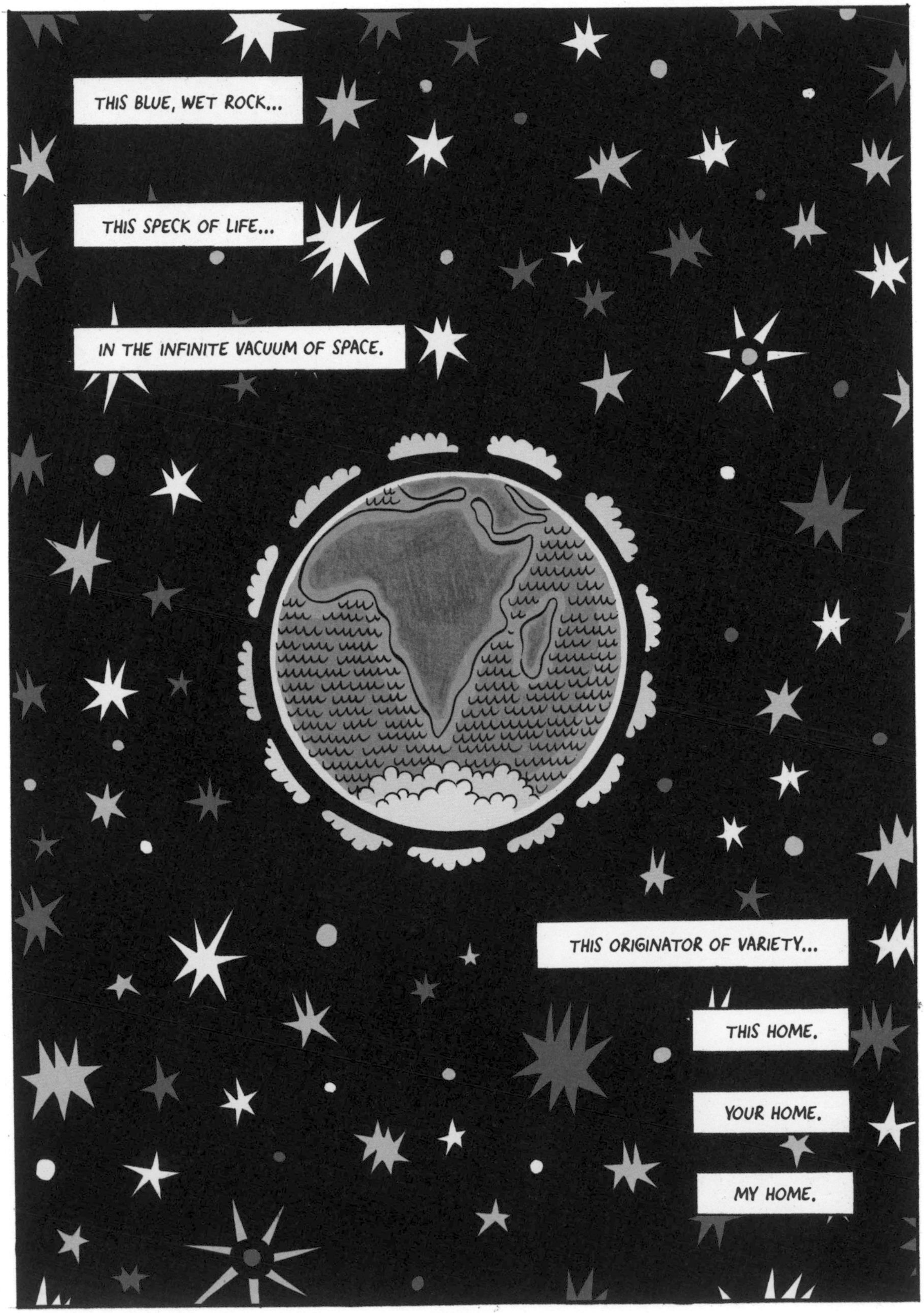
THIS BLUE, WET ROCK...
THIS SPECK OF LIFE...
IN THE INFINITE VACUUM OF SPACE.
THIS ORIGINATOR OF VARIETY...
THIS HOME.
YOUR HOME.
MY HOME.

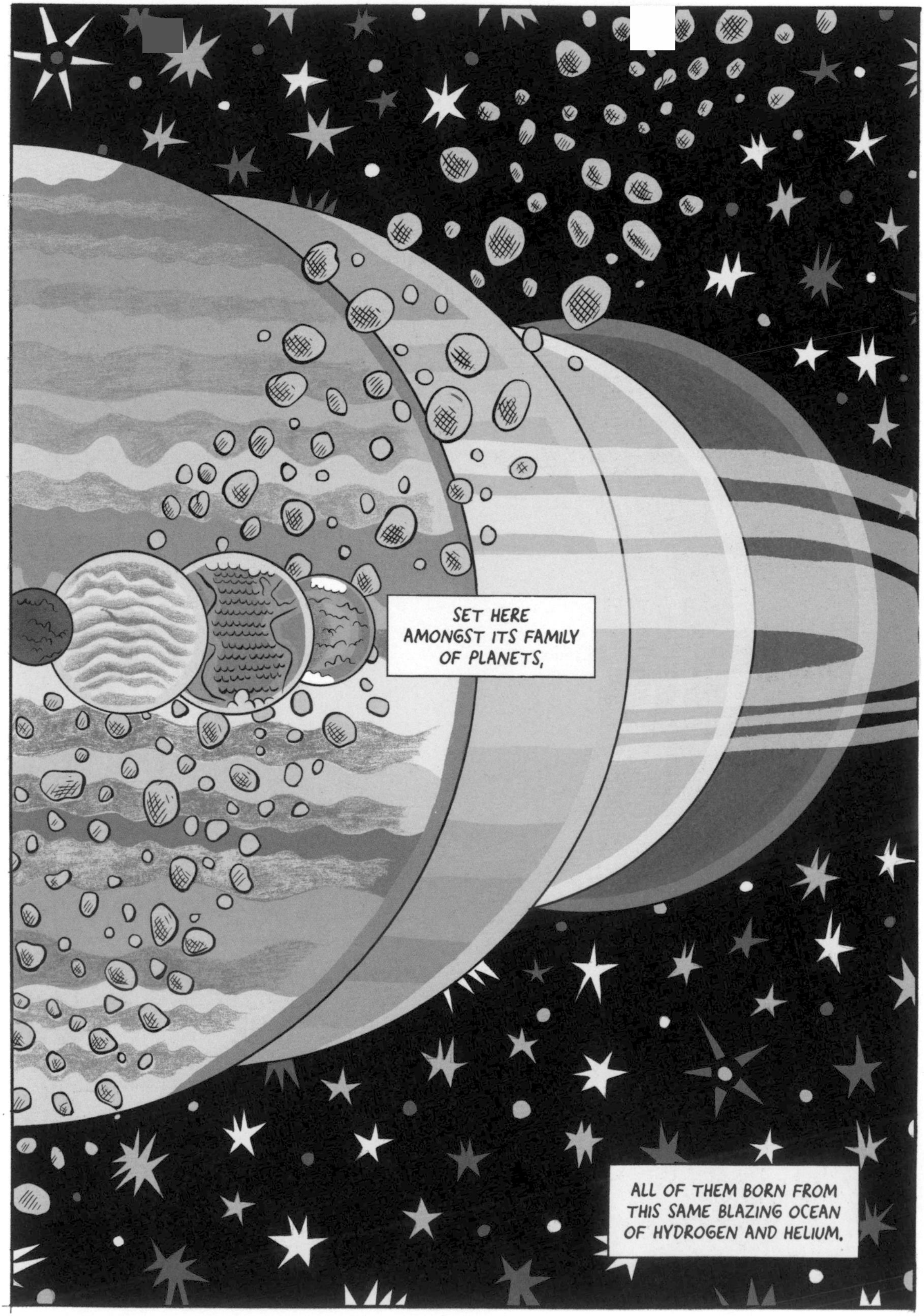
SET HERE AMONGST ITS FAMILY OF PLANETS,
ALL OF THEM BORN FROM THIS SAME BLAZING OCEAN OF HYDROGEN AND HELIUM.

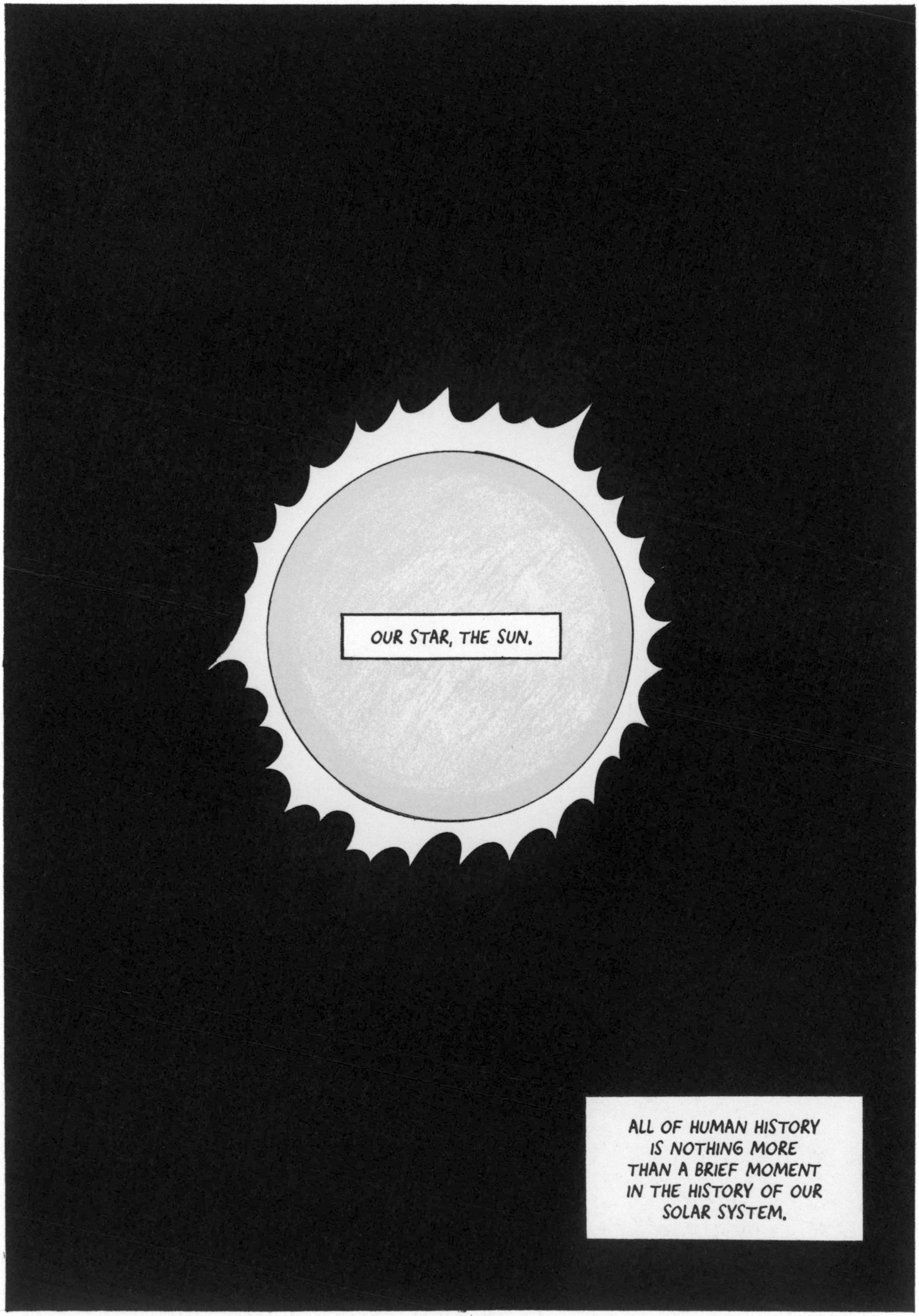
OUR STAR, THE SUN.
ALL OF HUMAN HISTORY IS NOTHING MORE THAN A BRIEF MOMENT IN THE HISTORY OF OUR SOLAR SYSTEM.

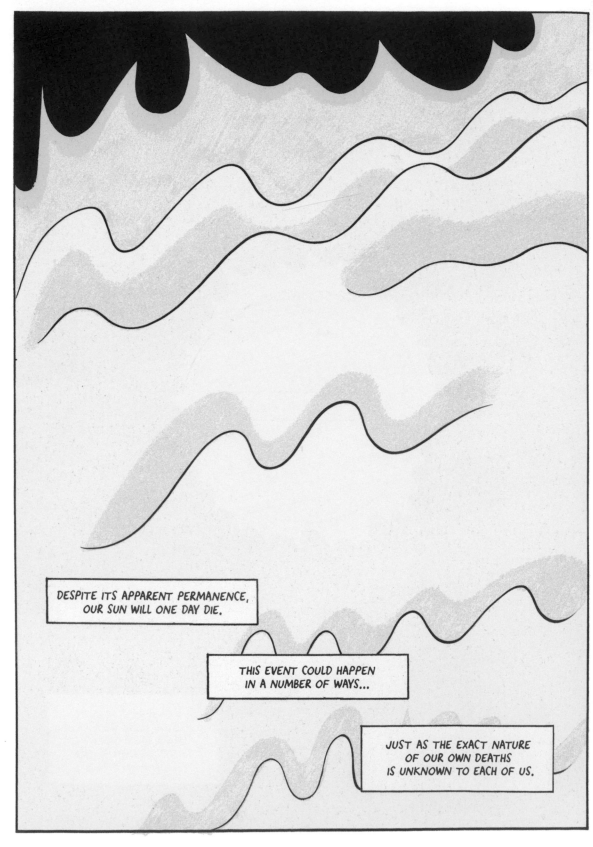

DESPITE ITS APPARENT PERMANENCE, OUR SUN WILL ONE DAY DIE.

THIS EVENT COULD HAPPEN IN A NUMBER OF WAYS...

JUST AS THE EXACT NATURE OF OUR OWN DEATHS IS UNKNOWN TO EACH OF US.

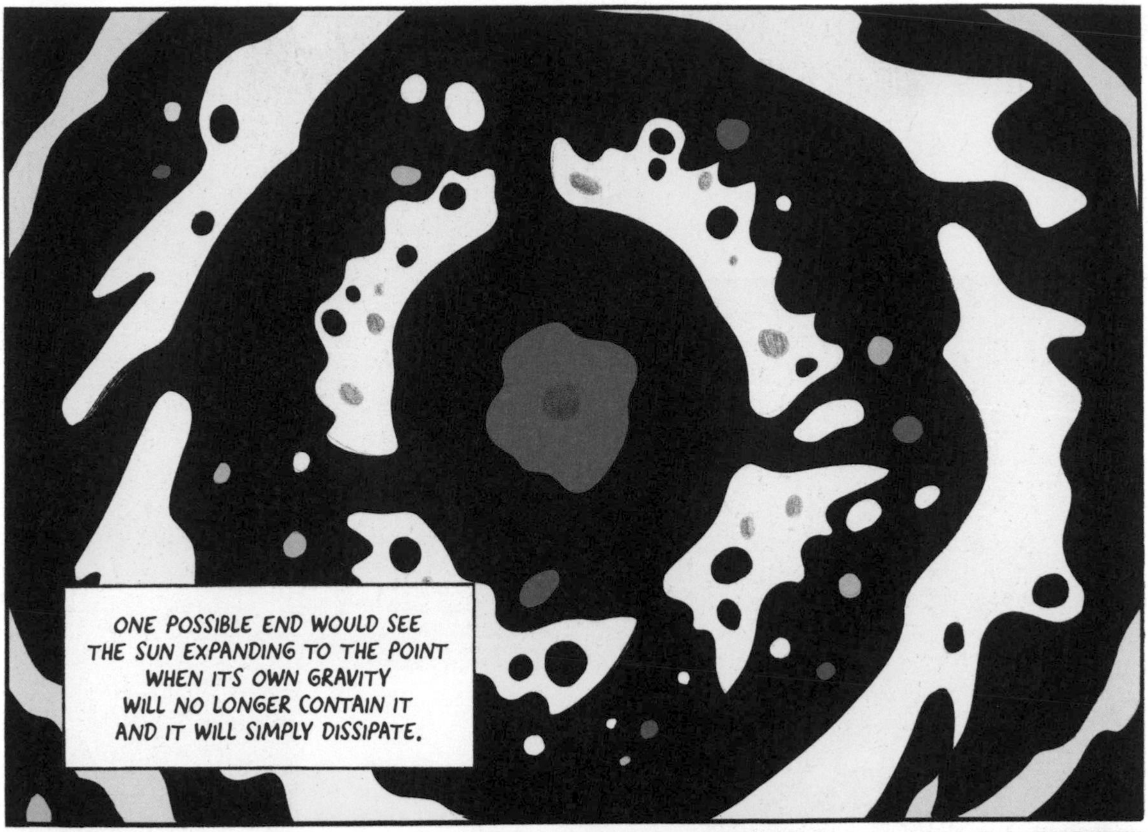
ONE POSSIBLE END WOULD SEE THE SUN EXPANDING TO THE POINT WHEN ITS OWN GRAVITY WILL NO LONGER CONTAIN IT AND IT WILL SIMPLY DISSIPATE.

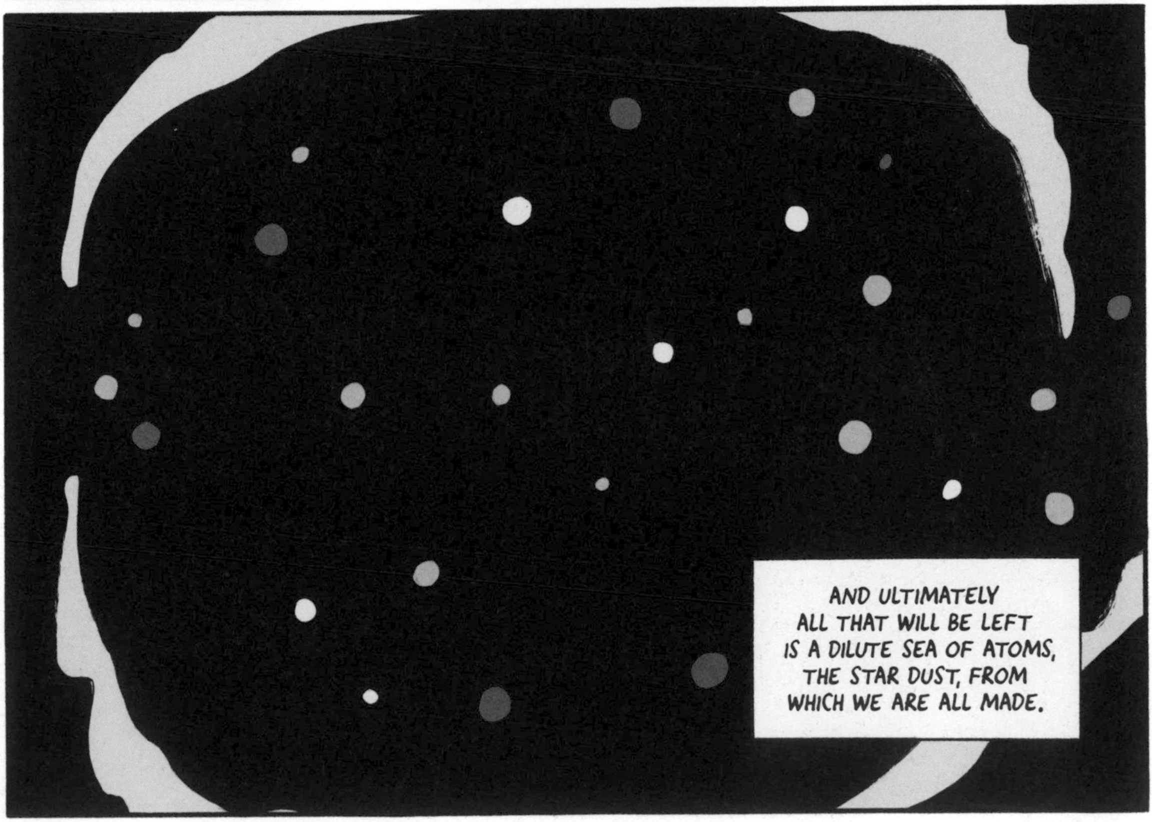
AND ULTIMATELY ALL THAT WILL BE LEFT IS A DILUTE SEA OF ATOMS, THE STAR DUST, FROM WHICH WE ARE ALL MADE.

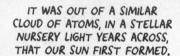

IT WAS OUT OF A SIMILAR CLOUD OF ATOMS, IN A STELLAR NURSERY LIGHT YEARS ACROSS, THAT OUR SUN FIRST FORMED.

THAT MOMENTOUS EVENT HAPPENED IN THE UTTER SILENCE OF THE UNIVERSE. NO ONE WITNESSED IT, NO ONE CONCEIVED OR THOUGHT ABOUT IT.

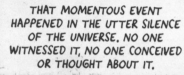

BUT TO THIS VANTAGE POINT IN TIME, 4.6 BILLION YEARS LATER, OUR DREAMING MINDS HAVE SENT A TIME TRAVELLER...

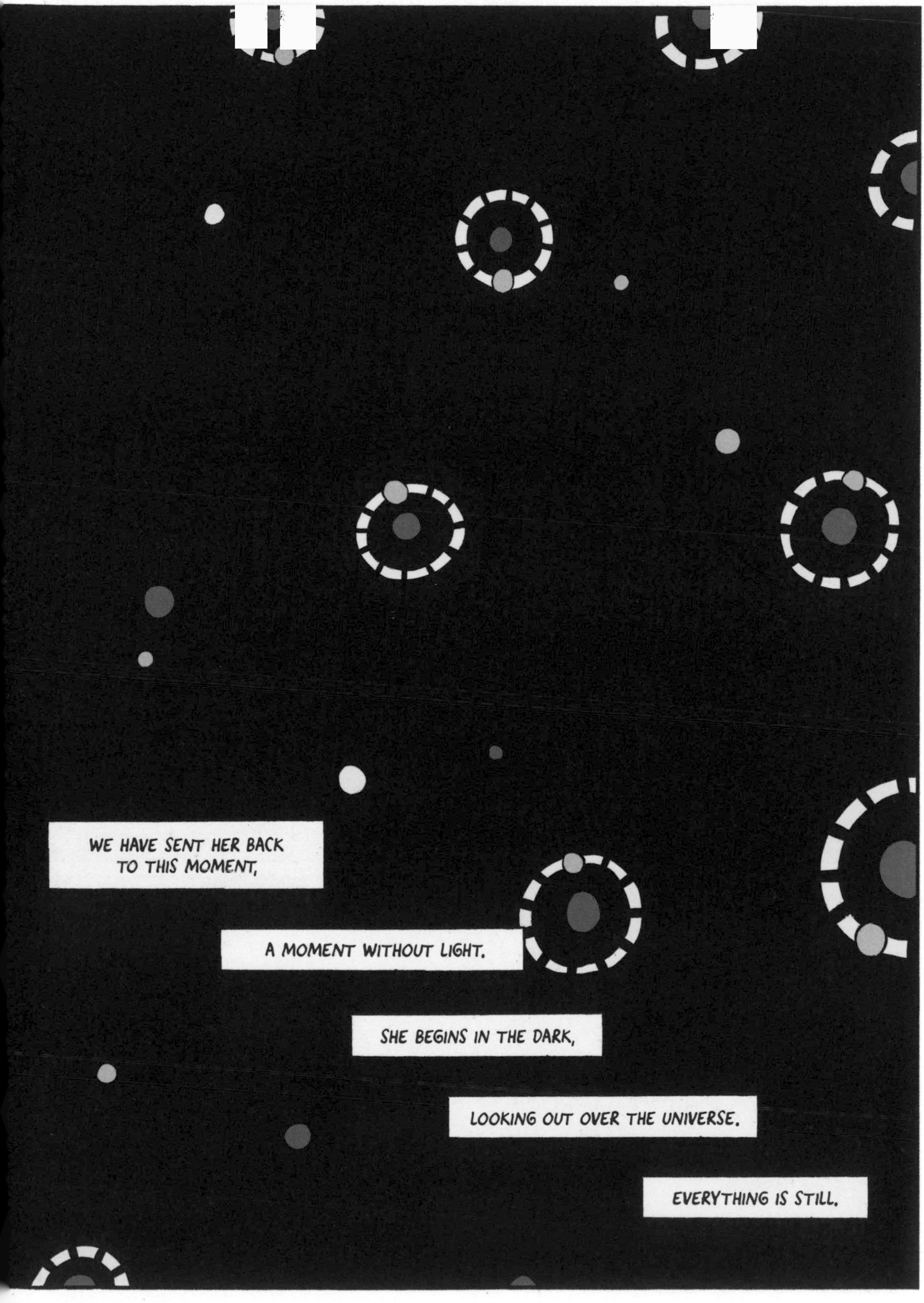
WE HAVE SENT HER BACK TO THIS MOMENT,
A MOMENT WITHOUT LIGHT.
SHE BEGINS IN THE DARK,
LOOKING OUT OVER THE UNIVERSE.
EVERYTHING IS STILL.

SHE IS THERE...

OUR EMISSARY IS THERE...

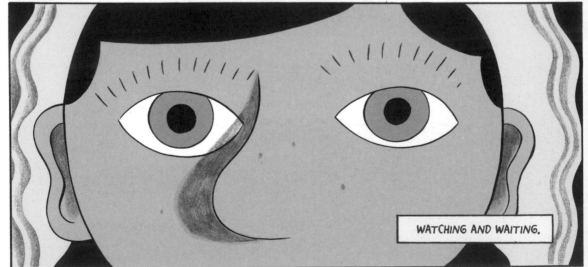

WATCHING AND WAITING.

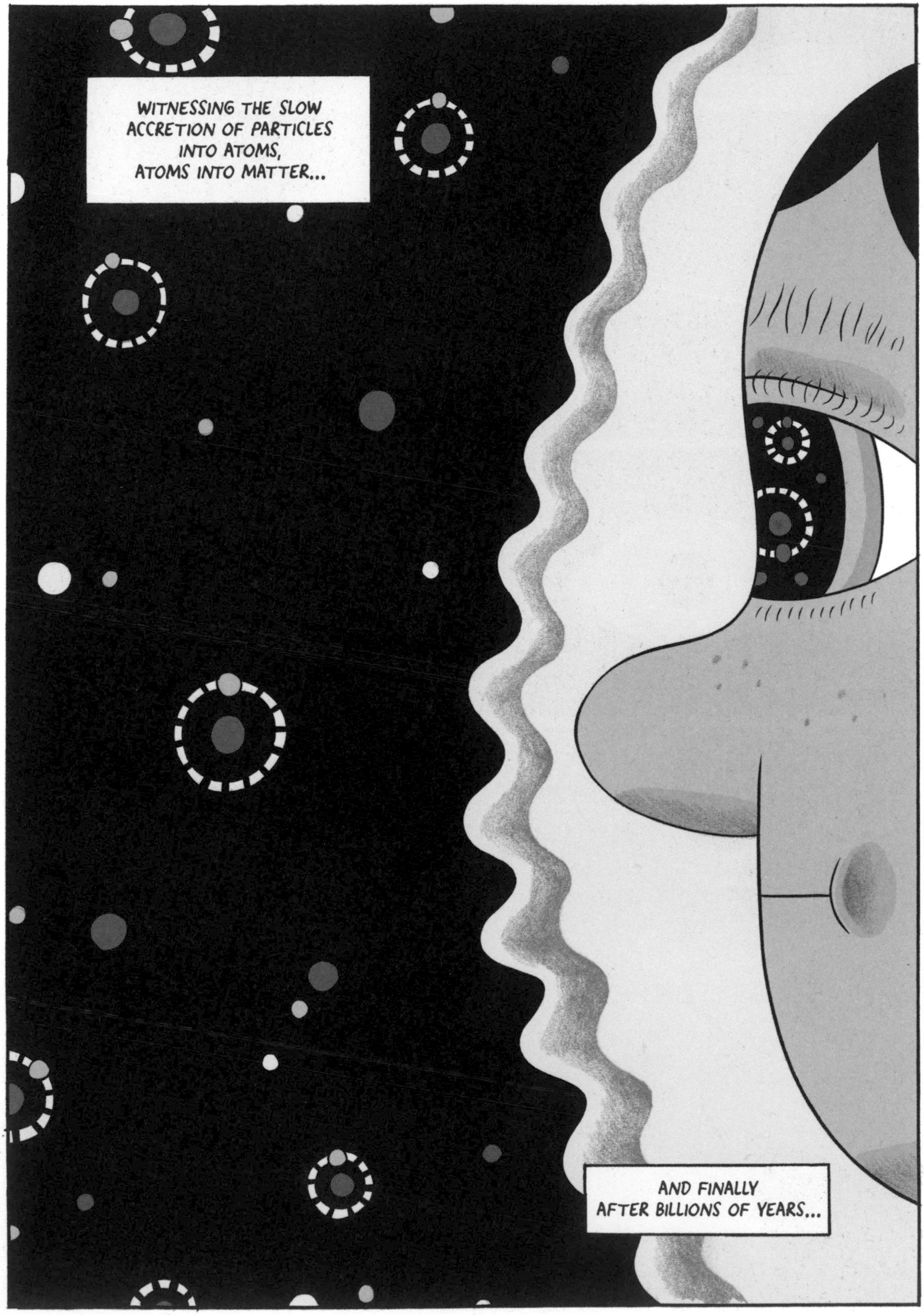
WITNESSING THE SLOW ACCRETION OF PARTICLES INTO ATOMS, ATOMS INTO MATTER...
AND FINALLY AFTER BILLIONS OF YEARS...

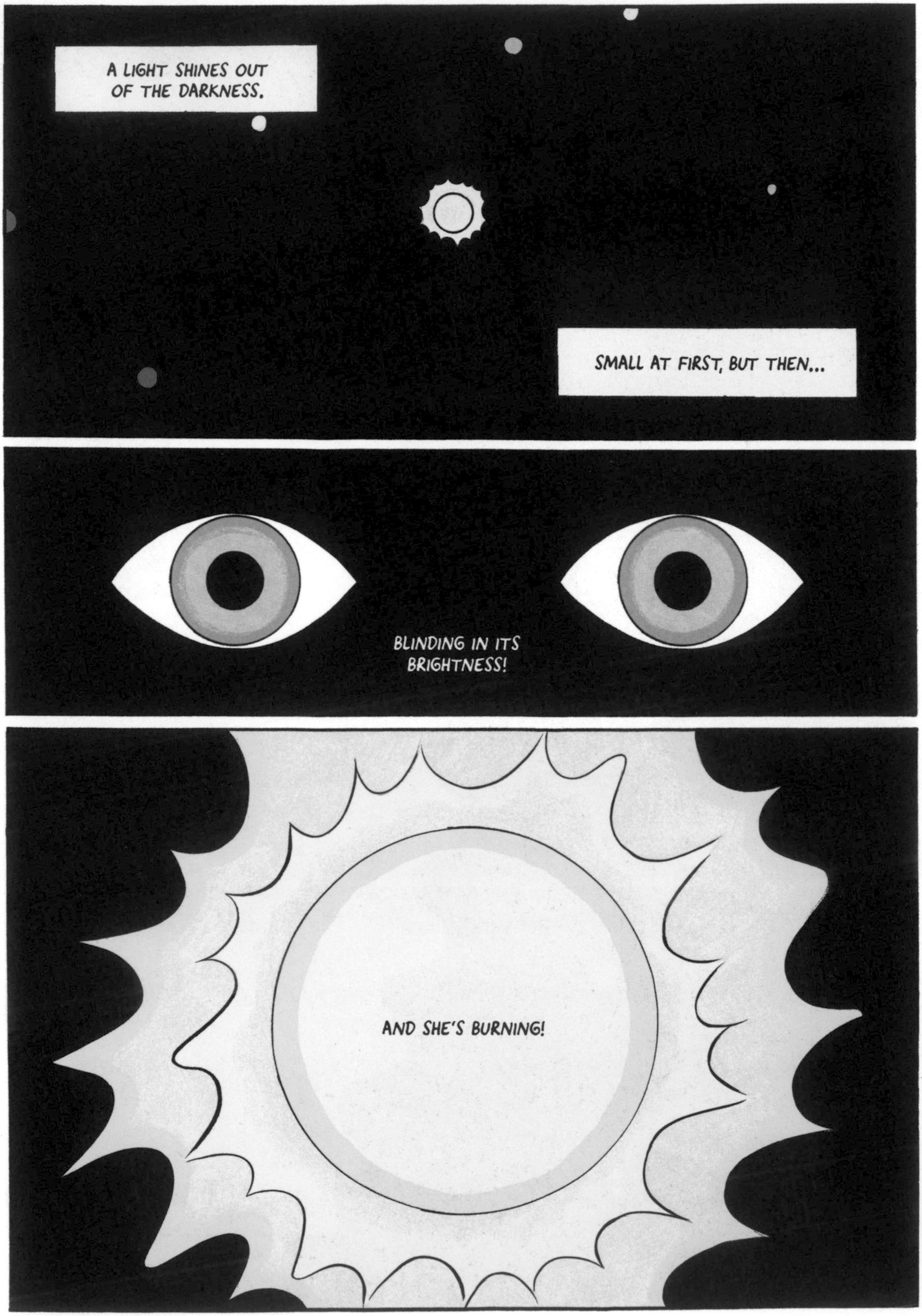
A LIGHT SHINES OUT OF THE DARKNESS.
SMALL AT FIRST, BUT THEN...
BLINDING IN ITS BRIGHTNESS!
AND SHE'S BURNING!

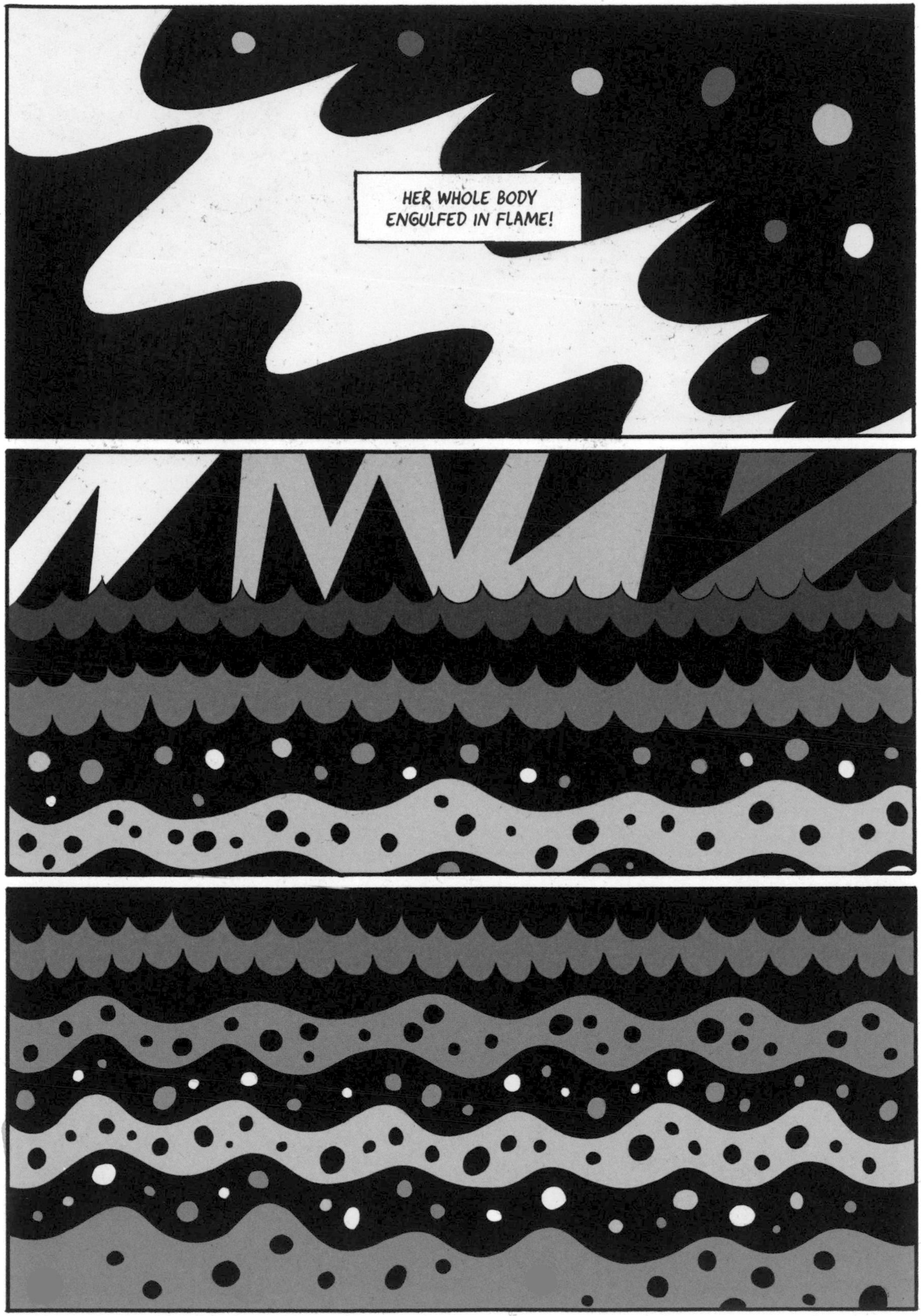
HER WHOLE BODY
ENGULFED IN FLAME!

SHE IS SUCKED DOWN, DOWN, DOWN,

DOWN INTO THE DARK,

INTO THE COOL, CALM DEPTHS...

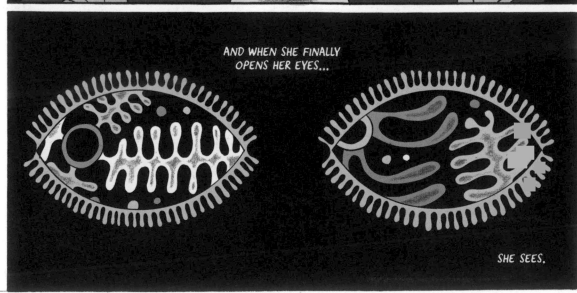

AND WHEN SHE FINALLY OPENS HER EYES...

SHE SEES.

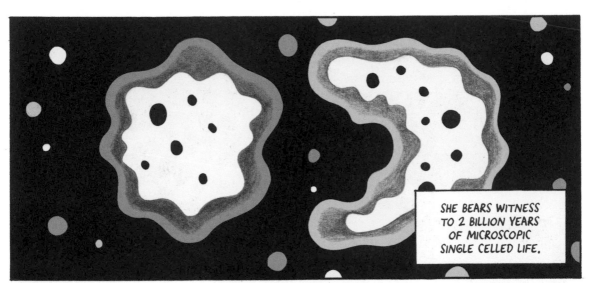

SHE BEARS WITNESS TO 2 BILLION YEARS OF MICROSCOPIC SINGLE CELLED LIFE.

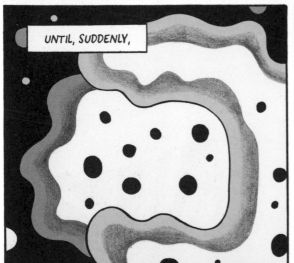

UNTIL, SUDDENLY,

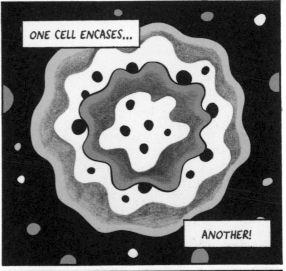

ONE CELL ENCASES...

ANOTHER!

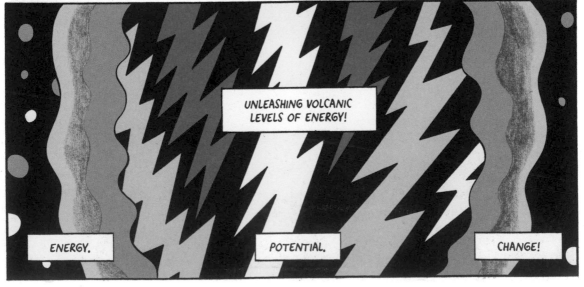

UNLEASHING VOLCANIC LEVELS OF ENERGY!

ENERGY.

POTENTIAL.

CHANGE!

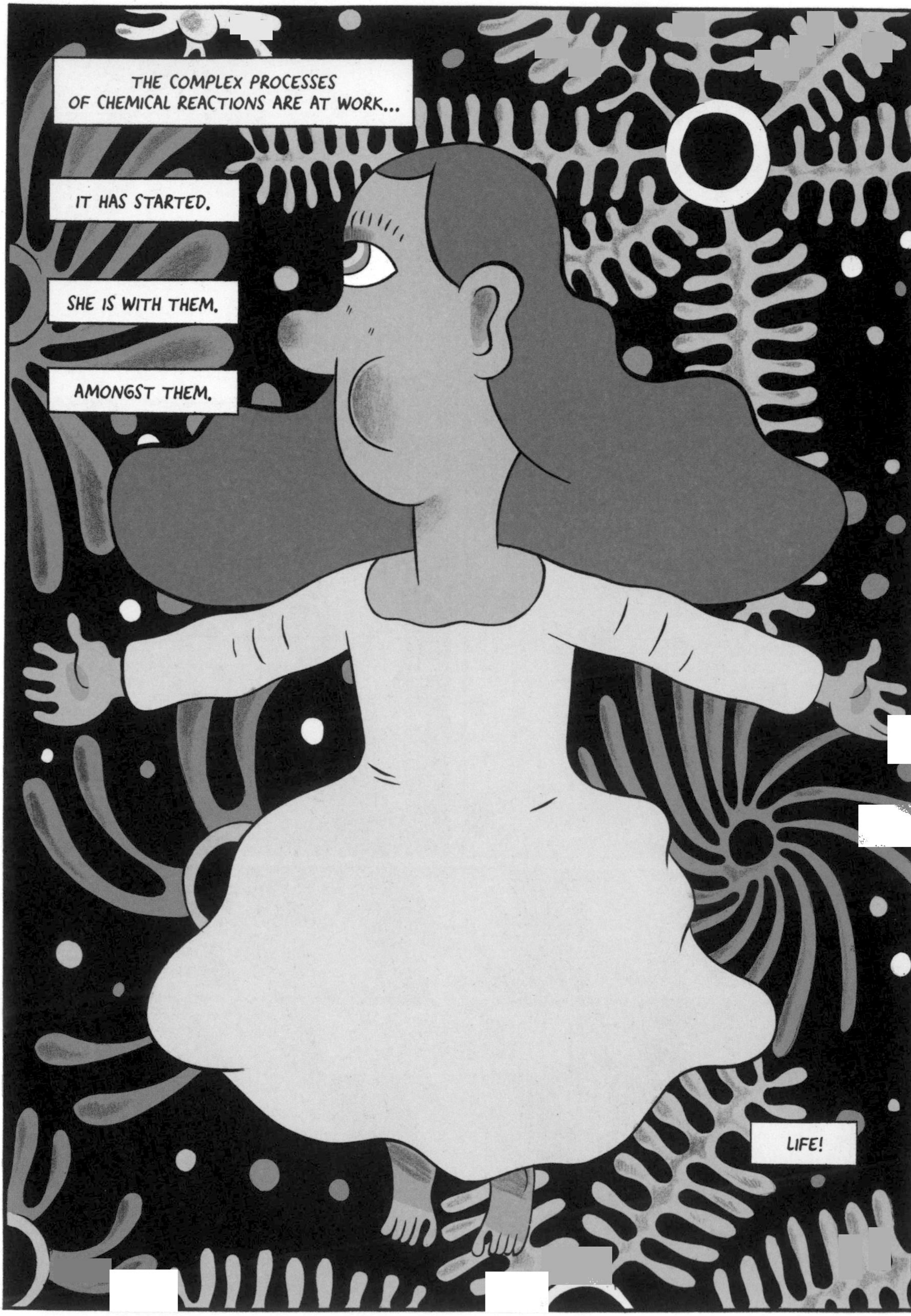
THE COMPLEX PROCESSES OF CHEMICAL REACTIONS ARE AT WORK...
IT HAS STARTED.
SHE IS WITH THEM.
AMONGST THEM.
LIFE!

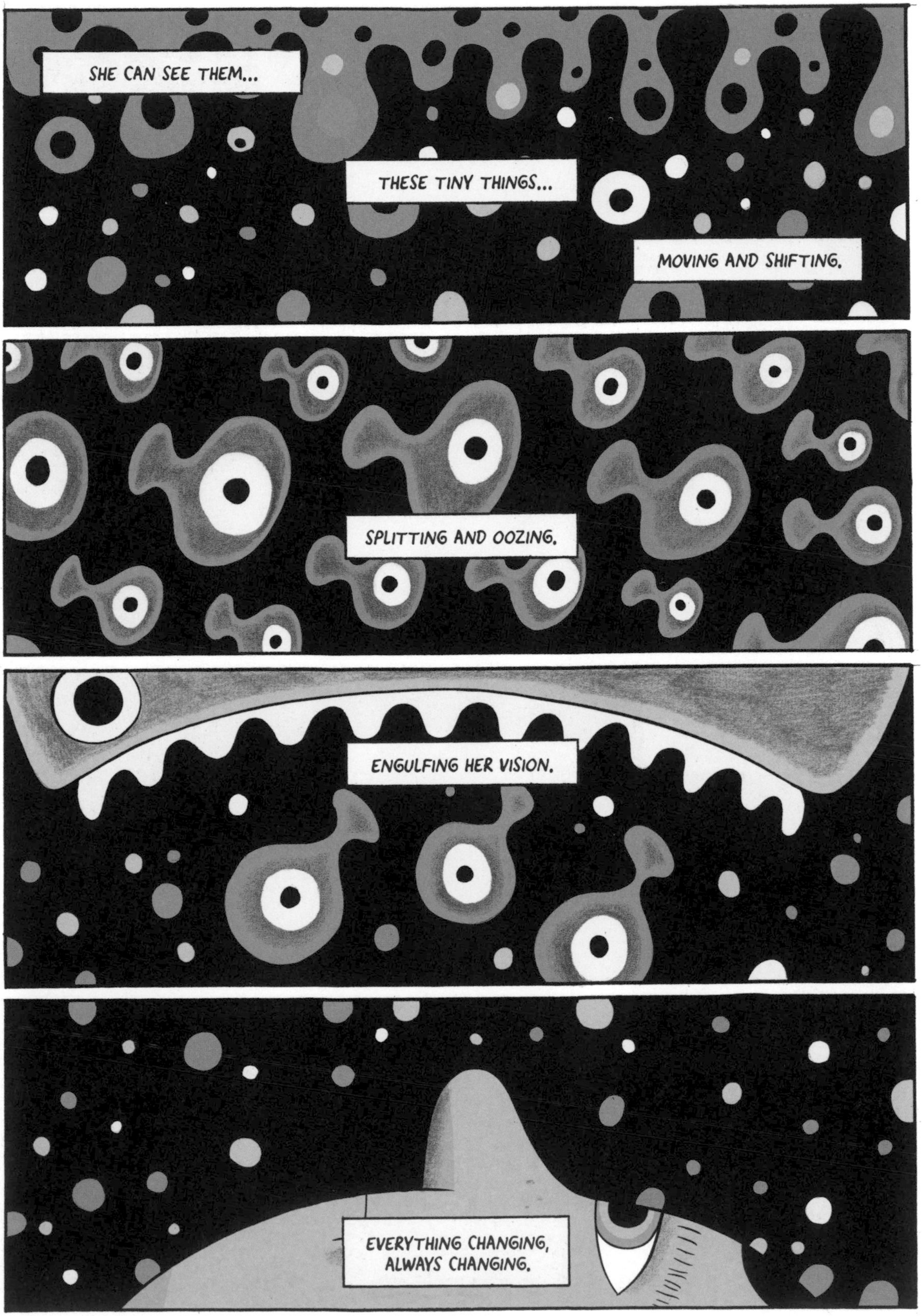
SHE CAN SEE THEM...
THESE TINY THINGS...
MOVING AND SHIFTING.
SPLITTING AND OOZING.
ENGULFING HER VISION.
EVERYTHING CHANGING, ALWAYS CHANGING.

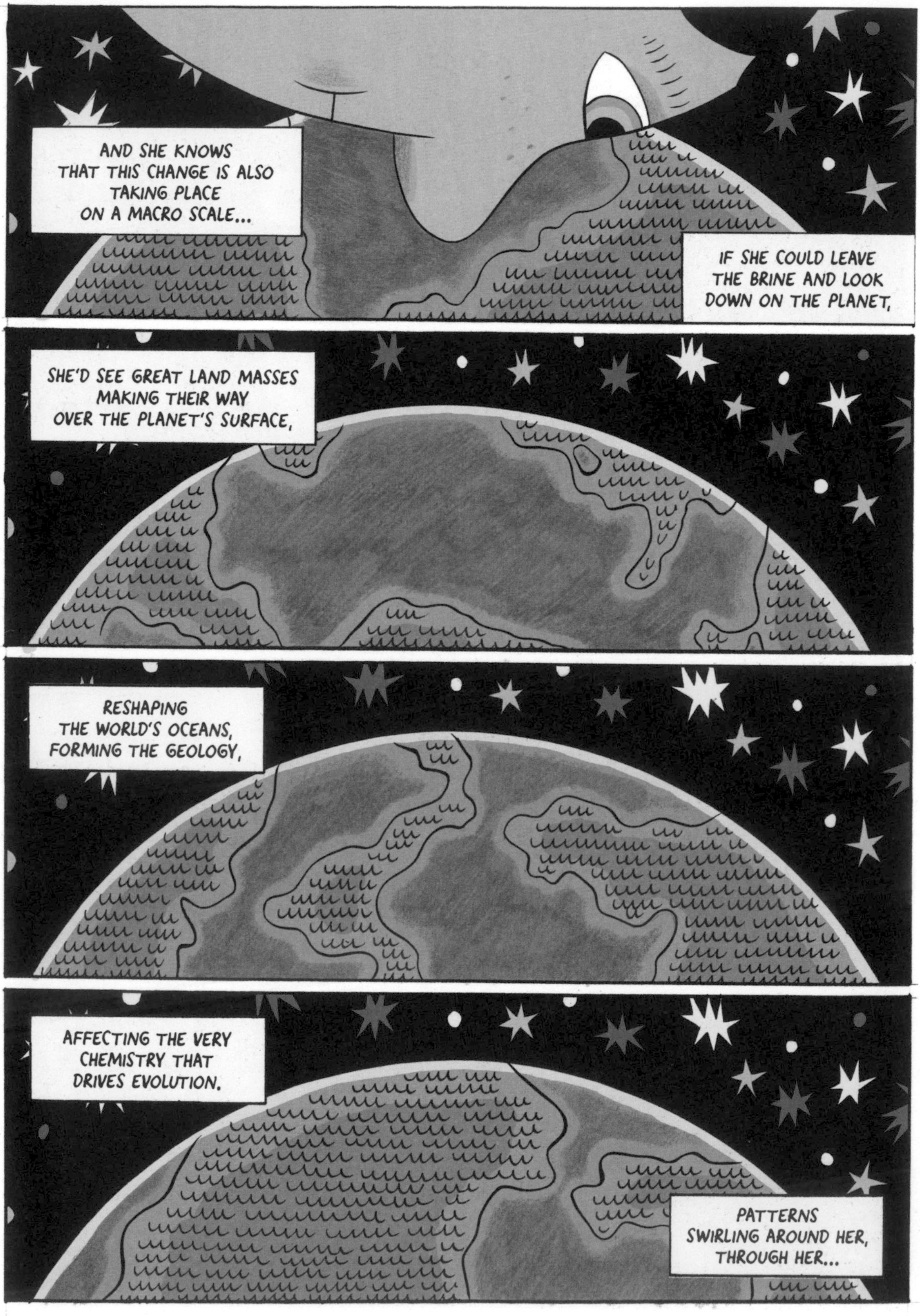
AND SHE KNOWS THAT THIS CHANGE IS ALSO TAKING PLACE ON A MACRO SCALE...
IF SHE COULD LEAVE THE BRINE AND LOOK DOWN ON THE PLANET,
SHE'D SEE GREAT LAND MASSES MAKING THEIR WAY OVER THE PLANET'S SURFACE,
RESHAPING THE WORLD'S OCEANS, FORMING THE GEOLOGY,
AFFECTING THE VERY CHEMISTRY THAT DRIVES EVOLUTION.
PATTERNS SWIRLING AROUND HER, THROUGH HER...

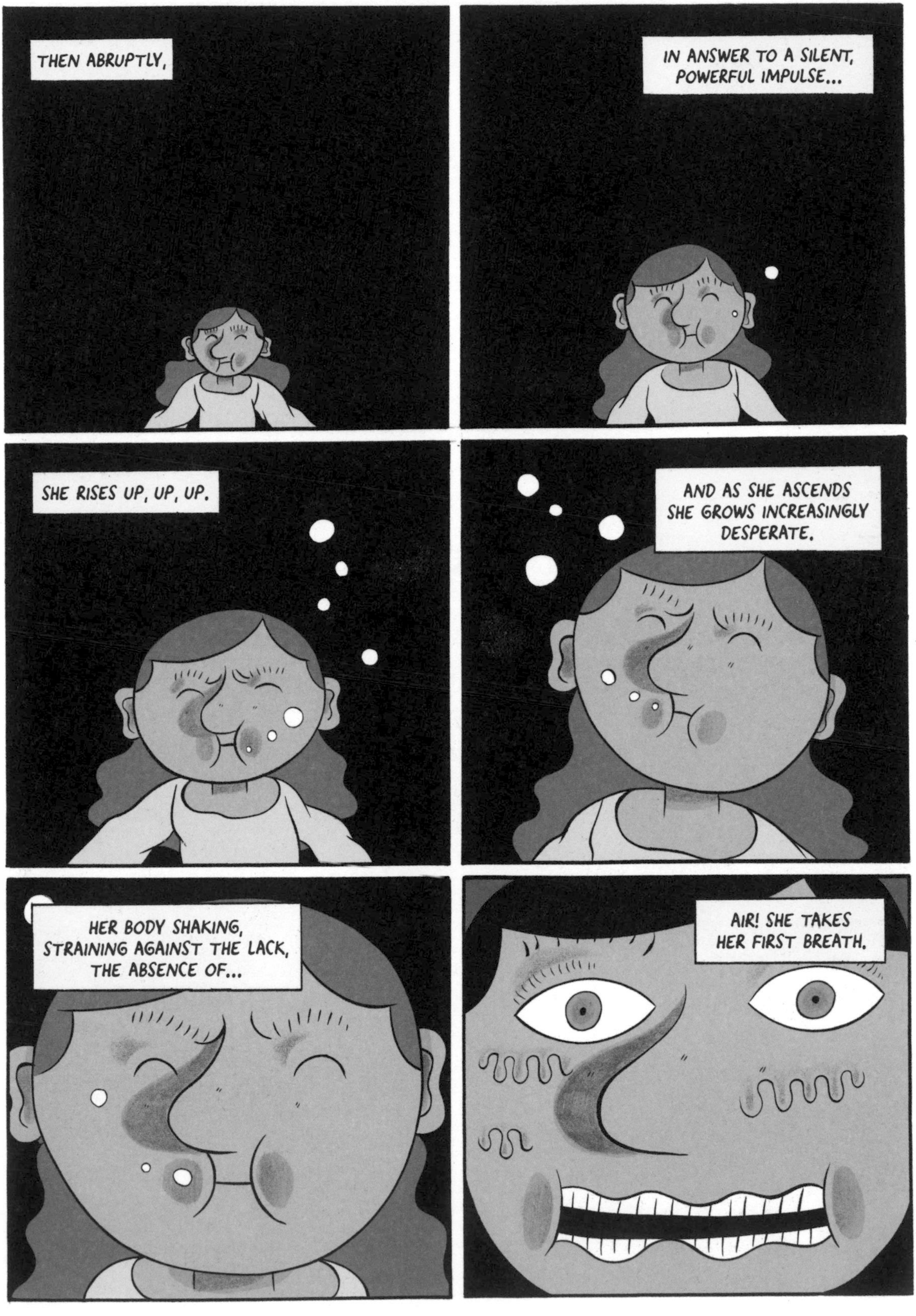
THEN ABRUPTLY,
IN ANSWER TO A SILENT, POWERFUL IMPULSE...
SHE RISES UP, UP, UP.
AND AS SHE ASCENDS SHE GROWS INCREASINGLY DESPERATE.
HER BODY SHAKING, STRAINING AGAINST THE LACK, THE ABSENCE OF...
AIR! SHE TAKES HER FIRST BREATH.

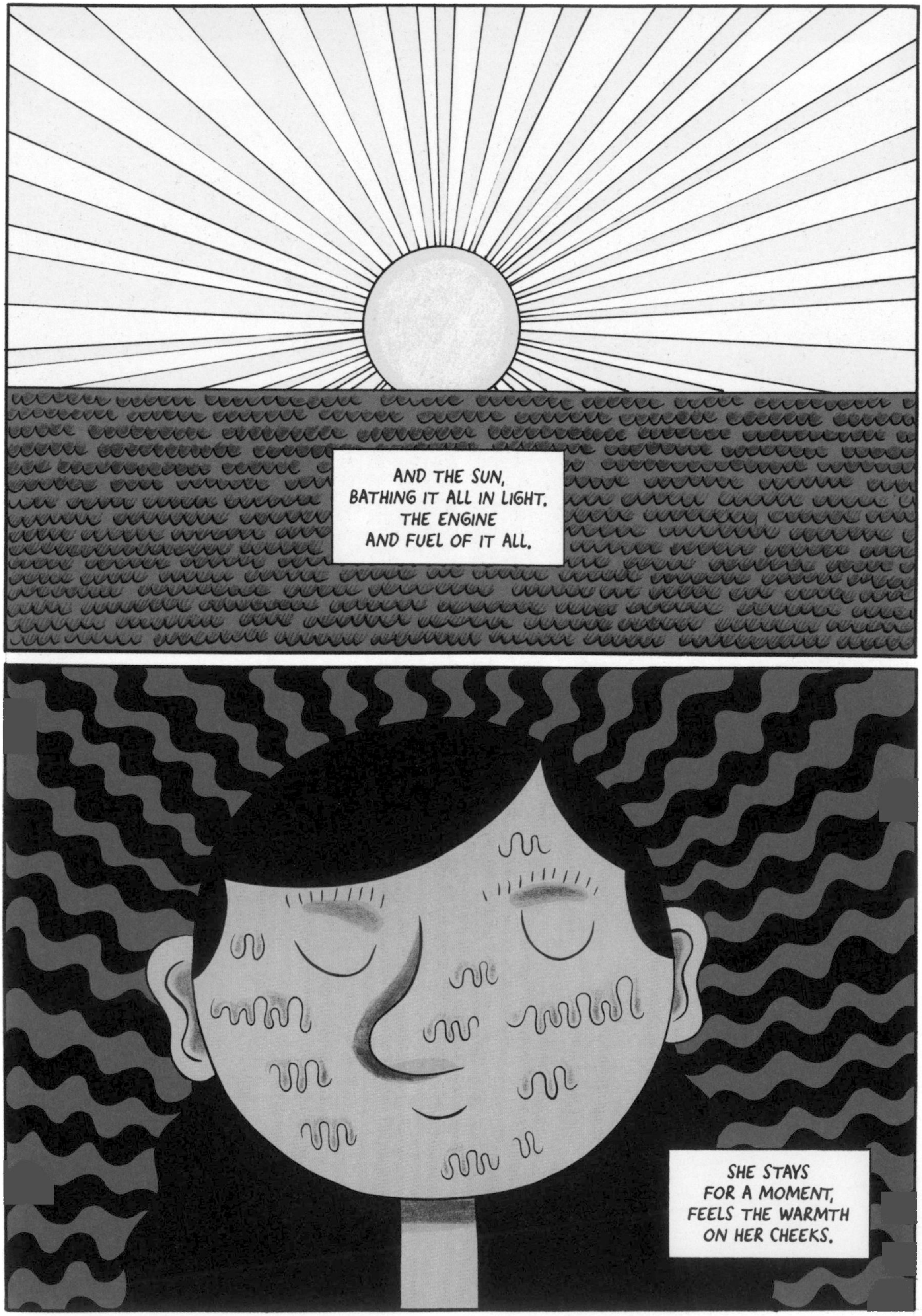
AND THE SUN, BATHING IT ALL IN LIGHT. THE ENGINE AND FUEL OF IT ALL.
SHE STAYS FOR A MOMENT, FEELS THE WARMTH ON HER CHEEKS.

1,910,000
years later...

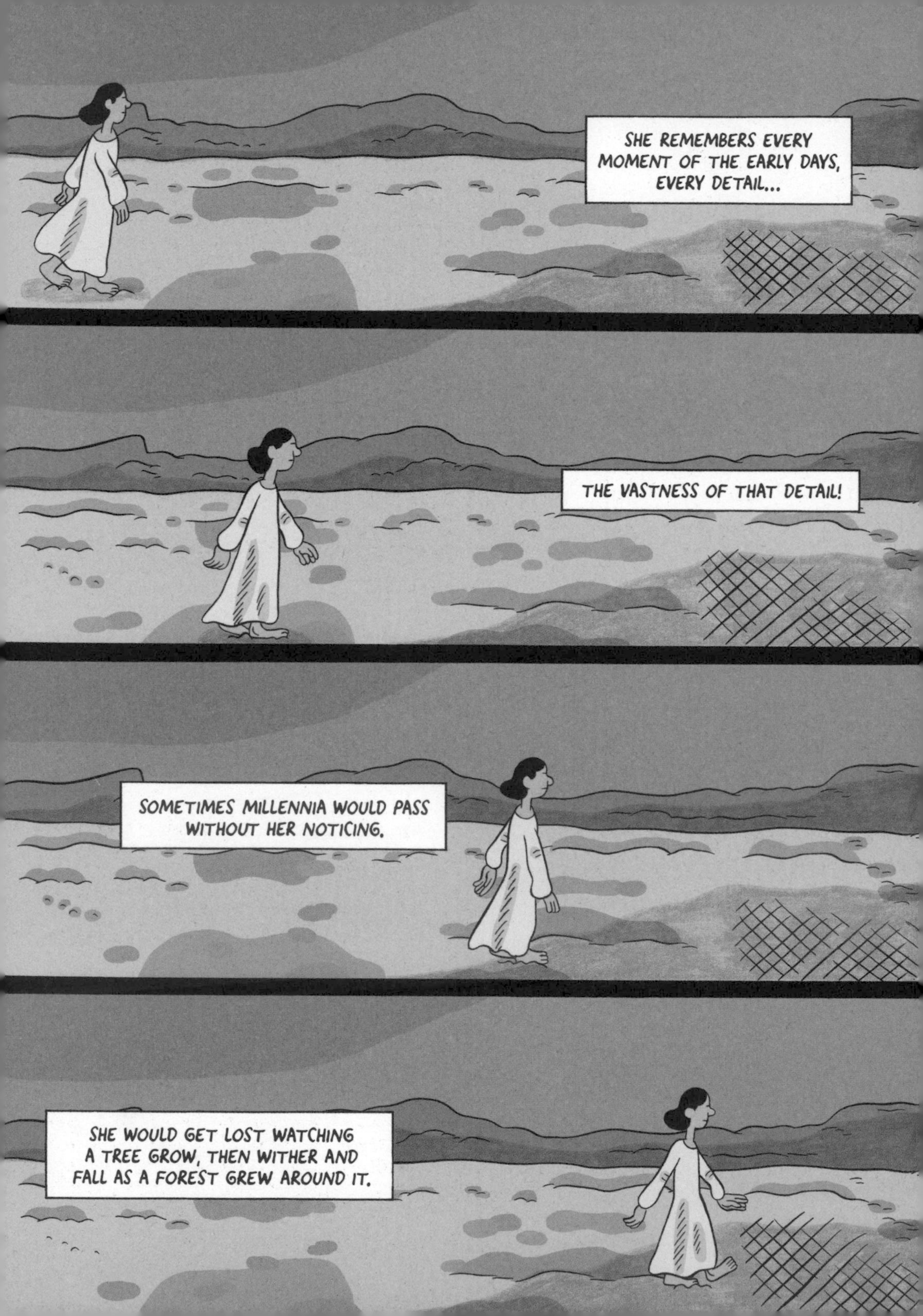
SHE REMEMBERS EVERY MOMENT OF THE EARLY DAYS, EVERY DETAIL...
THE VASTNESS OF THAT DETAIL!
SOMETIMES MILLENNIA WOULD PASS WITHOUT HER NOTICING.
SHE WOULD GET LOST WATCHING A TREE GROW, THEN WITHER AND FALL AS A FOREST GREW AROUND IT.

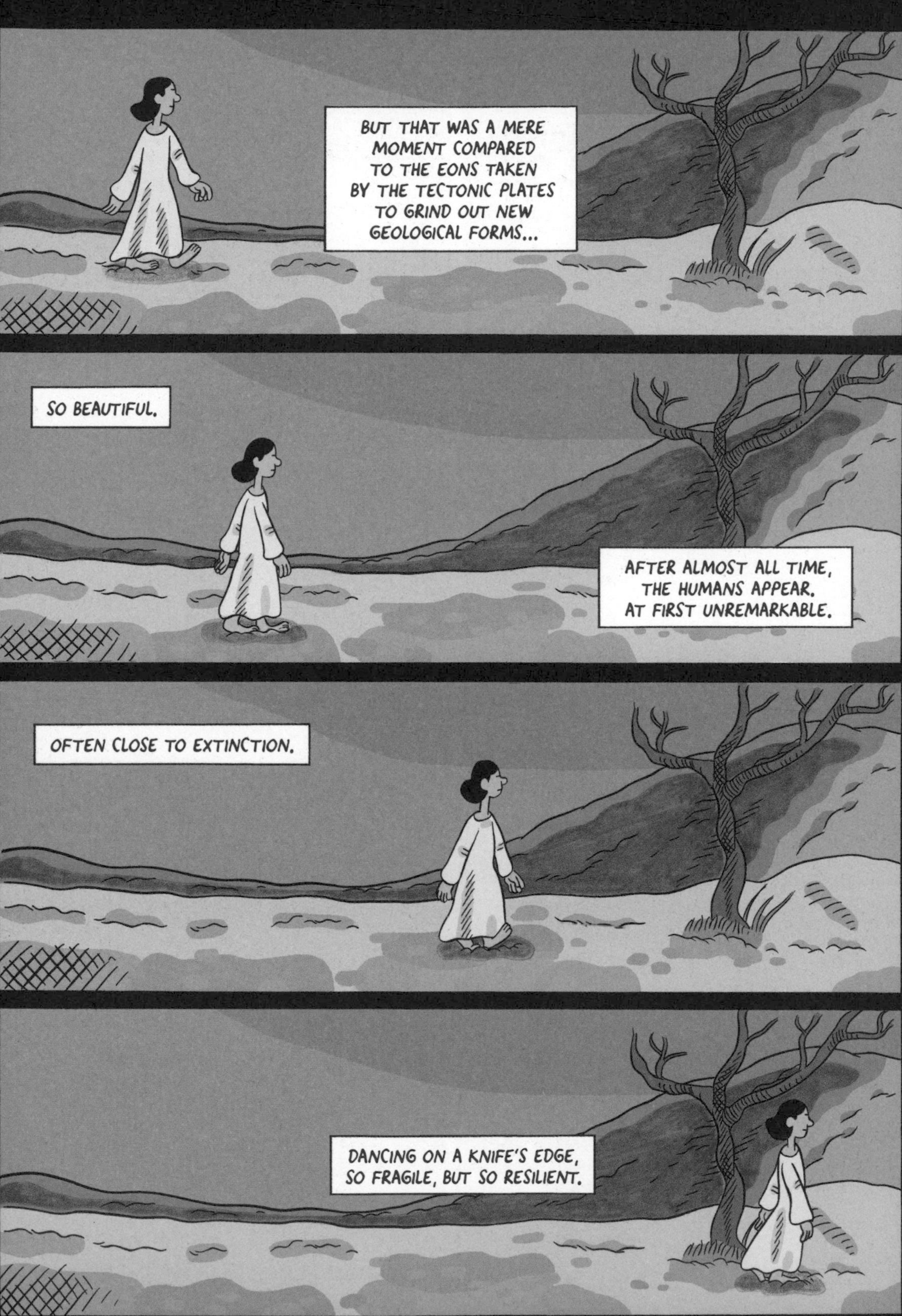
BUT THAT WAS A MERE MOMENT COMPARED TO THE EONS TAKEN BY THE TECTONIC PLATES TO GRIND OUT NEW GEOLOGICAL FORMS...
SO BEAUTIFUL.
AFTER ALMOST ALL TIME, THE HUMANS APPEAR. AT FIRST UNREMARKABLE.
OFTEN CLOSE TO EXTINCTION.
DANCING ON A KNIFE'S EDGE, SO FRAGILE, BUT SO RESILIENT.

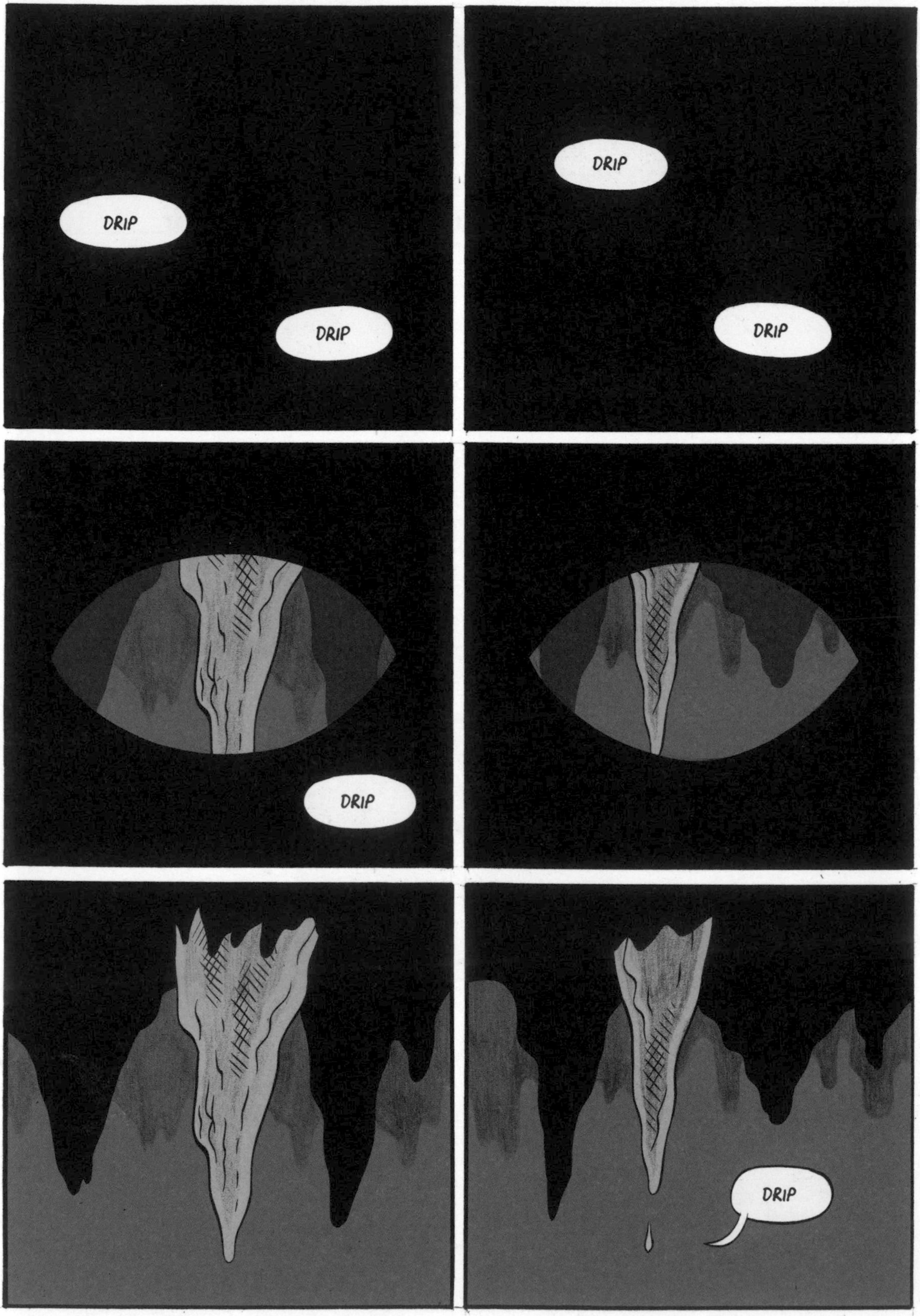
DRIP
DRIP
DRIP
DRIP
DRIP
DRIP

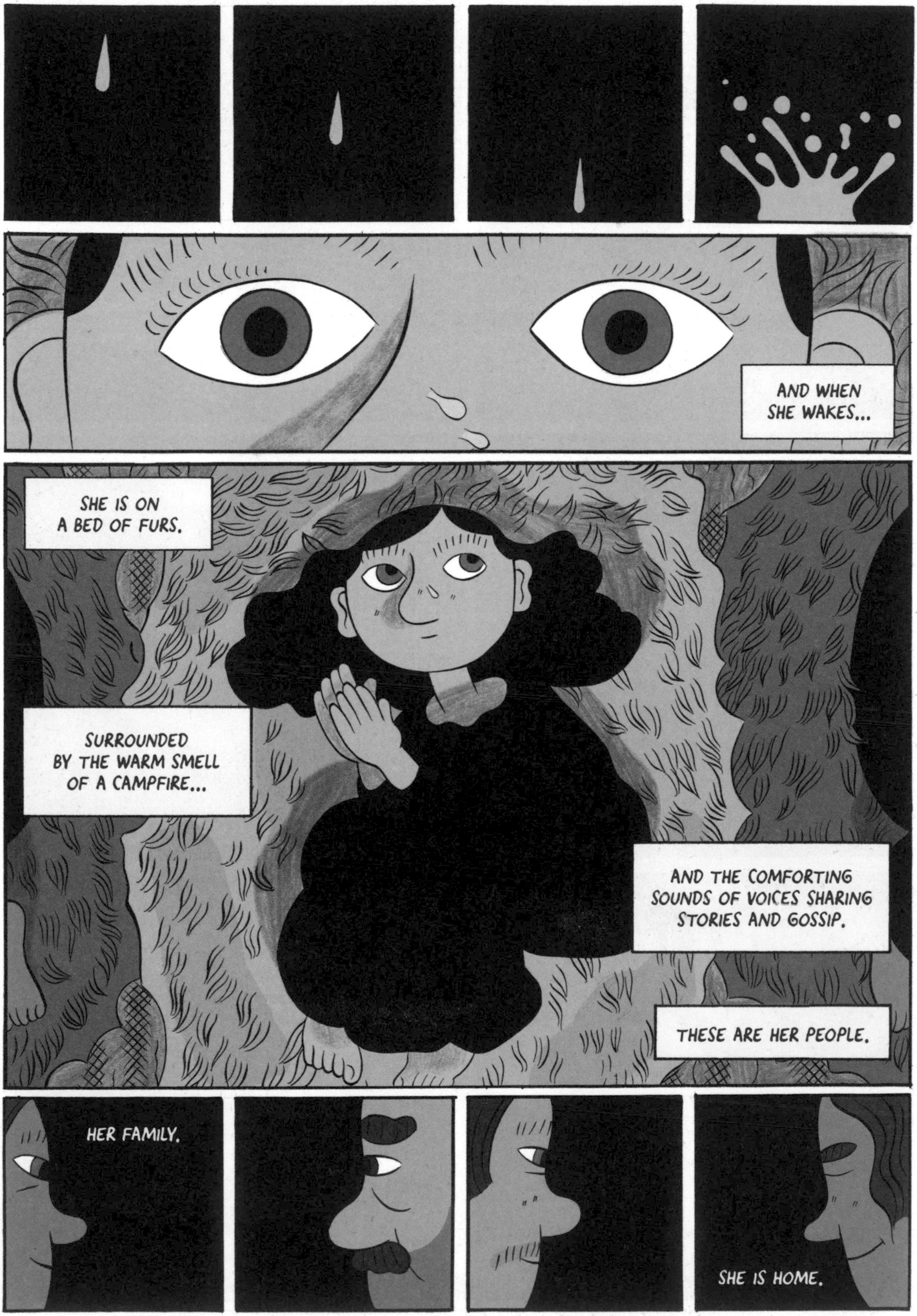
AND WHEN SHE WAKES...
SHE IS ON A BED OF FURS.
SURROUNDED BY THE WARM SMELL OF A CAMPFIRE...
AND THE COMFORTING SOUNDS OF VOICES SHARING STORIES AND GOSSIP.
THESE ARE HER PEOPLE.
HER FAMILY.
SHE IS HOME.

SHE GETS HERSELF UP, EAGER TO JOIN IN...

GNAW

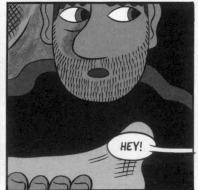

HEY!

TAKE THIS.

I WANT YOU TO CARVE THIS INTO SOMETHING.

IT'S IMPORTANT.

AND IT'LL BE FOR ALL OF US.

HMM, THIS IS A NICE BIT OF IVORY...

IT'D MAKE A GOOD CUP...

NO! THIS ISN'T FOR SOME CUP!

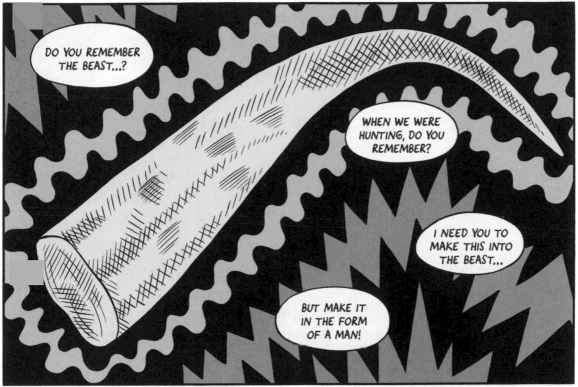

DO YOU REMEMBER THE BEAST...?

WHEN WE WERE HUNTING, DO YOU REMEMBER?

I NEED YOU TO MAKE THIS INTO THE BEAST...

BUT MAKE IT IN THE FORM OF A MAN!

I REMEMBER IT. I REMEMBER THE BEAST.

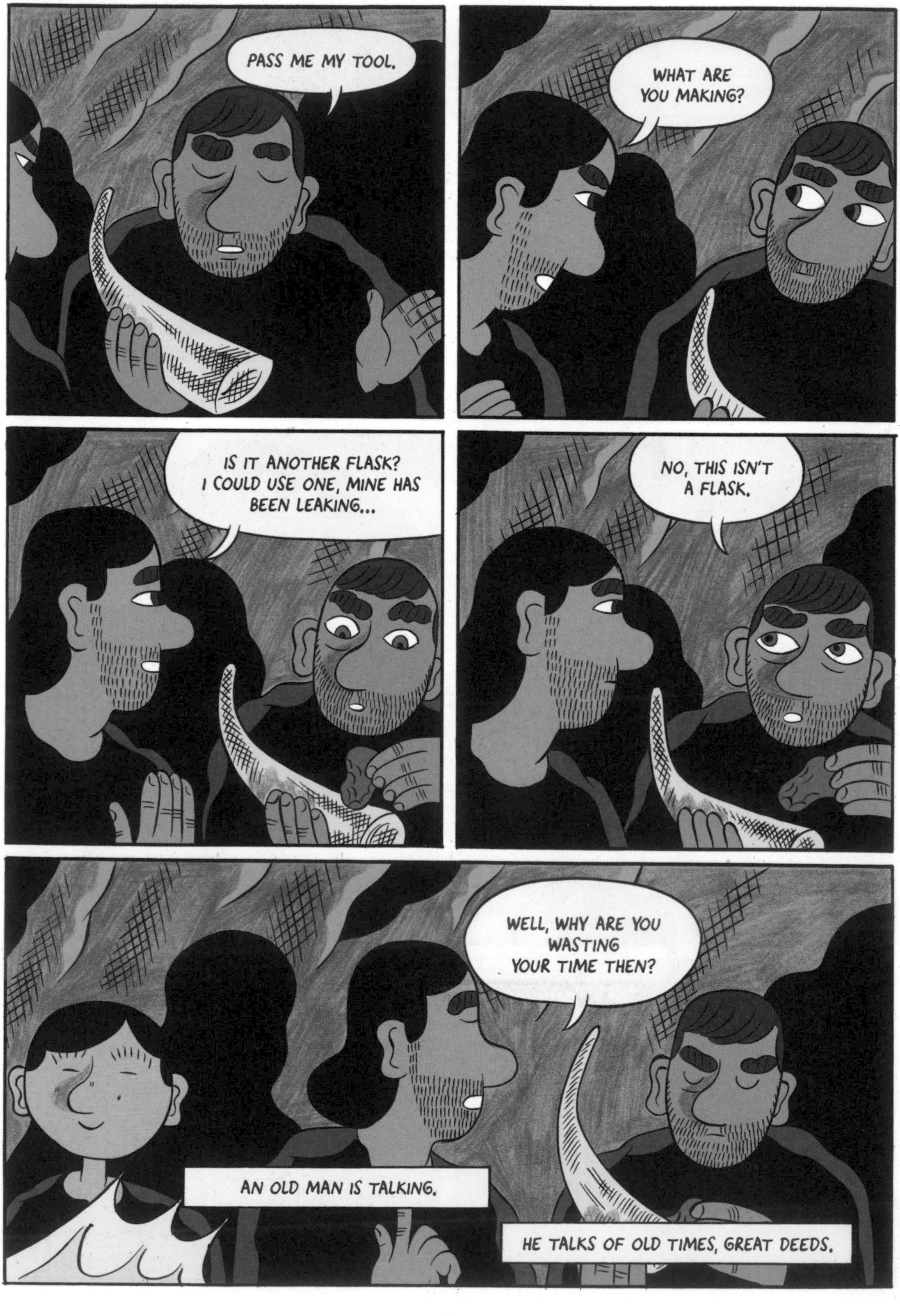
PASS ME MY TOOL.
WHAT ARE YOU MAKING?
IS IT ANOTHER FLASK? I COULD USE ONE, MINE HAS BEEN LEAKING...
NO, THIS ISN'T A FLASK.
WELL, WHY ARE YOU WASTING YOUR TIME THEN?
AN OLD MAN IS TALKING.
HE TALKS OF OLD TIMES, GREAT DEEDS.

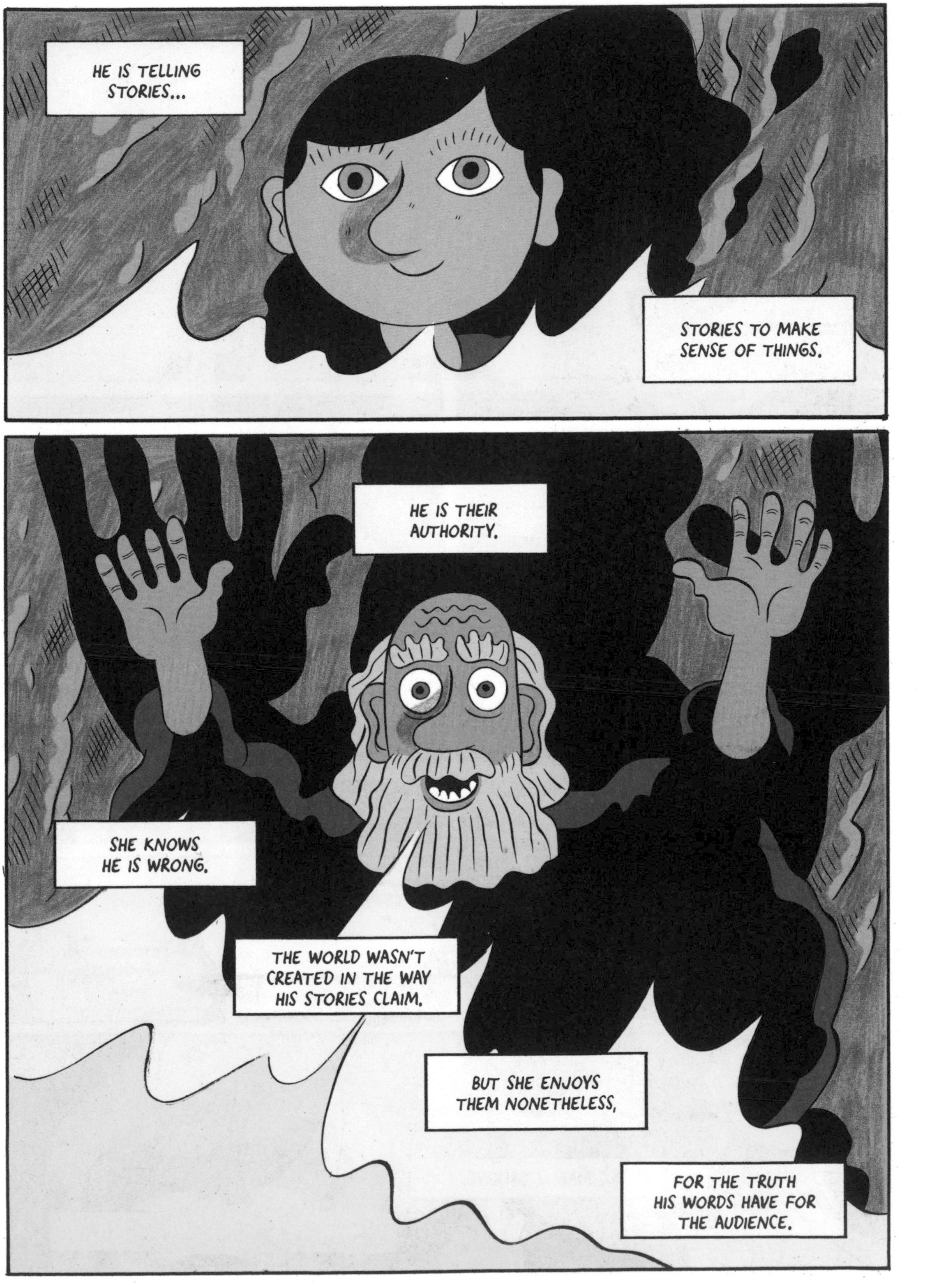
HE IS TELLING STORIES...
STORIES TO MAKE SENSE OF THINGS.
HE IS THEIR AUTHORITY.
SHE KNOWS HE IS WRONG.
THE WORLD WASN'T CREATED IN THE WAY HIS STORIES CLAIM.
BUT SHE ENJOYS THEM NONETHELESS,
FOR THE TRUTH HIS WORDS HAVE FOR THE AUDIENCE.

AND SHE ENJOYS WATCHING HIM.

SEEING THE WILD LIGHT IN HIS EYES.

HE IS A PERFORMER, AN ARTIST.

SHE LOOKS AROUND AT THE AUDIENCE,

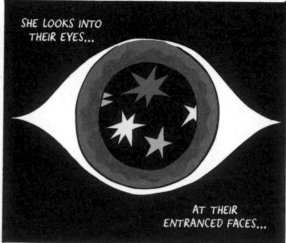

SHE LOOKS INTO THEIR EYES...

AT THEIR ENTRANCED FACES...

THEIR MINDS

DIVINING MEANING

FROM THE

STORYTELLER'S WORDS.

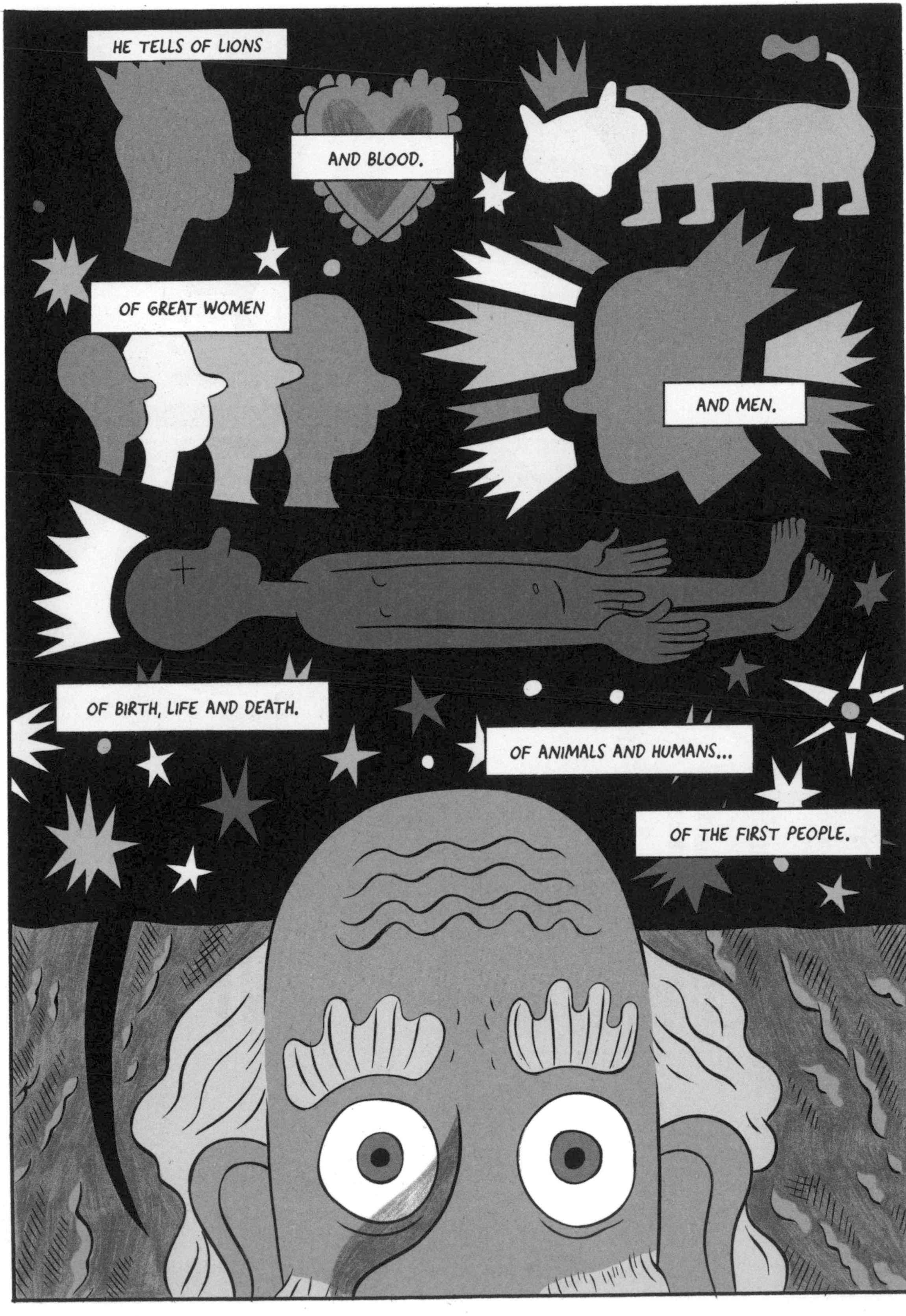
HE TELLS OF LIONS
AND BLOOD.
OF GREAT WOMEN
AND MEN.
OF BIRTH, LIFE AND DEATH.
OF ANIMALS AND HUMANS...
OF THE FIRST PEOPLE.

HE USES HIS HANDS AS HE TALKS, CREATING SHAPES, LIGHT AND DARK.

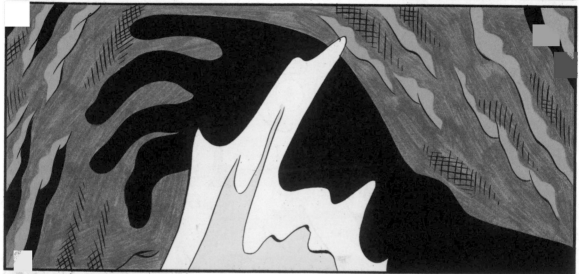

OBSCURING AND REVEALING THE MARKINGS ON THE WALLS.

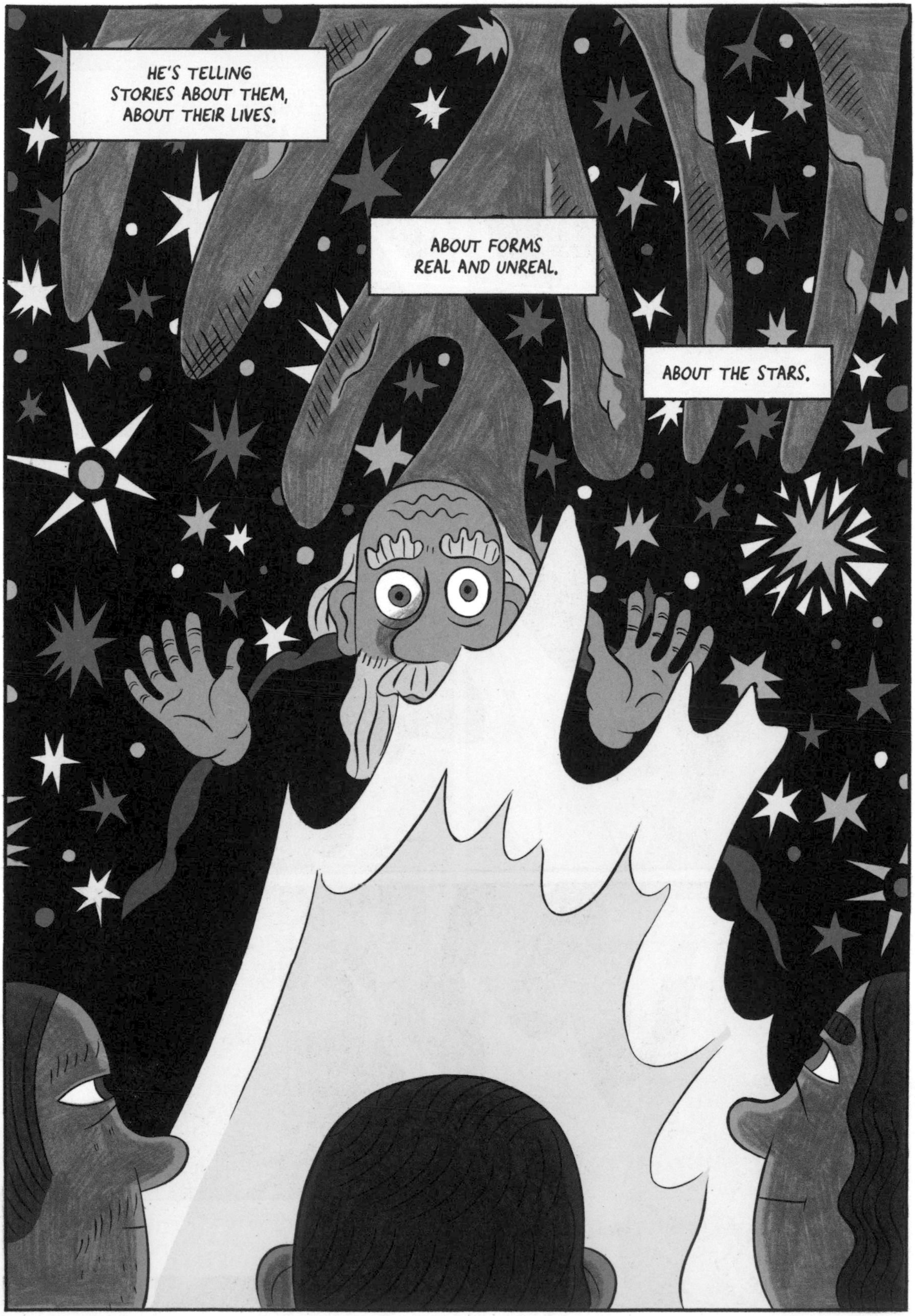
HE'S TELLING STORIES ABOUT THEM, ABOUT THEIR LIVES.
ABOUT FORMS REAL AND UNREAL.
ABOUT THE STARS.

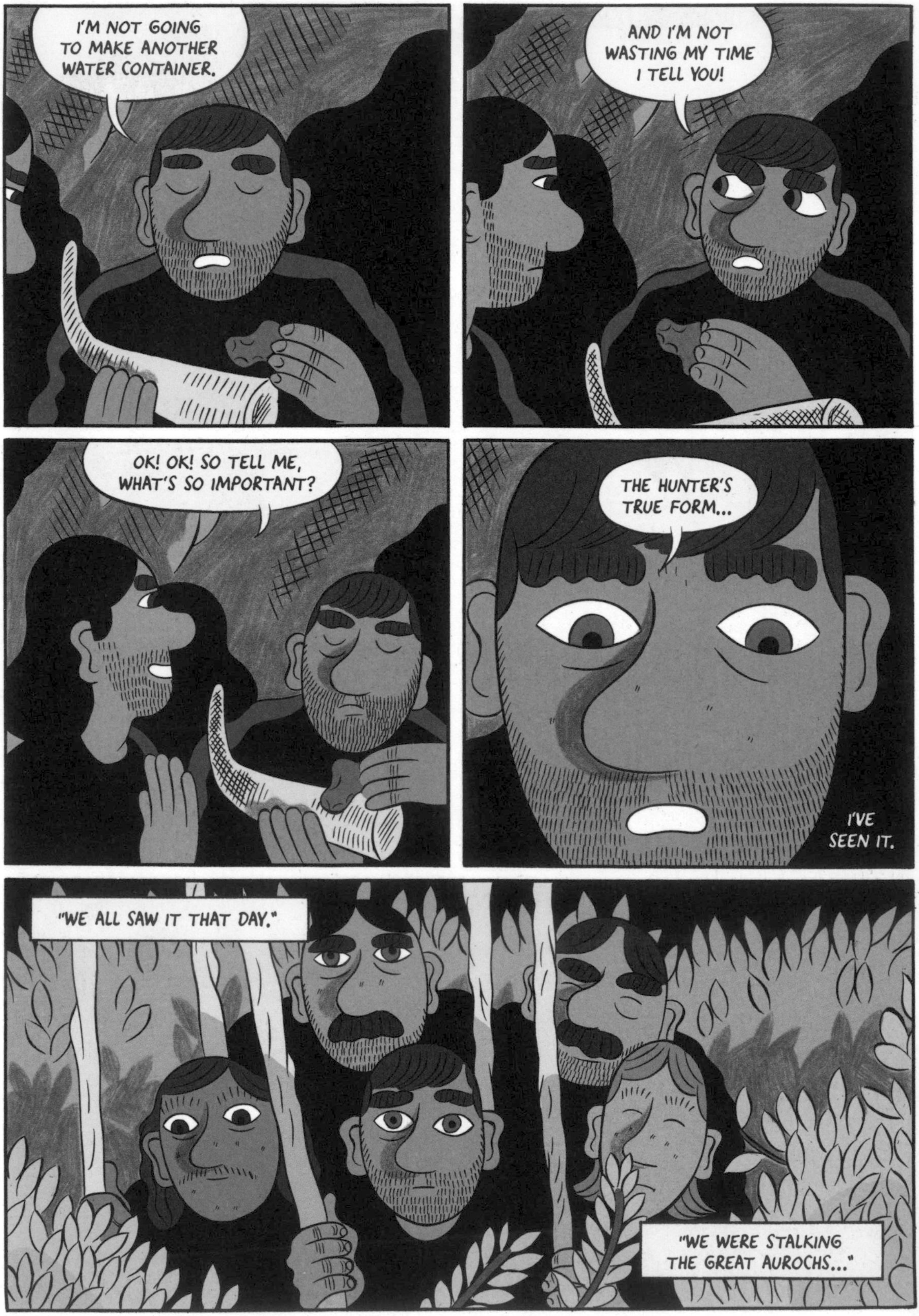
I'M NOT GOING TO MAKE ANOTHER WATER CONTAINER.
AND I'M NOT WASTING MY TIME I TELL YOU!
OK! OK! SO TELL ME, WHAT'S SO IMPORTANT?
THE HUNTER'S TRUE FORM...
I'VE SEEN IT.
"WE ALL SAW IT THAT DAY."
"WE WERE STALKING THE GREAT AUROCHS..."

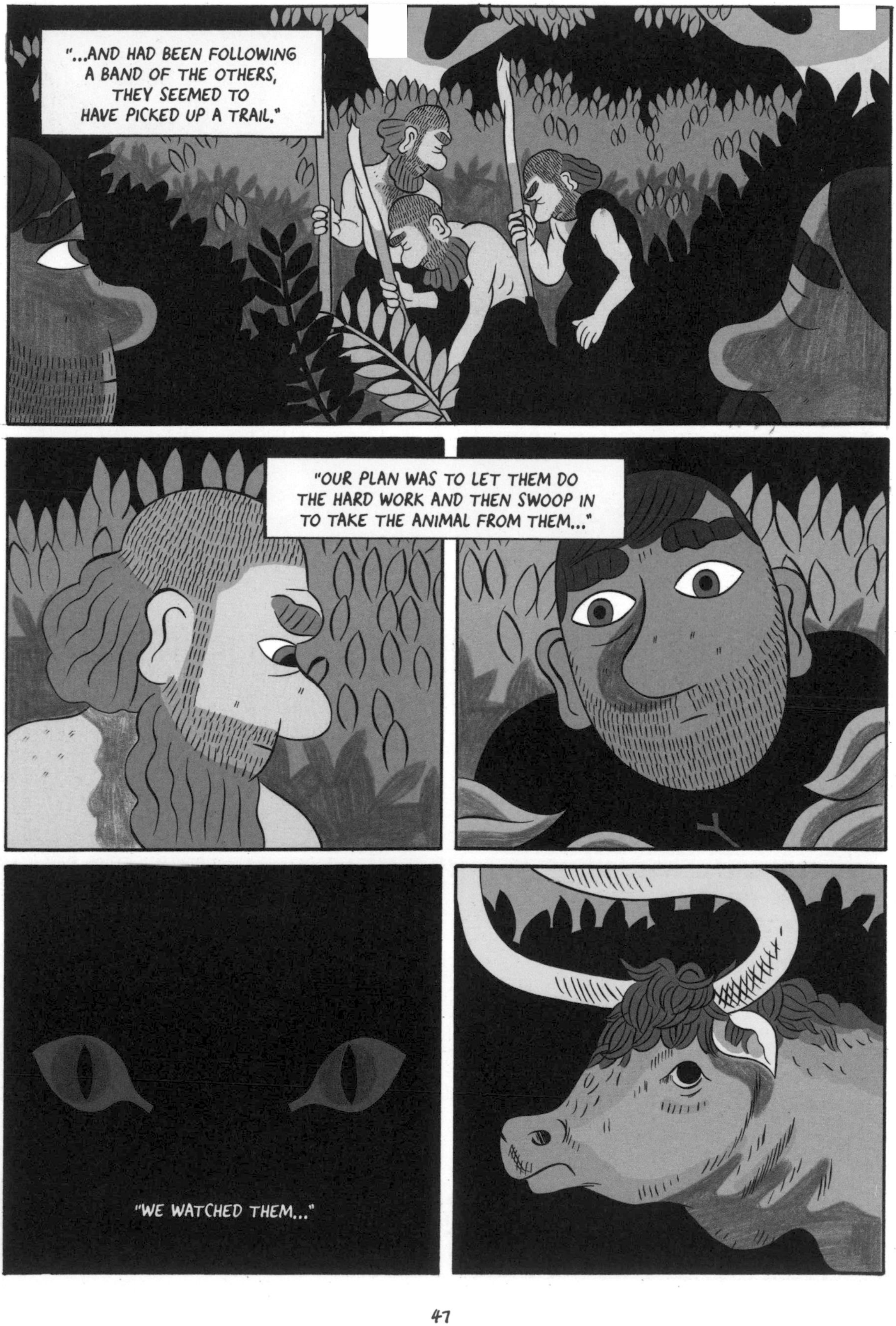
"...AND HAD BEEN FOLLOWING A BAND OF THE OTHERS, THEY SEEMED TO HAVE PICKED UP A TRAIL."
"OUR PLAN WAS TO LET THEM DO THE HARD WORK AND THEN SWOOP IN TO TAKE THE ANIMAL FROM THEM..."
"WE WATCHED THEM..."

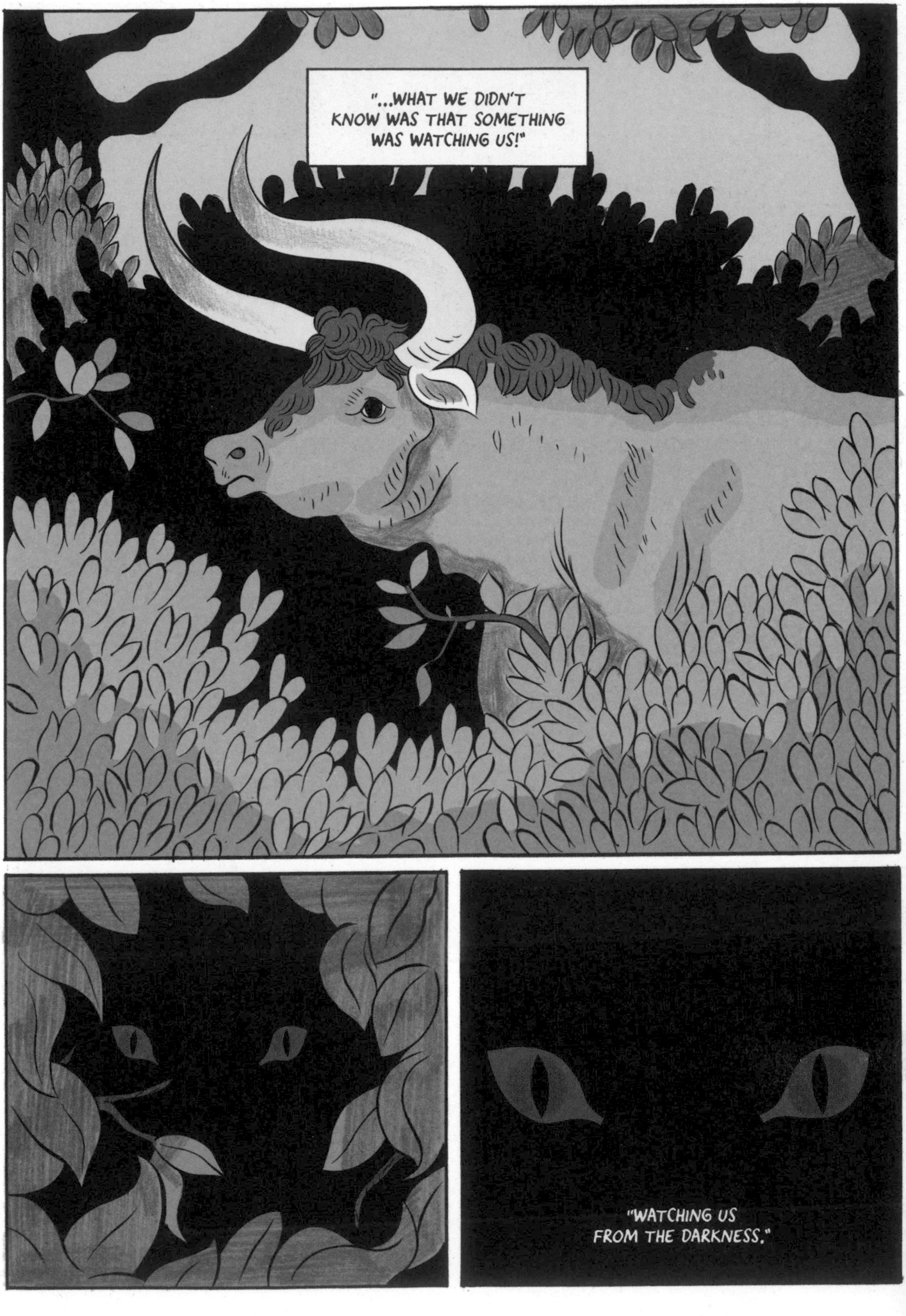
"...WHAT WE DIDN'T KNOW WAS THAT SOMETHING WAS WATCHING US!"
"WATCHING US FROM THE DARKNESS."

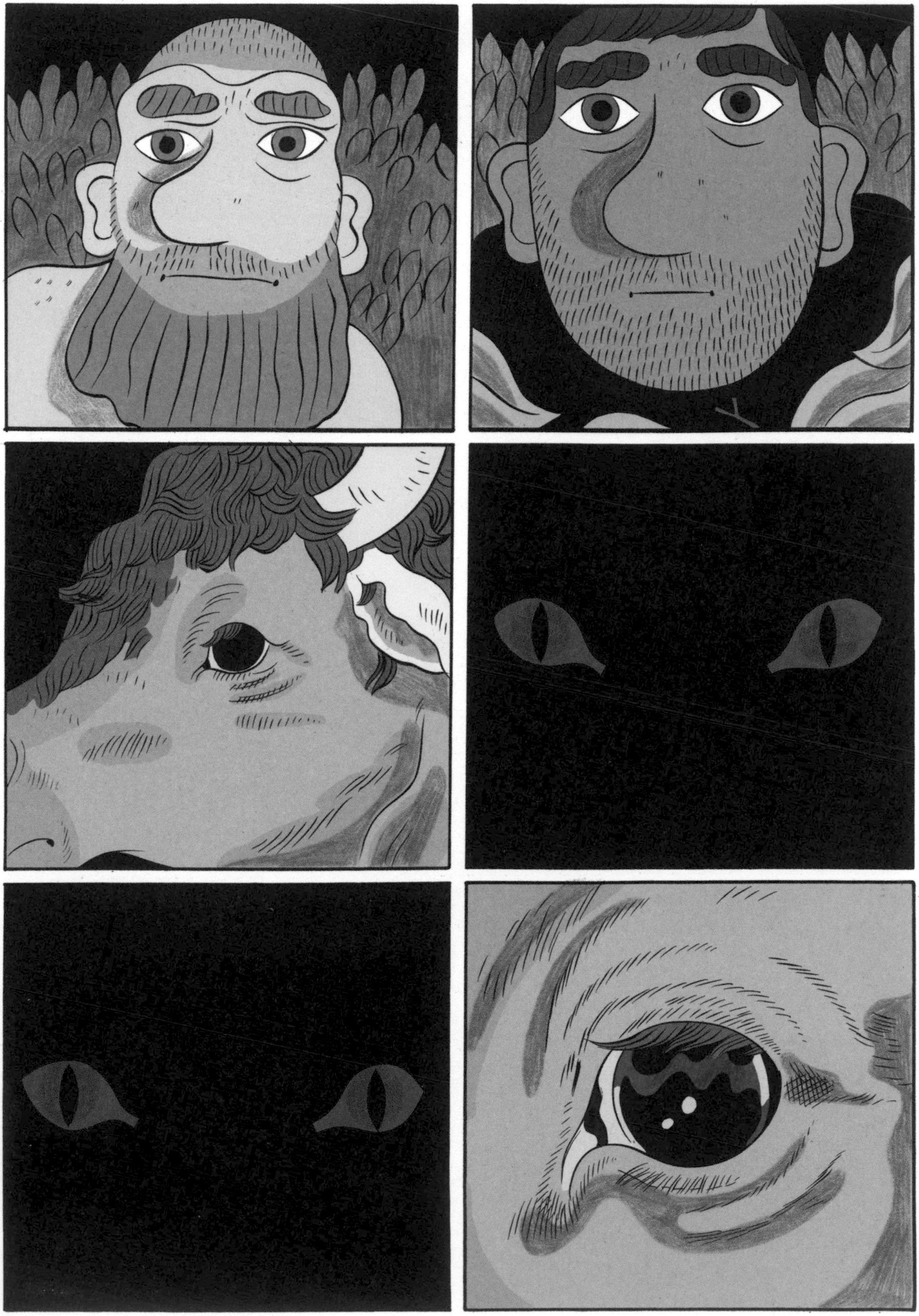

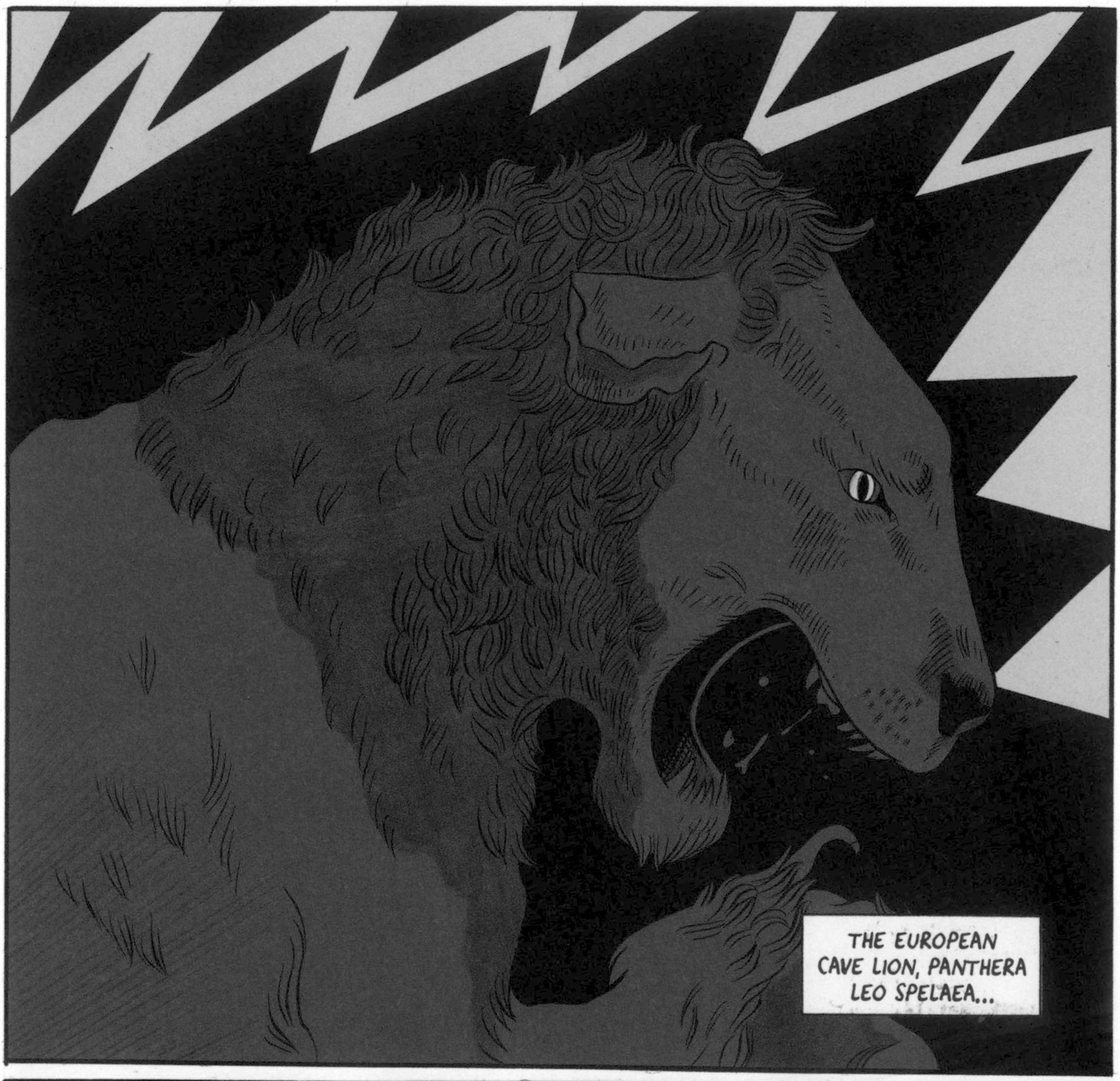
THE EUROPEAN CAVE LION, PANTHERA LEO SPELAEA...

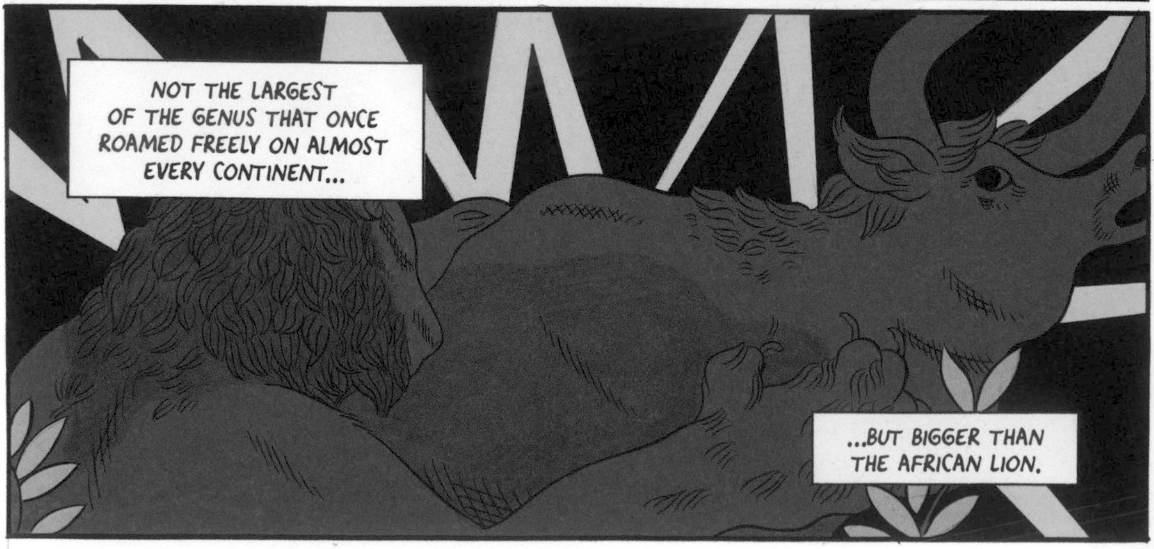
NOT THE LARGEST OF THE GENUS THAT ONCE ROAMED FREELY ON ALMOST EVERY CONTINENT...
...BUT BIGGER THAN THE AFRICAN LION.

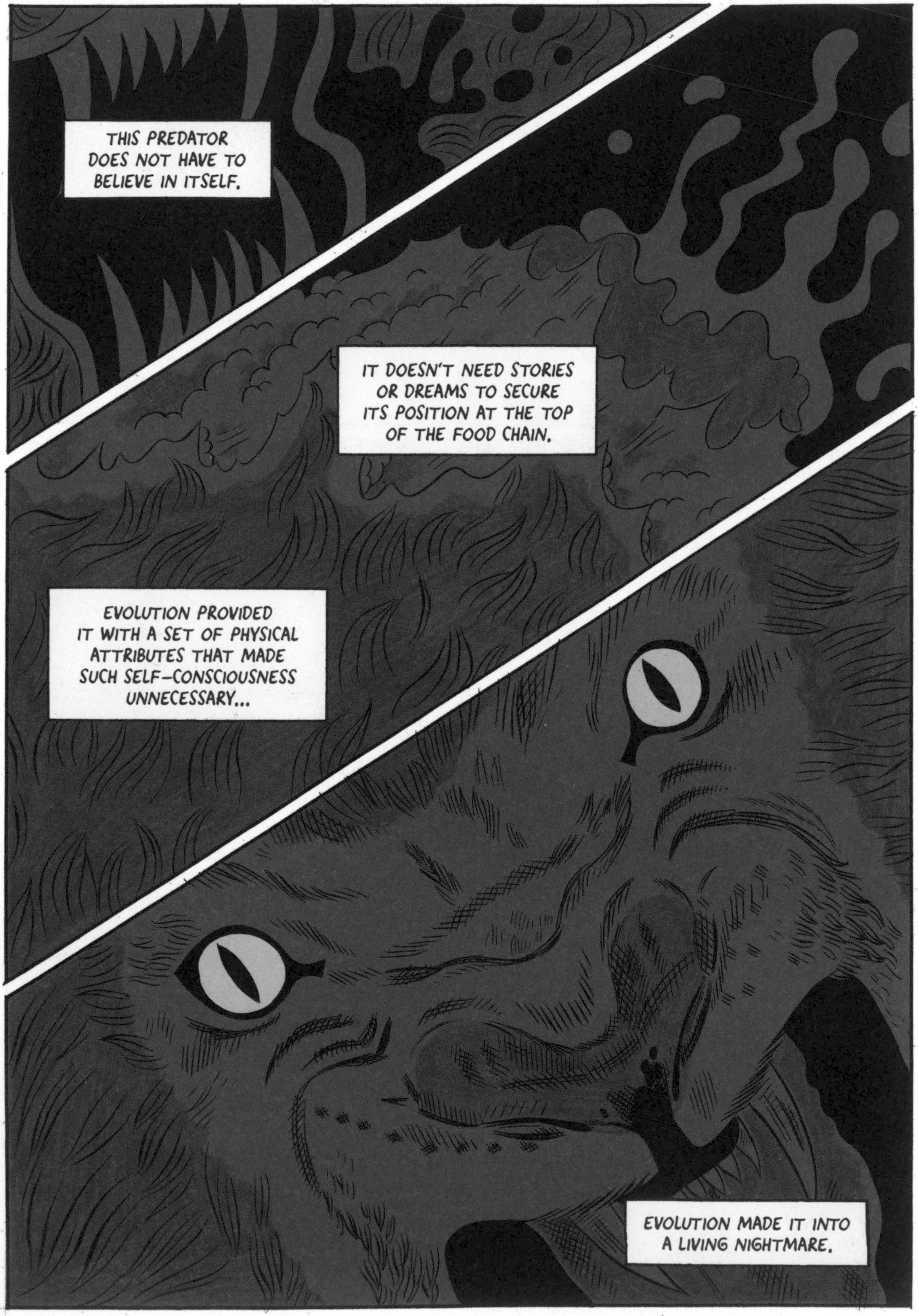
THIS PREDATOR DOES NOT HAVE TO BELIEVE IN ITSELF.
IT DOESN'T NEED STORIES OR DREAMS TO SECURE ITS POSITION AT THE TOP OF THE FOOD CHAIN.
EVOLUTION PROVIDED IT WITH A SET OF PHYSICAL ATTRIBUTES THAT MADE SUCH SELF-CONSCIOUSNESS UNNECESSARY...
EVOLUTION MADE IT INTO A LIVING NIGHTMARE.

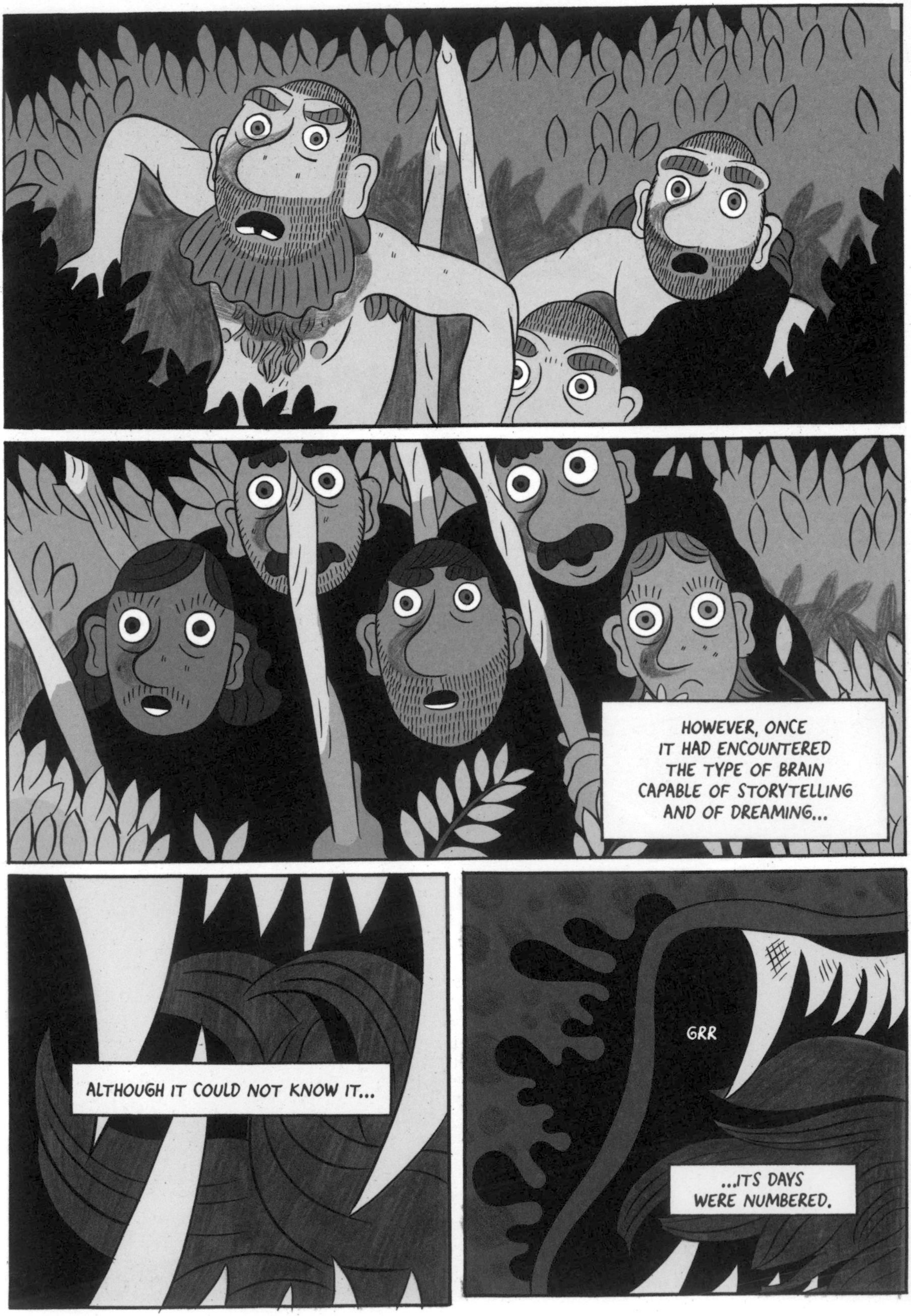
HOWEVER, ONCE IT HAD ENCOUNTERED THE TYPE OF BRAIN CAPABLE OF STORYTELLING AND OF DREAMING...
ALTHOUGH IT COULD NOT KNOW IT...
GRR
...ITS DAYS WERE NUMBERED.

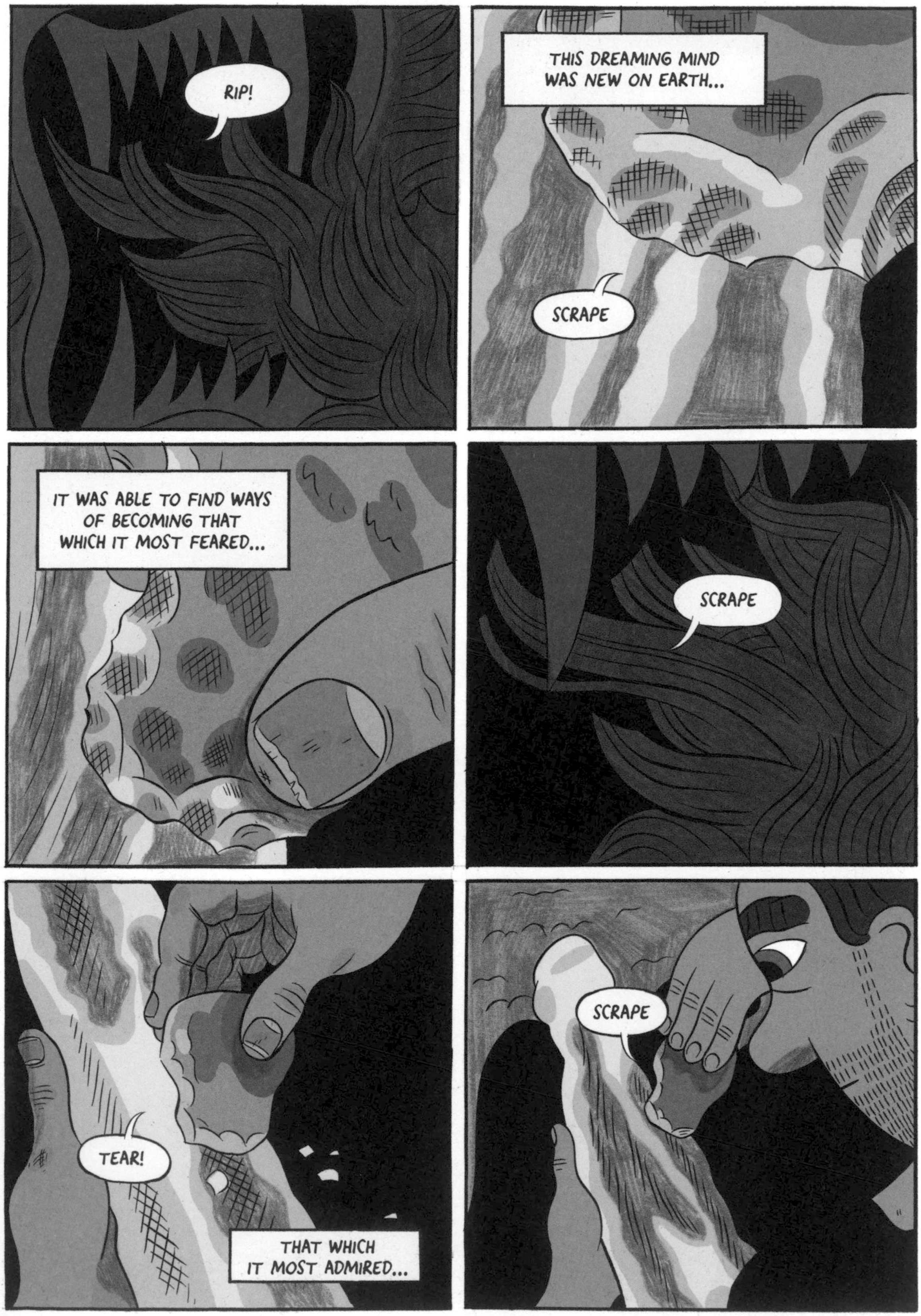
RIP!
THIS DREAMING MIND WAS NEW ON EARTH...
SCRAPE
IT WAS ABLE TO FIND WAYS OF BECOMING THAT WHICH IT MOST FEARED...
SCRAPE
TEAR!
THAT WHICH IT MOST ADMIRED...
SCRAPE

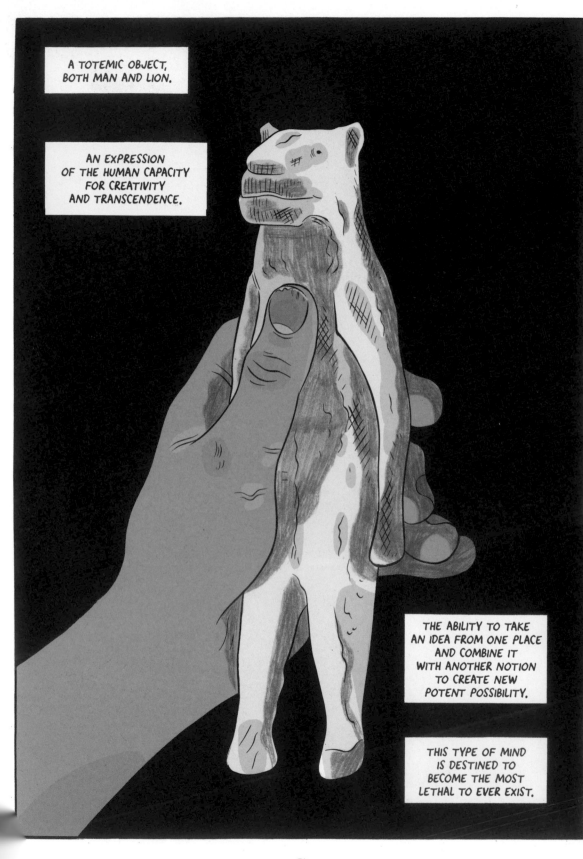

A TOTEMIC OBJECT, BOTH MAN AND LION.

AN EXPRESSION OF THE HUMAN CAPACITY FOR CREATIVITY AND TRANSCENDENCE.

THE ABILITY TO TAKE AN IDEA FROM ONE PLACE AND COMBINE IT WITH ANOTHER NOTION TO CREATE NEW POTENT POSSIBILITY.

THIS TYPE OF MIND IS DESTINED TO BECOME THE MOST LETHAL TO EVER EXIST.

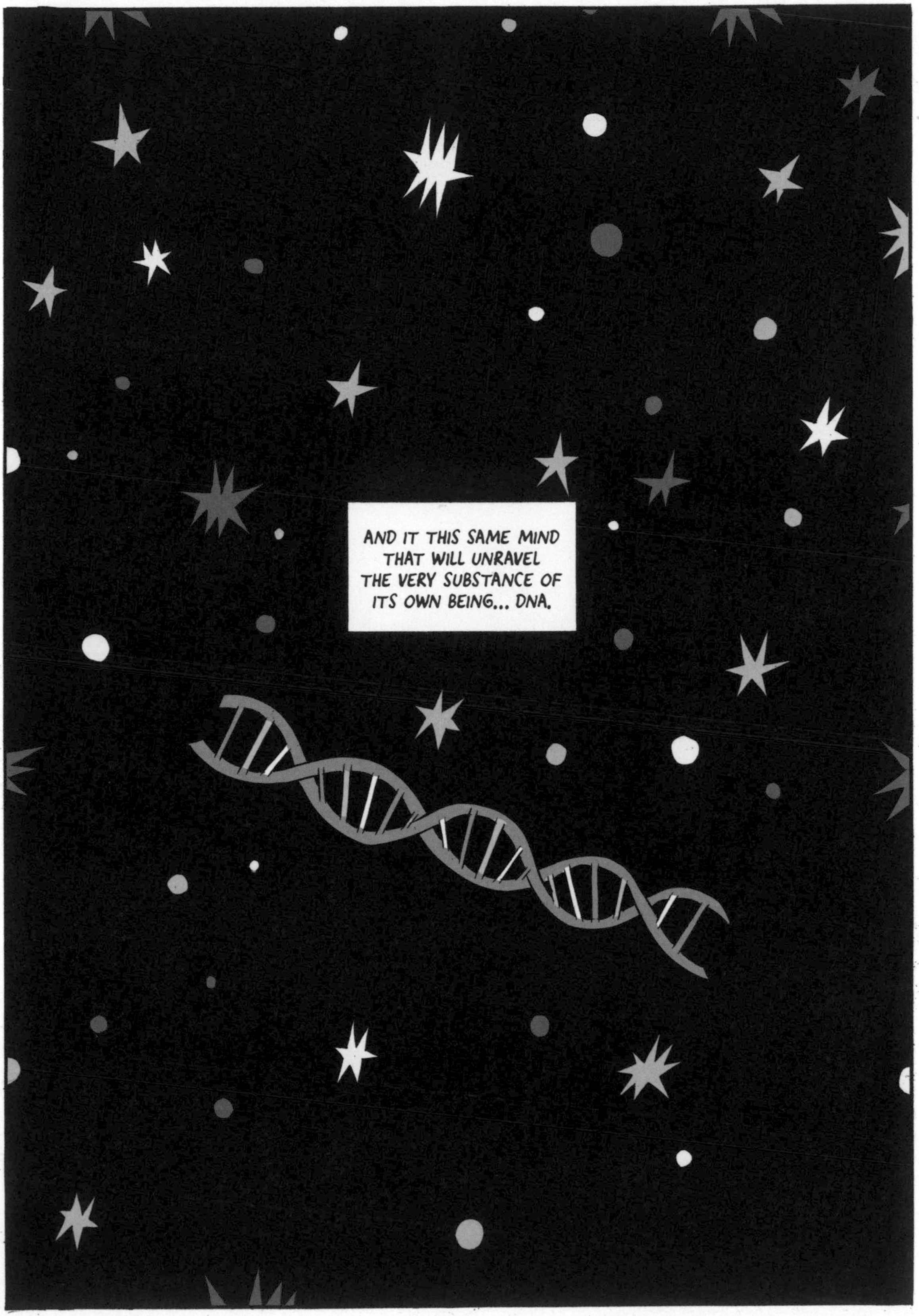
AND IT THIS SAME MIND THAT WILL UNRAVEL THE VERY SUBSTANCE OF ITS OWN BEING... DNA.

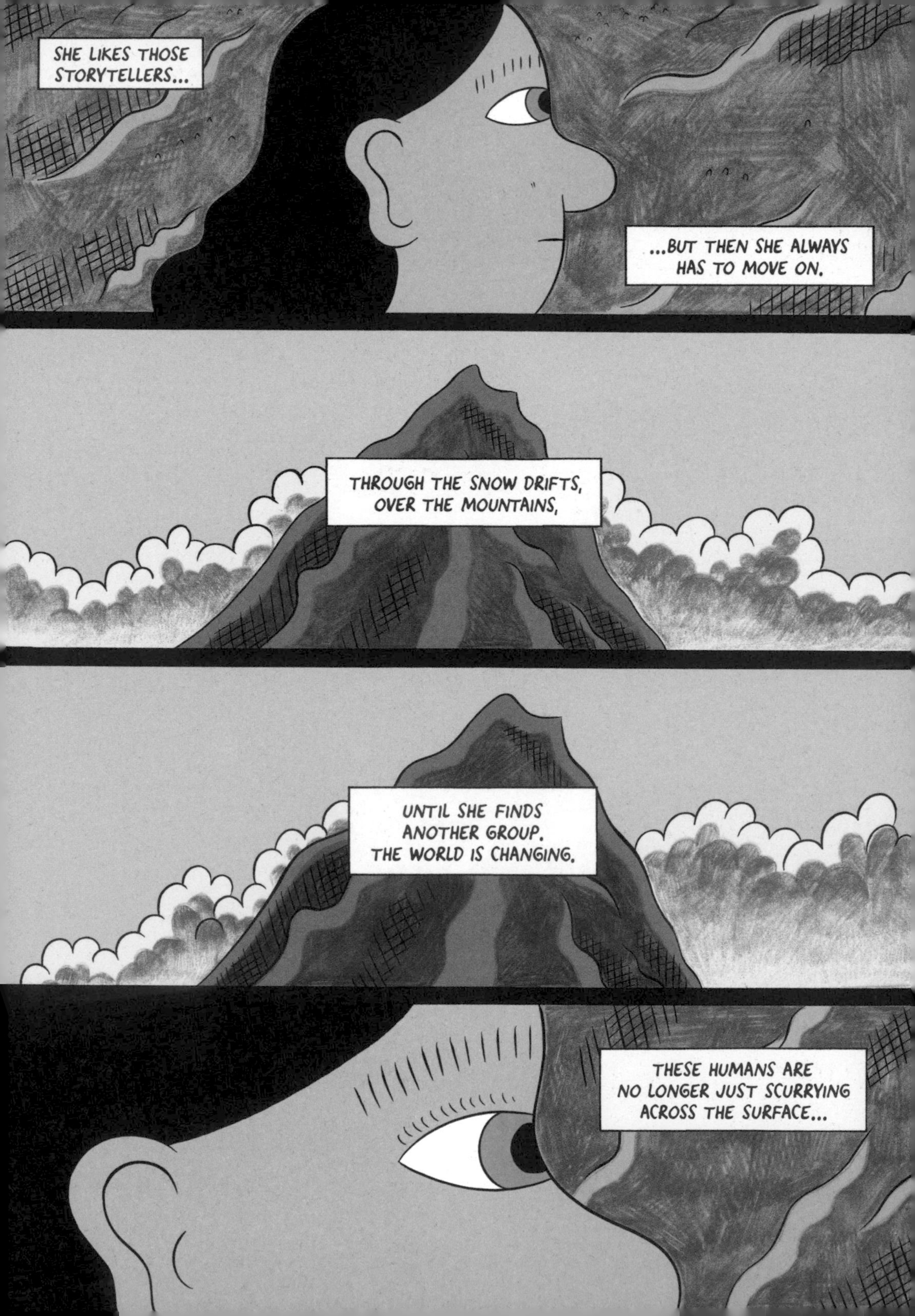
SHE LIKES THOSE STORYTELLERS...
...BUT THEN SHE ALWAYS HAS TO MOVE ON.
THROUGH THE SNOW DRIFTS, OVER THE MOUNTAINS,
UNTIL SHE FINDS ANOTHER GROUP. THE WORLD IS CHANGING.
THESE HUMANS ARE NO LONGER JUST SCURRYING ACROSS THE SURFACE...

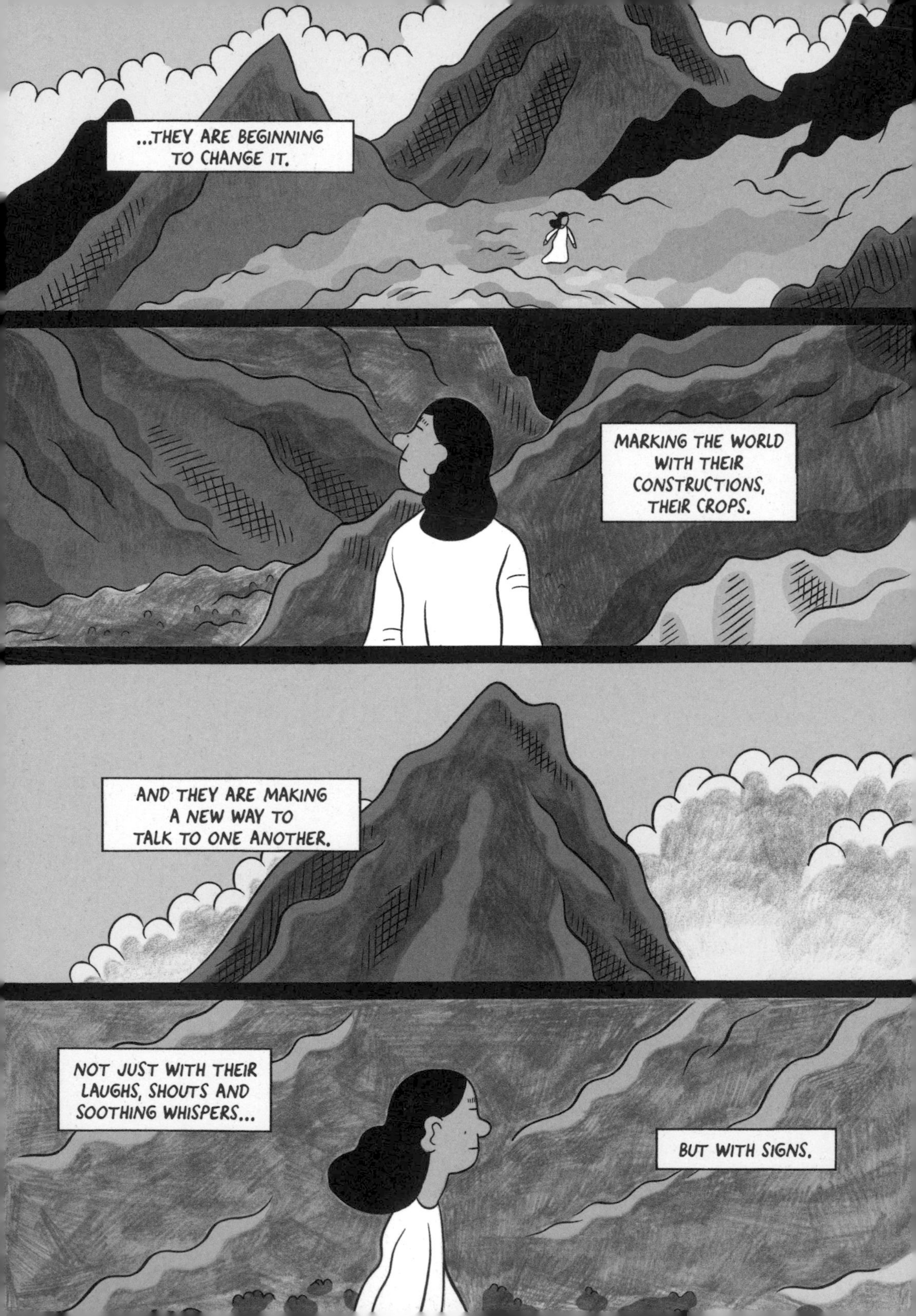
...THEY ARE BEGINNING TO CHANGE IT.
MARKING THE WORLD WITH THEIR CONSTRUCTIONS, THEIR CROPS.
AND THEY ARE MAKING A NEW WAY TO TALK TO ONE ANOTHER.
NOT JUST WITH THEIR LAUGHS, SHOUTS AND SOOTHING WHISPERS...
BUT WITH SIGNS.

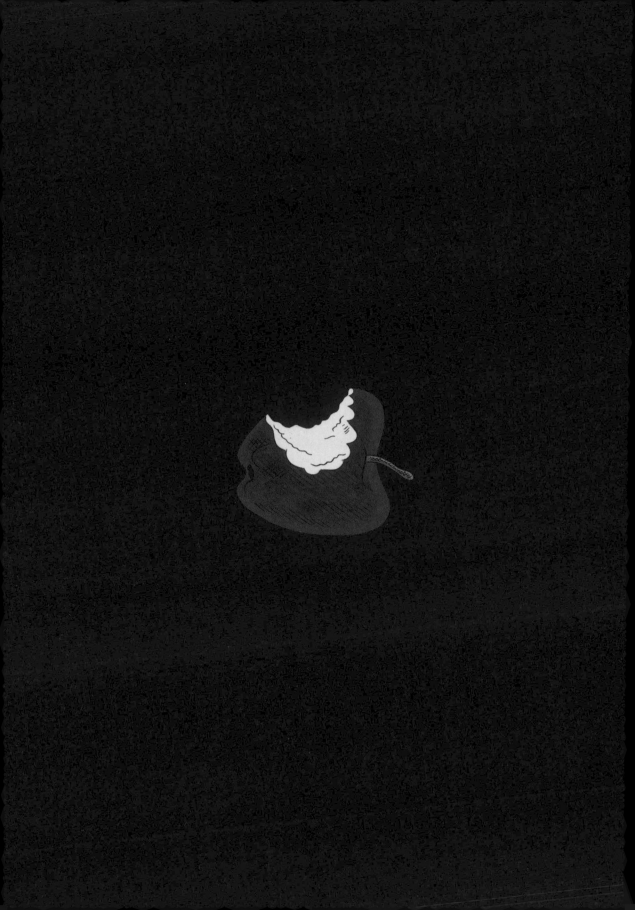

34,000
years later...

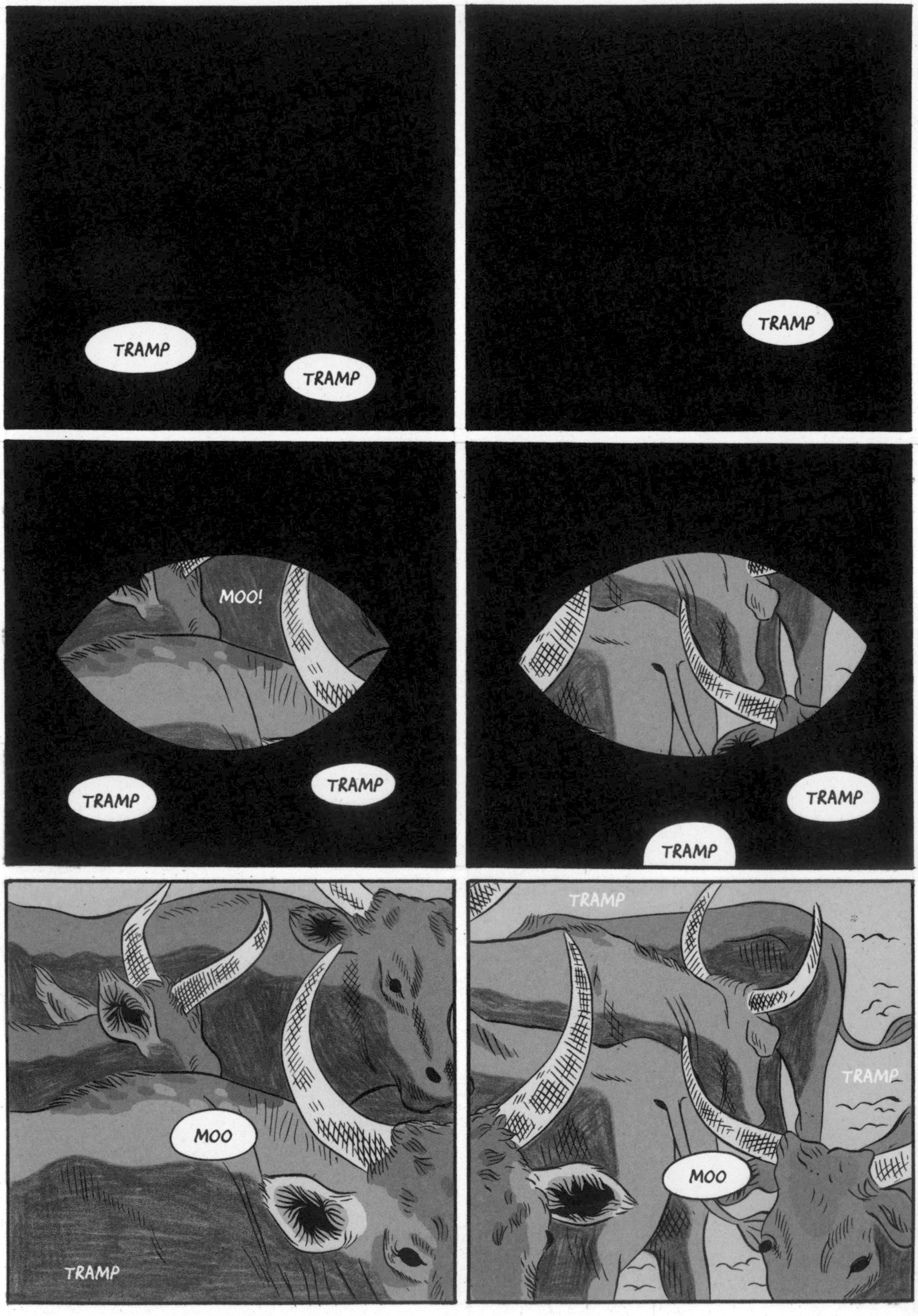

TRAMP
TRAMP
TRAMP
MOO!
TRAMP
TRAMP
TRAMP
TRAMP
MOO
TRAMP
TRAMP
TRAMP
MOO

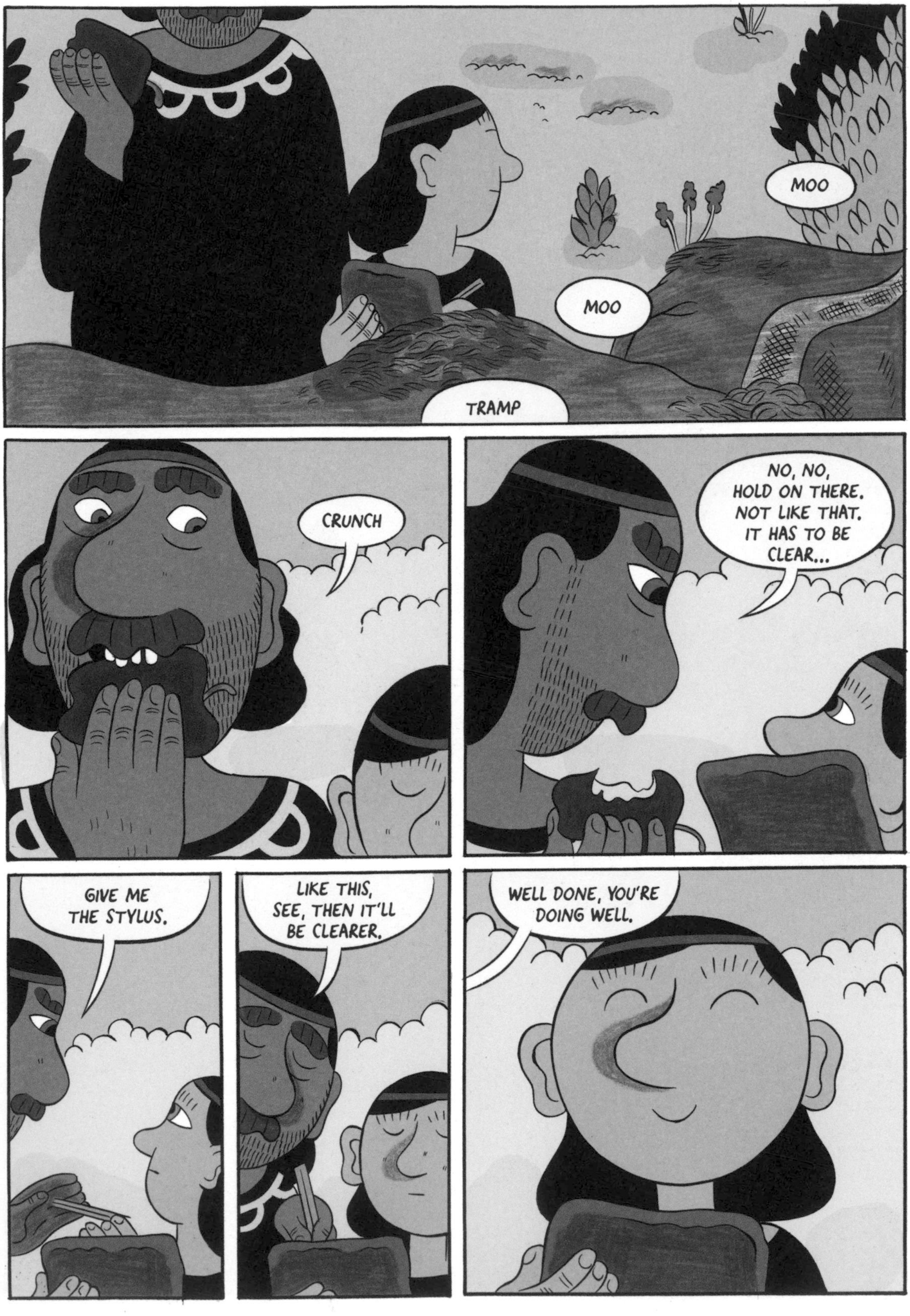
MOO
MOO
TRAMP
CRUNCH
NO, NO, HOLD ON THERE. NOT LIKE THAT. IT HAS TO BE CLEAR...
GIVE ME THE STYLUS.
LIKE THIS, SEE, THEN IT'LL BE CLEARER.
WELL DONE, YOU'RE DOING WELL.

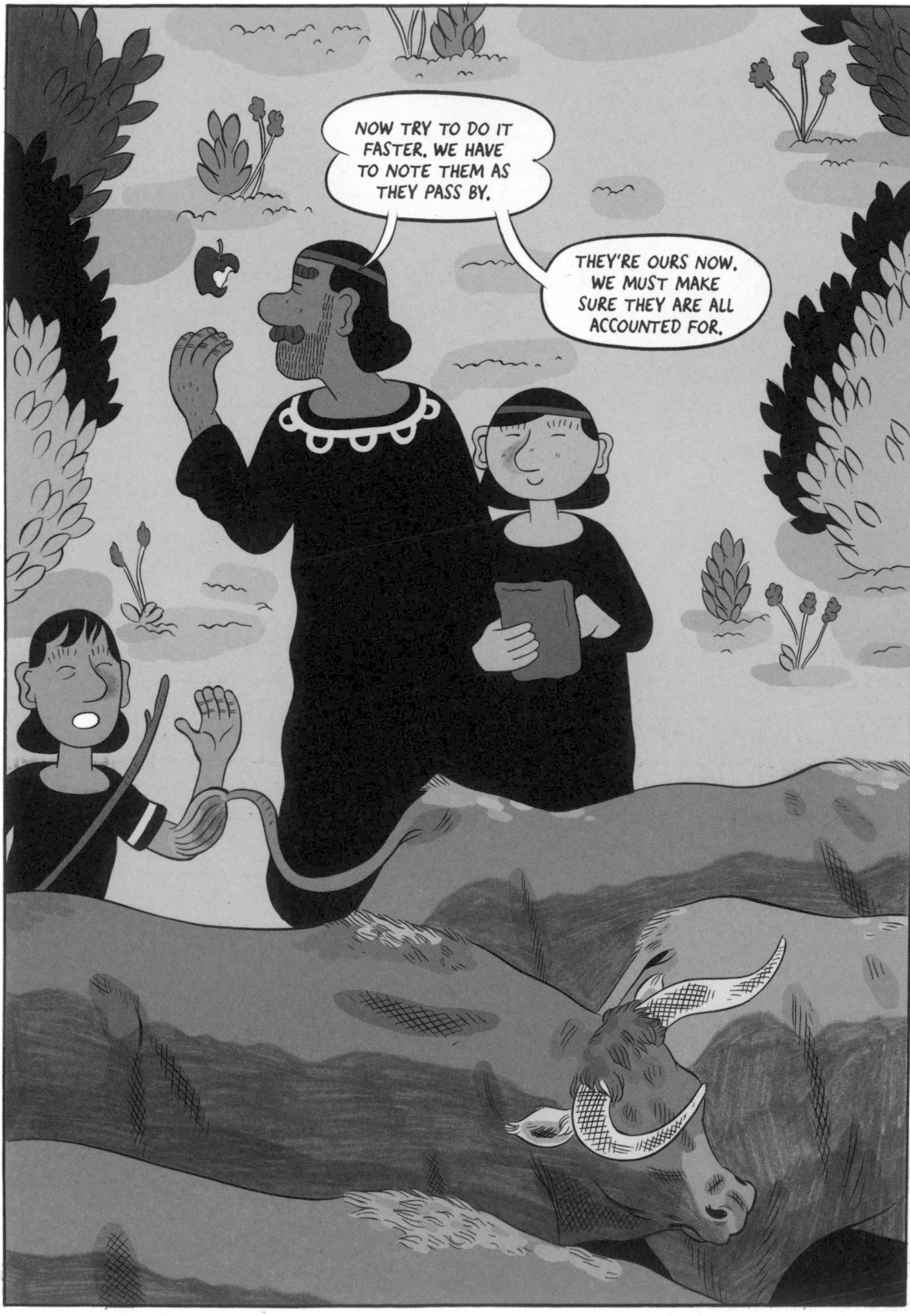
NOW TRY TO DO IT FASTER. WE HAVE TO NOTE THEM AS THEY PASS BY.
THEY'RE OURS NOW. WE MUST MAKE SURE THEY ARE ALL ACCOUNTED FOR.

ALMOST ALL FARMED APPLES ORIGINATE FROM ONE SPECIES...

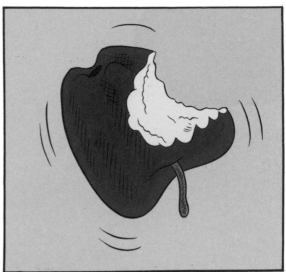

...A WILD APPLE IN THE TIEN SHAN MOUNTAINS, IN WHAT WILL ONE DAY BE KAZAKHSTAN.

MOO

MOOO

MOVE ON!

MOO

FROM THERE IT TRAVELS WEST ALONG THE SILK ROAD.

AT FIRST THE SEEDS ARE MOVED THROUGH BIRD DROPPINGS...

THOSE SEEDS GERMINATE AND GROW INTO FRUIT, WHICH IN TURN ARE TAKEN UP BY LARGER ANIMALS SUCH AS BEARS AND WILD HORSES.

THESE CREATURES SELECT THE LARGEST, SWEETEST FRUITS...

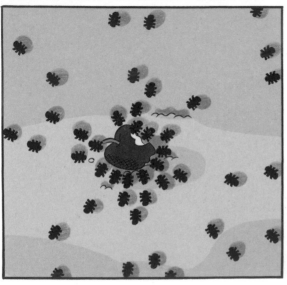

AND THE PLANT OBLIGES THEM BY GROWING LARGER APPLES.

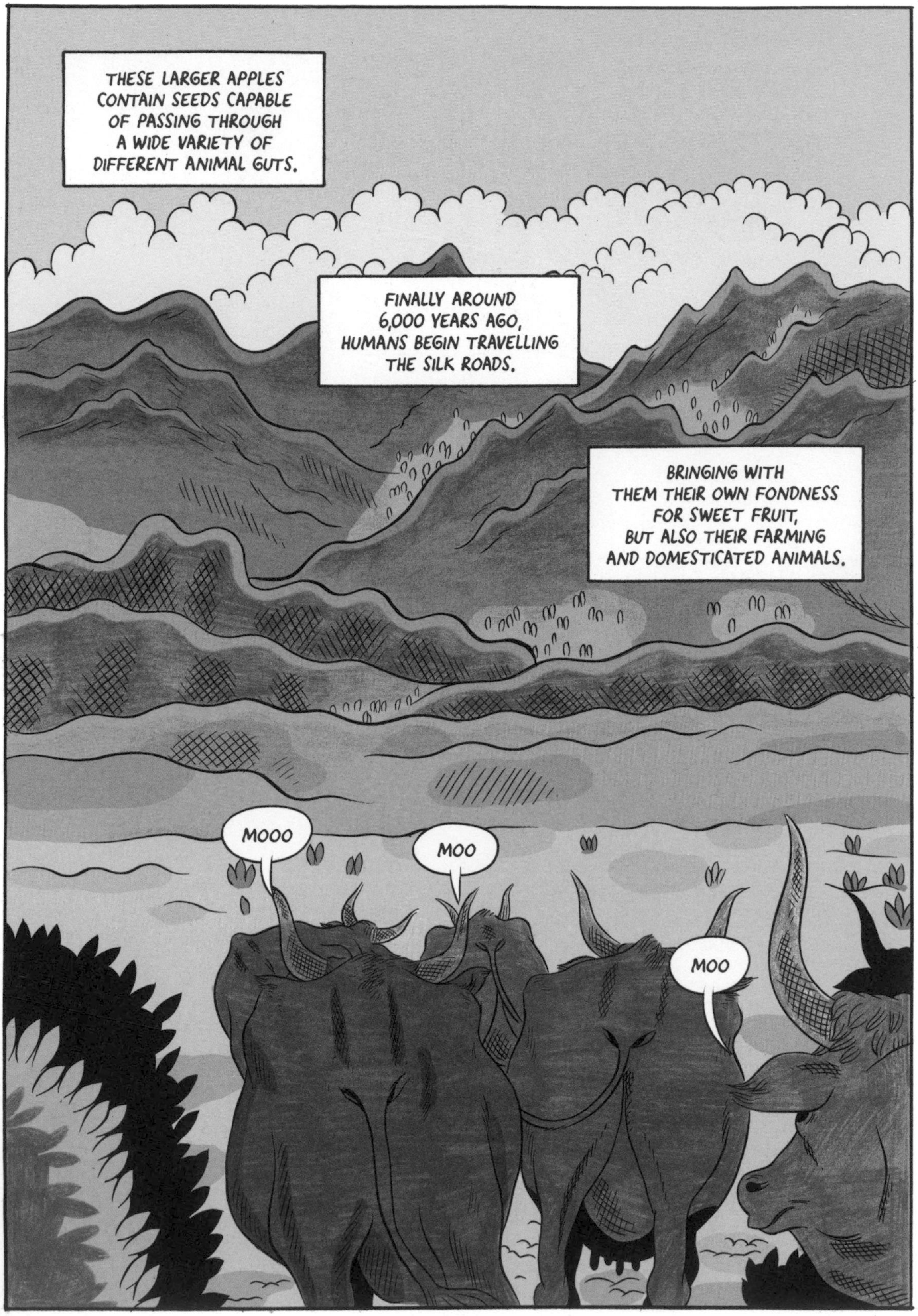

THESE LARGER APPLES CONTAIN SEEDS CAPABLE OF PASSING THROUGH A WIDE VARIETY OF DIFFERENT ANIMAL GUTS.
FINALLY AROUND 6,000 YEARS AGO, HUMANS BEGIN TRAVELLING THE SILK ROADS.
BRINGING WITH THEM THEIR OWN FONDNESS FOR SWEET FRUIT, BUT ALSO THEIR FARMING AND DOMESTICATED ANIMALS.
MOOO
MOO
MOO

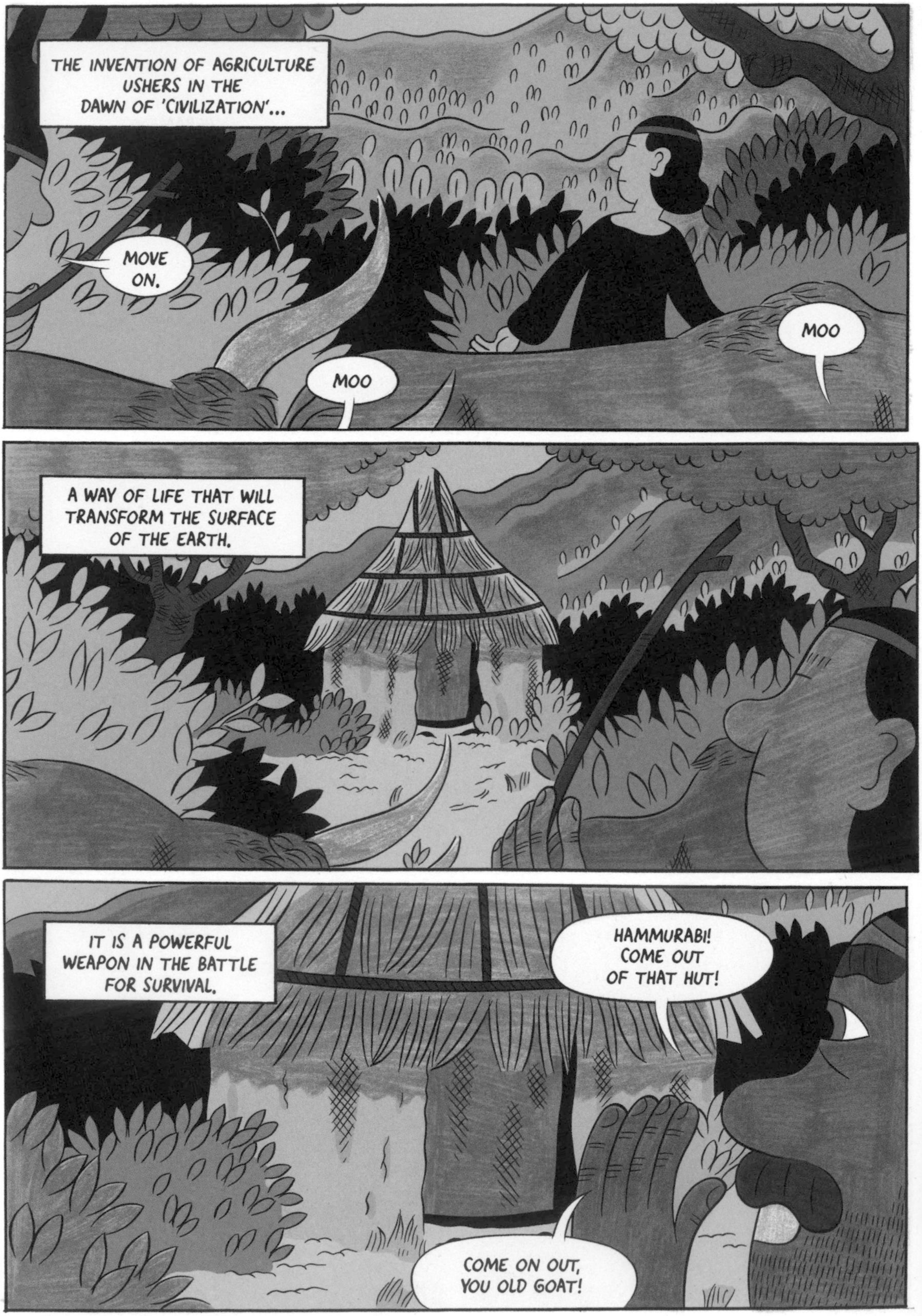
THE INVENTION OF AGRICULTURE USHERS IN THE DAWN OF 'CIVILIZATION'...
MOVE ON.
MOO
MOO
A WAY OF LIFE THAT WILL TRANSFORM THE SURFACE OF THE EARTH.
IT IS A POWERFUL WEAPON IN THE BATTLE FOR SURVIVAL.
HAMMURABI! COME OUT OF THAT HUT!
COME ON OUT, YOU OLD GOAT!

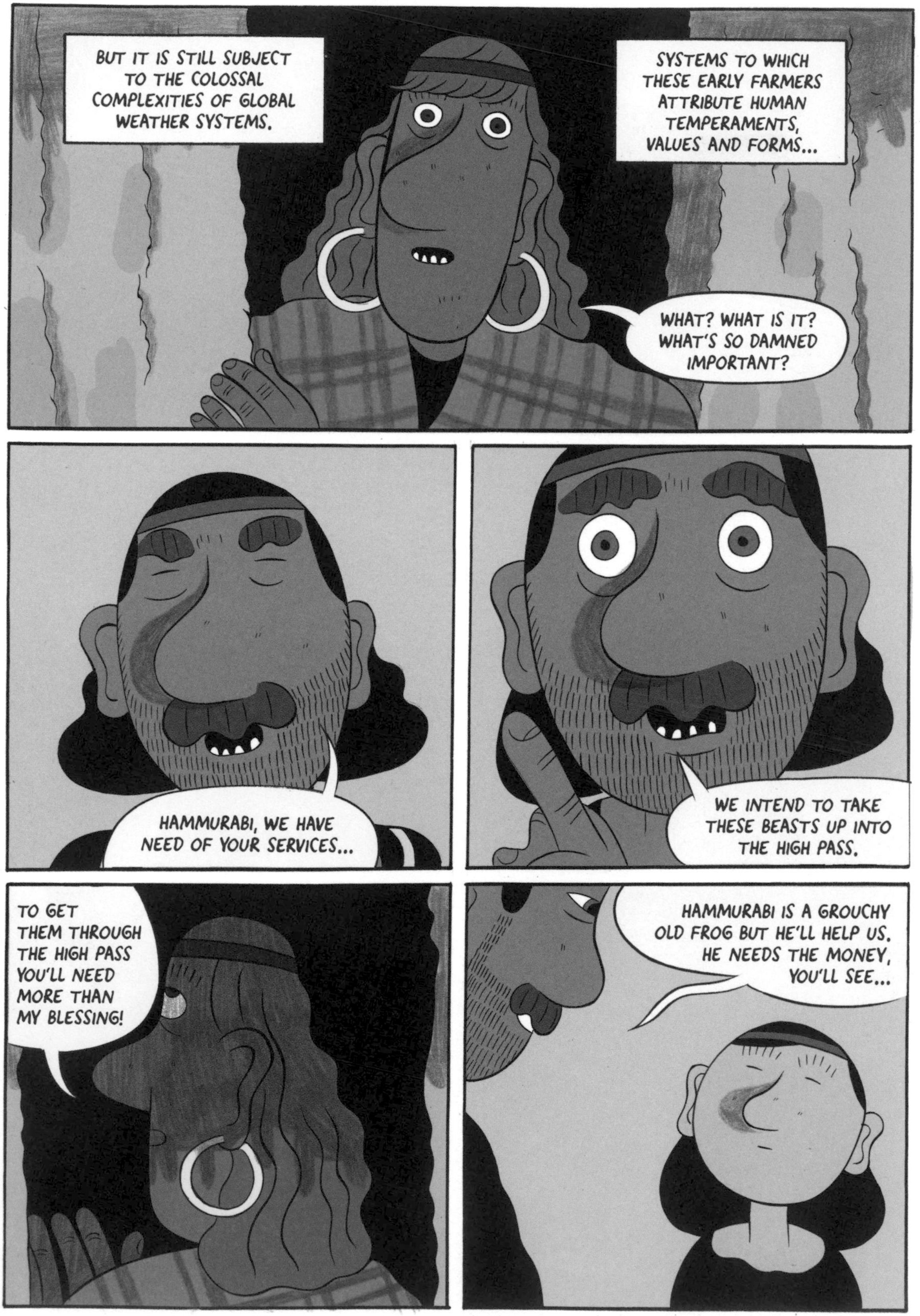
BUT IT IS STILL SUBJECT TO THE COLOSSAL COMPLEXITIES OF GLOBAL WEATHER SYSTEMS.
SYSTEMS TO WHICH THESE EARLY FARMERS ATTRIBUTE HUMAN TEMPERAMENTS, VALUES AND FORMS...
WHAT? WHAT IS IT? WHAT'S SO DAMNED IMPORTANT?
HAMMURABI, WE HAVE NEED OF YOUR SERVICES...
WE INTEND TO TAKE THESE BEASTS UP INTO THE HIGH PASS.
TO GET THEM THROUGH THE HIGH PASS YOU'LL NEED MORE THAN MY BLESSING!
HAMMURABI IS A GROUCHY OLD FROG BUT HE'LL HELP US. HE NEEDS THE MONEY, YOU'LL SEE...

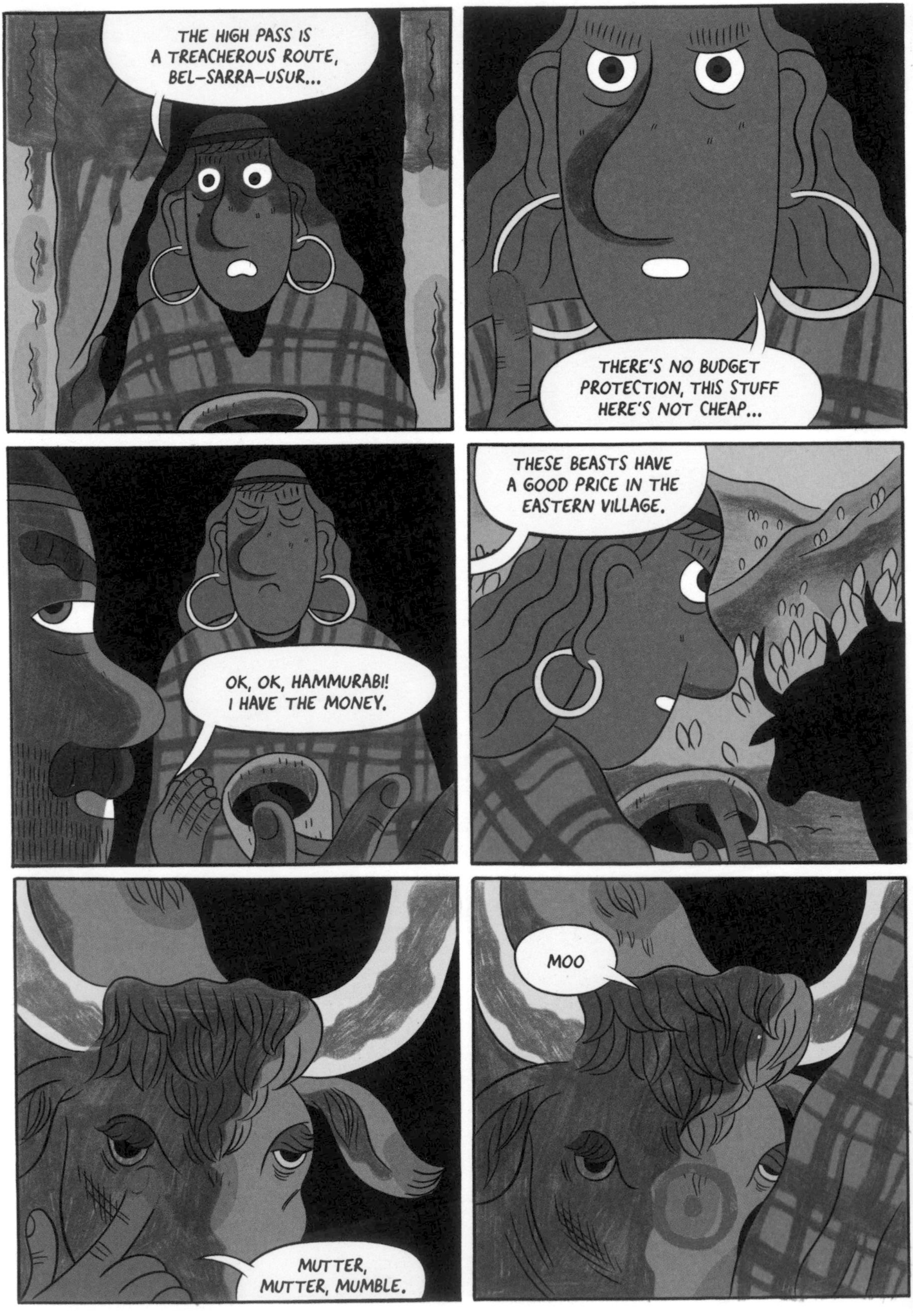
THE HIGH PASS IS A TREACHEROUS ROUTE, BEL-SARRA-USUR...
THERE'S NO BUDGET PROTECTION, THIS STUFF HERE'S NOT CHEAP...
OK, OK, HAMMURABI! I HAVE THE MONEY.
THESE BEASTS HAVE A GOOD PRICE IN THE EASTERN VILLAGE.
MUTTER, MUTTER, MUMBLE.
MOO

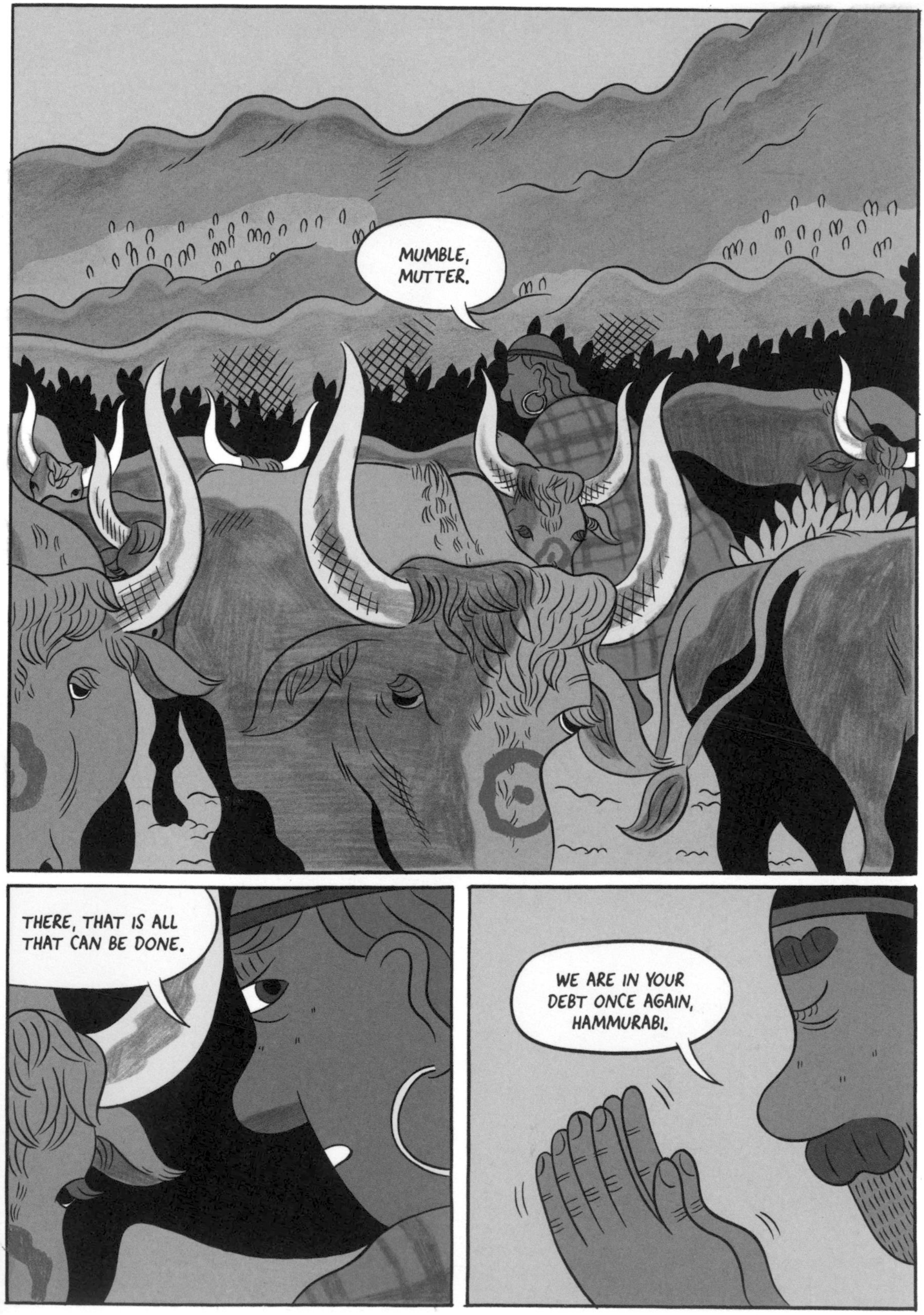
MUMBLE, MUTTER.
THERE, THAT IS ALL THAT CAN BE DONE.
WE ARE IN YOUR DEBT ONCE AGAIN, HAMMURABI.

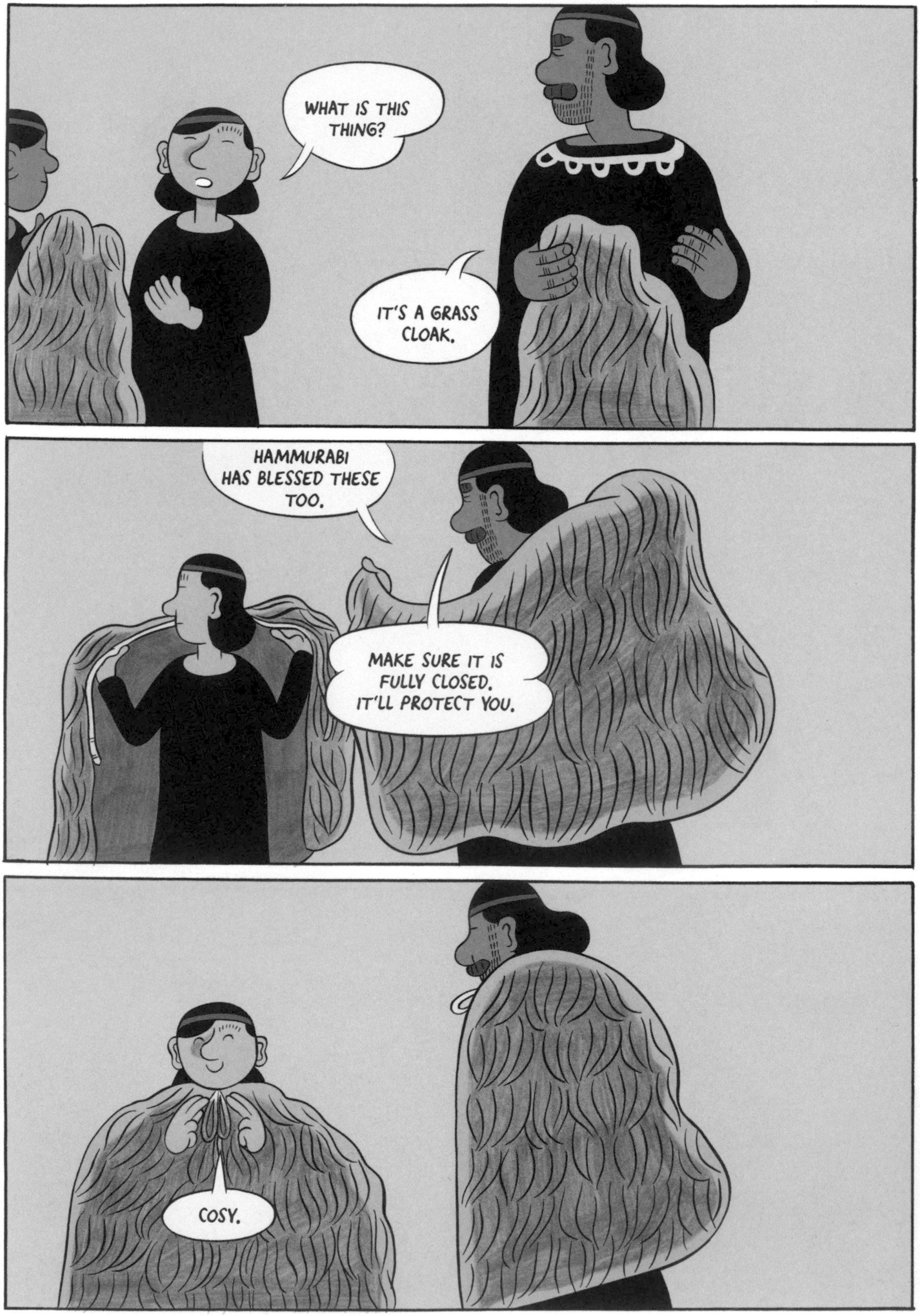
WHAT IS THIS THING?
IT'S A GRASS CLOAK.
HAMMURABI HAS BLESSED THESE TOO.
MAKE SURE IT IS FULLY CLOSED. IT'LL PROTECT YOU.
COSY.

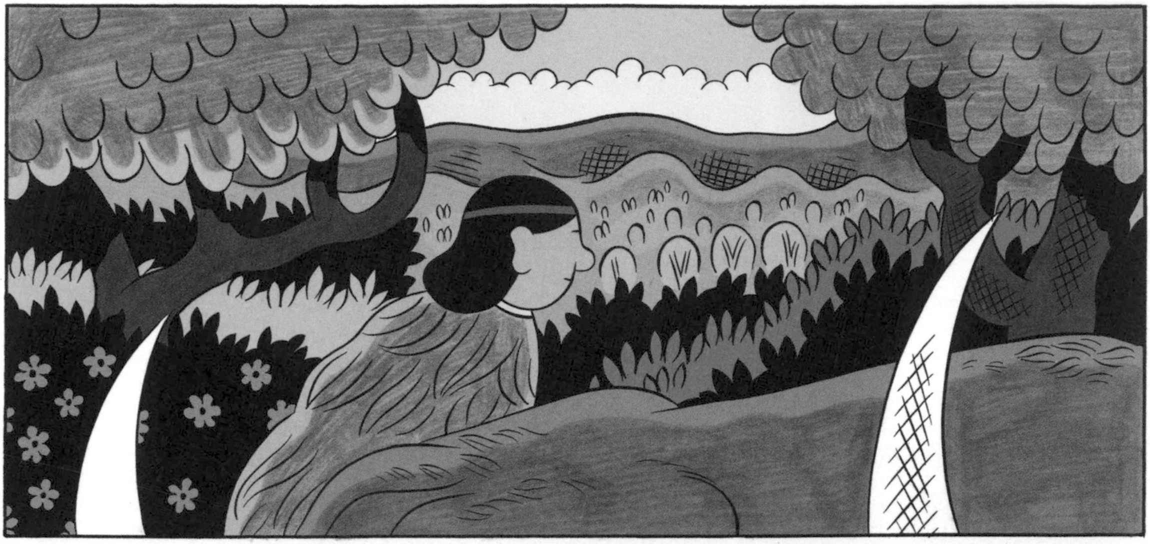

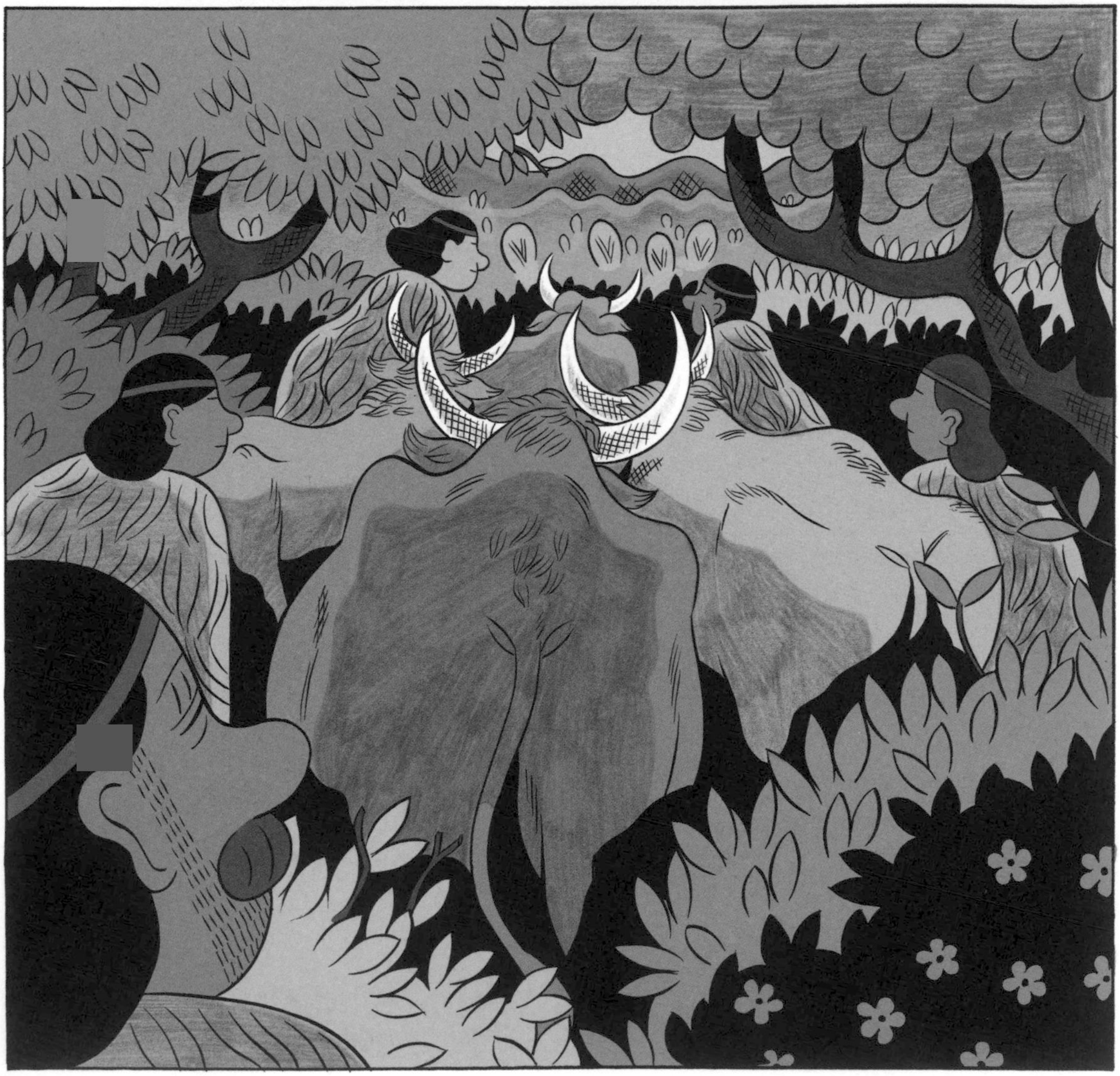

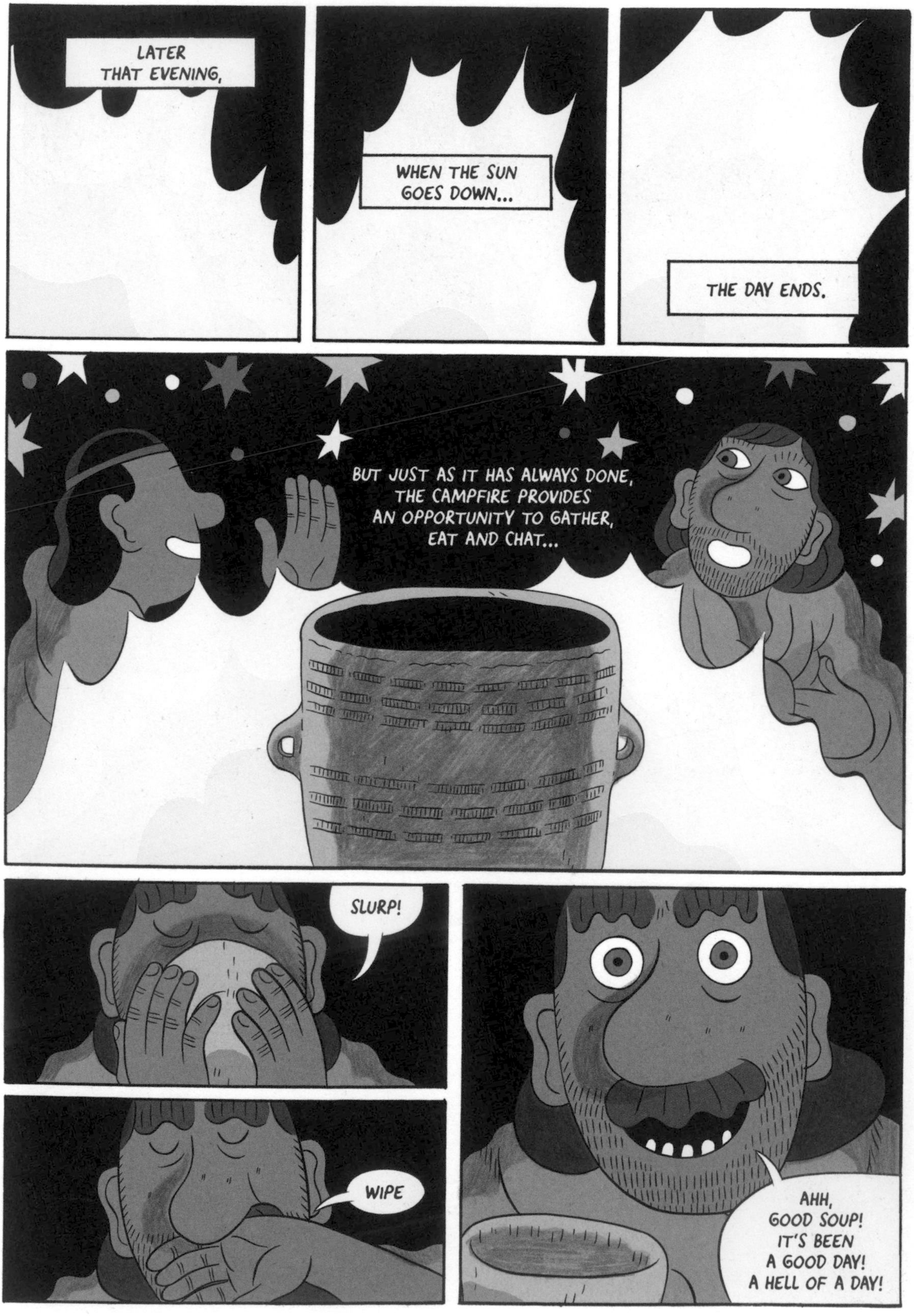
LATER THAT EVENING,
WHEN THE SUN GOES DOWN...
THE DAY ENDS.
BUT JUST AS IT HAS ALWAYS DONE, THE CAMPFIRE PROVIDES AN OPPORTUNITY TO GATHER, EAT AND CHAT...
SLURP!
WIPE
AHH, GOOD SOUP! IT'S BEEN A GOOD DAY! A HELL OF A DAY!

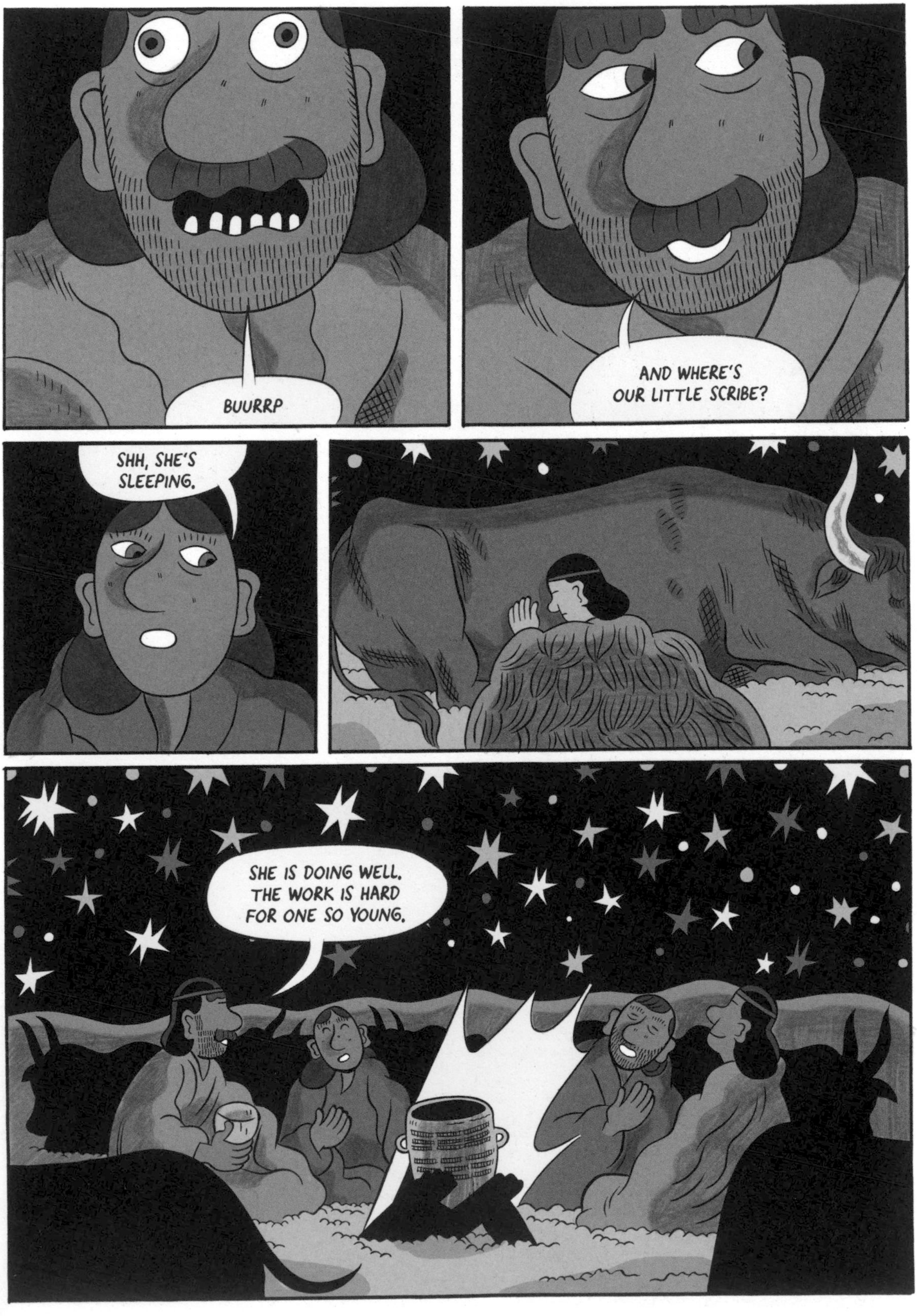
BUURRP
AND WHERE'S OUR LITTLE SCRIBE?
SHH, SHE'S SLEEPING.
SHE IS DOING WELL. THE WORK IS HARD FOR ONE SO YOUNG.

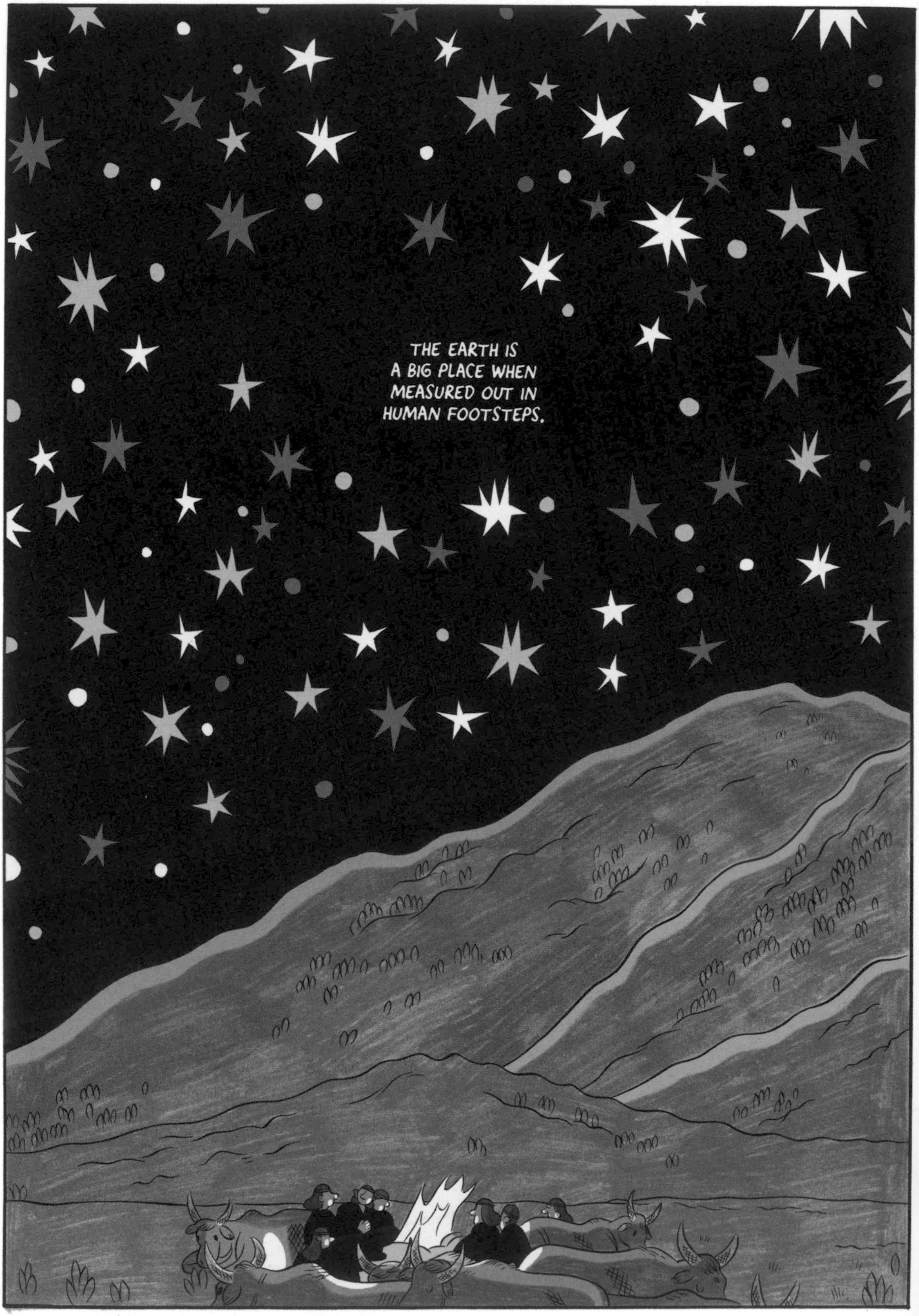
THE EARTH IS A BIG PLACE WHEN MEASURED OUT IN HUMAN FOOTSTEPS.

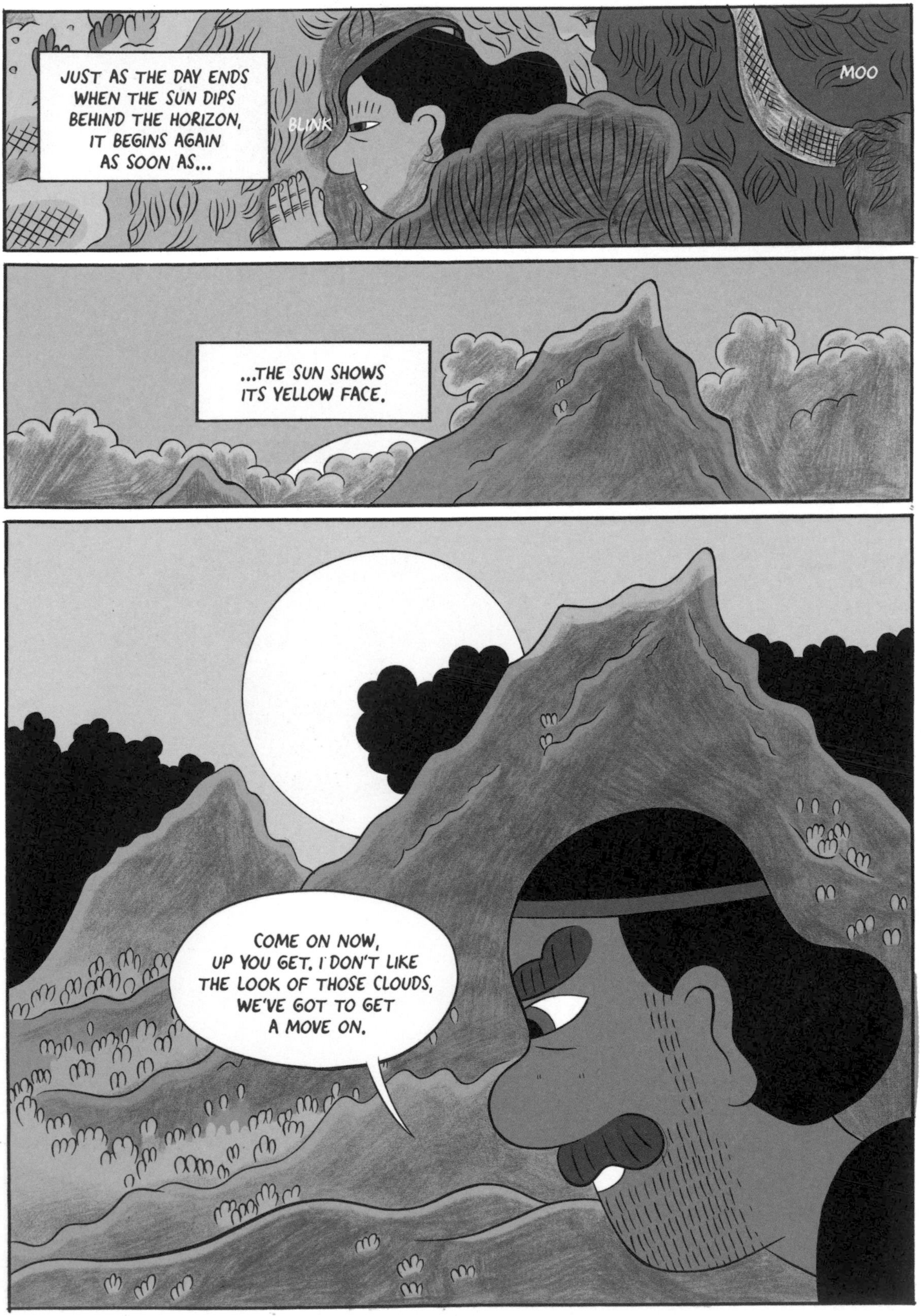
JUST AS THE DAY ENDS WHEN THE SUN DIPS BEHIND THE HORIZON, IT BEGINS AGAIN AS SOON AS...
BLINK
MOO
...THE SUN SHOWS ITS YELLOW FACE.
COME ON NOW, UP YOU GET. I DON'T LIKE THE LOOK OF THOSE CLOUDS, WE'VE GOT TO GET A MOVE ON.

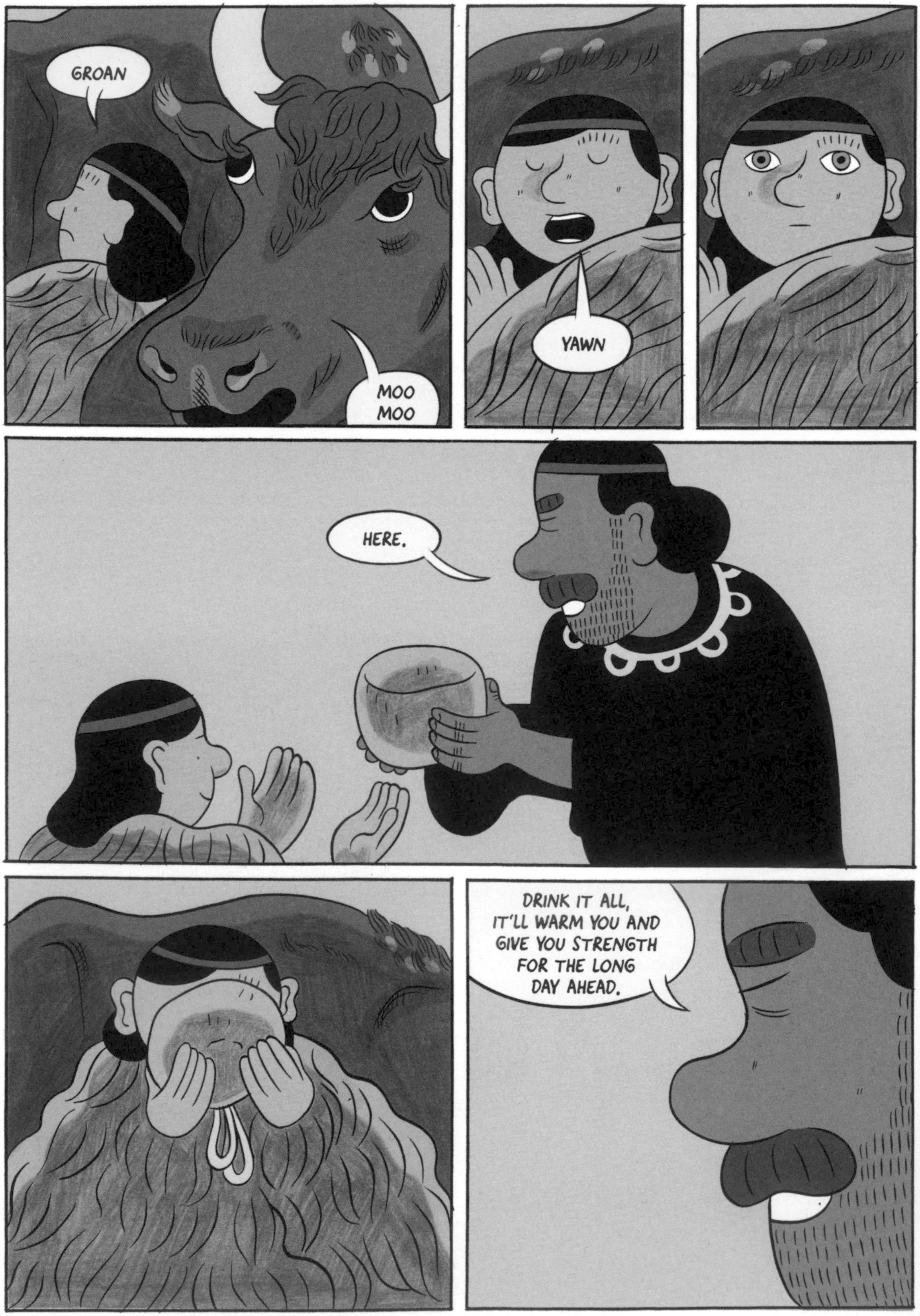
GROAN
MOO MOO
YAWN
HERE.
DRINK IT ALL, IT'LL WARM YOU AND GIVE YOU STRENGTH FOR THE LONG DAY AHEAD.

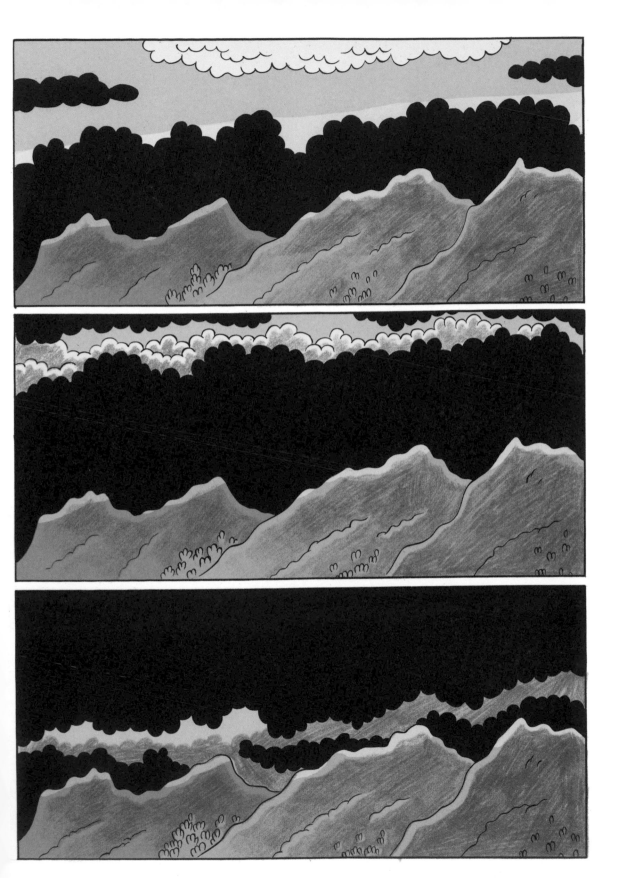

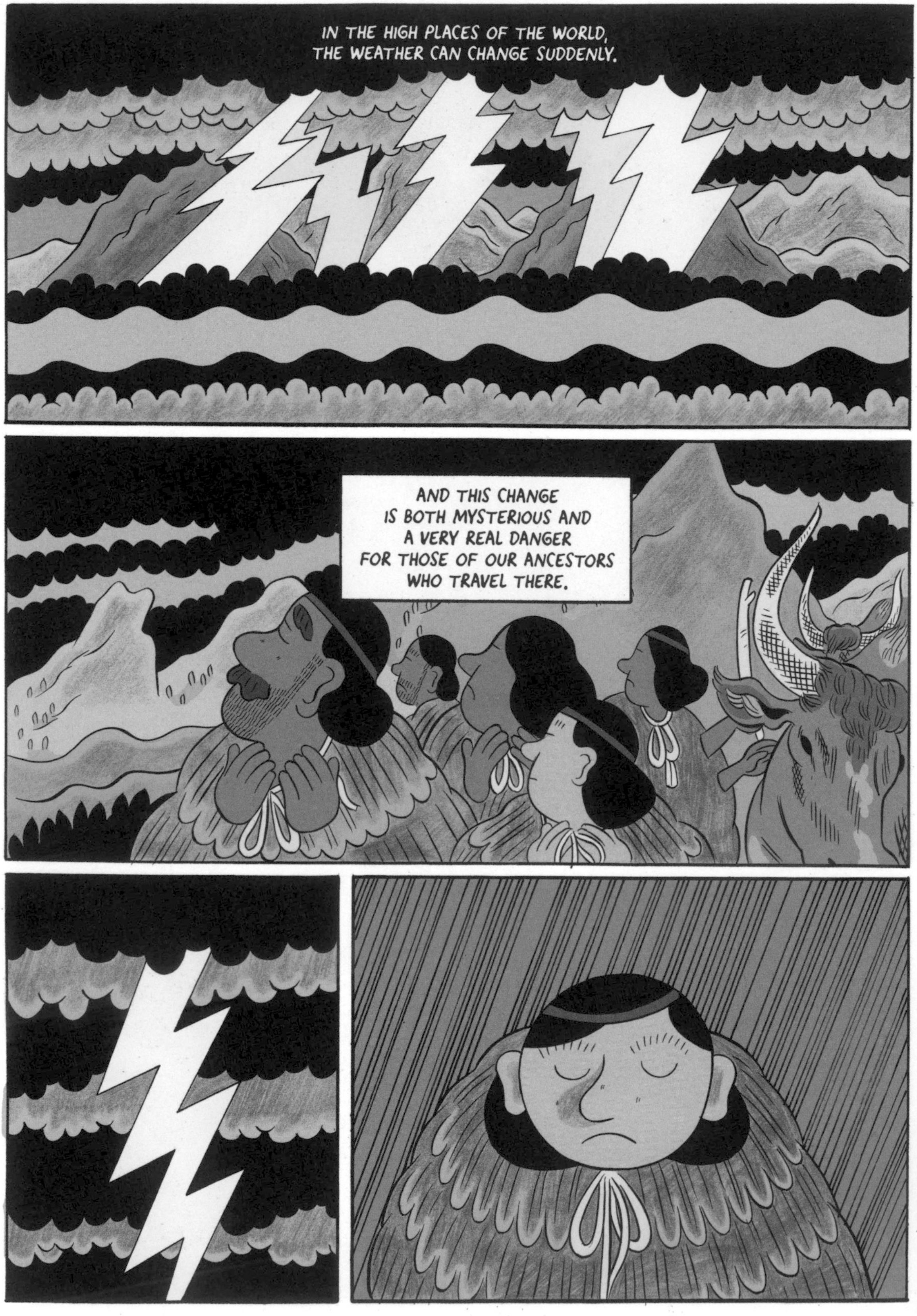
IN THE HIGH PLACES OF THE WORLD, THE WEATHER CAN CHANGE SUDDENLY.
AND THIS CHANGE IS BOTH MYSTERIOUS AND A VERY REAL DANGER FOR THOSE OF OUR ANCESTORS WHO TRAVEL THERE.

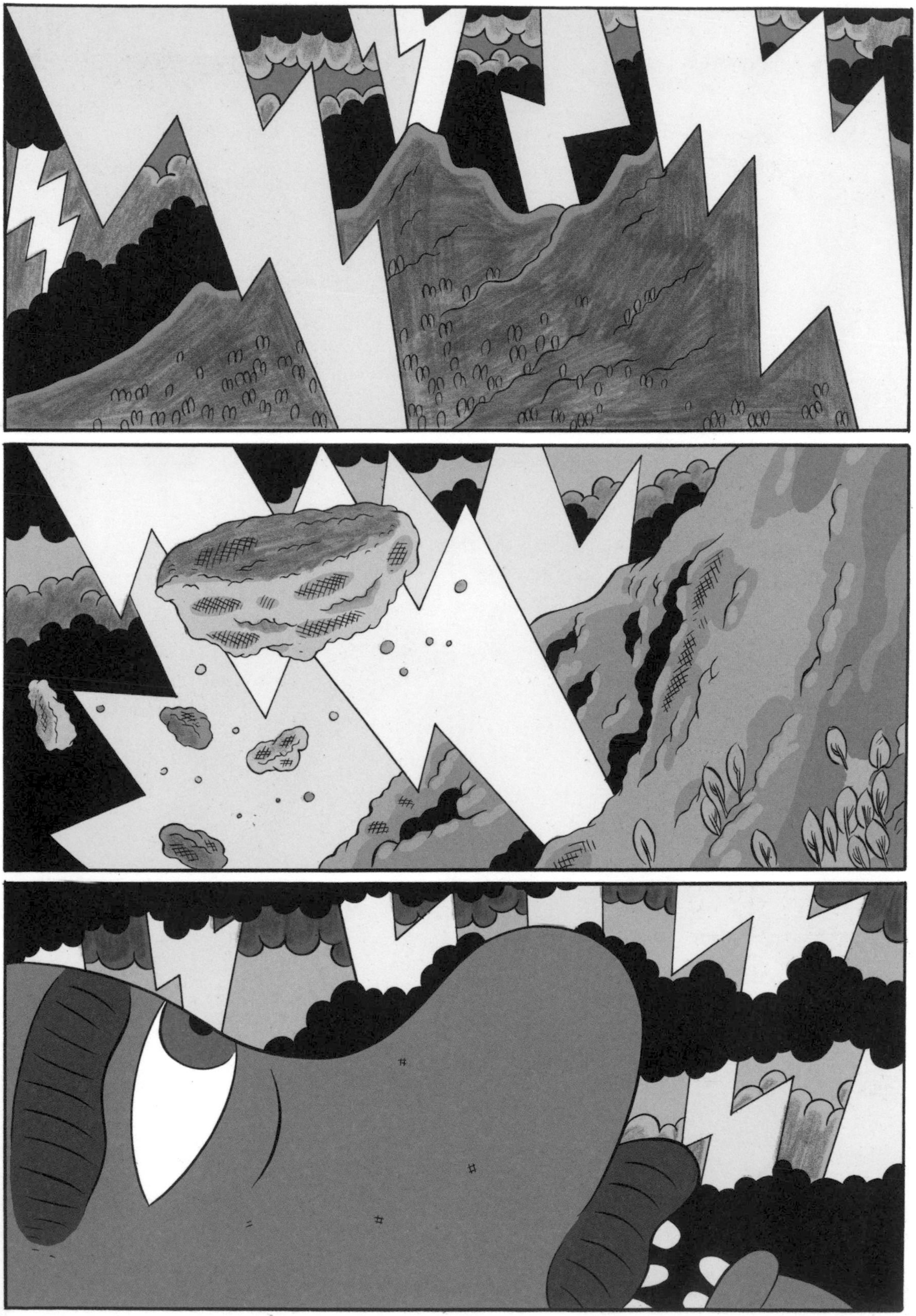

MOO!

MOO!

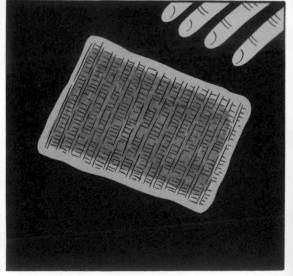

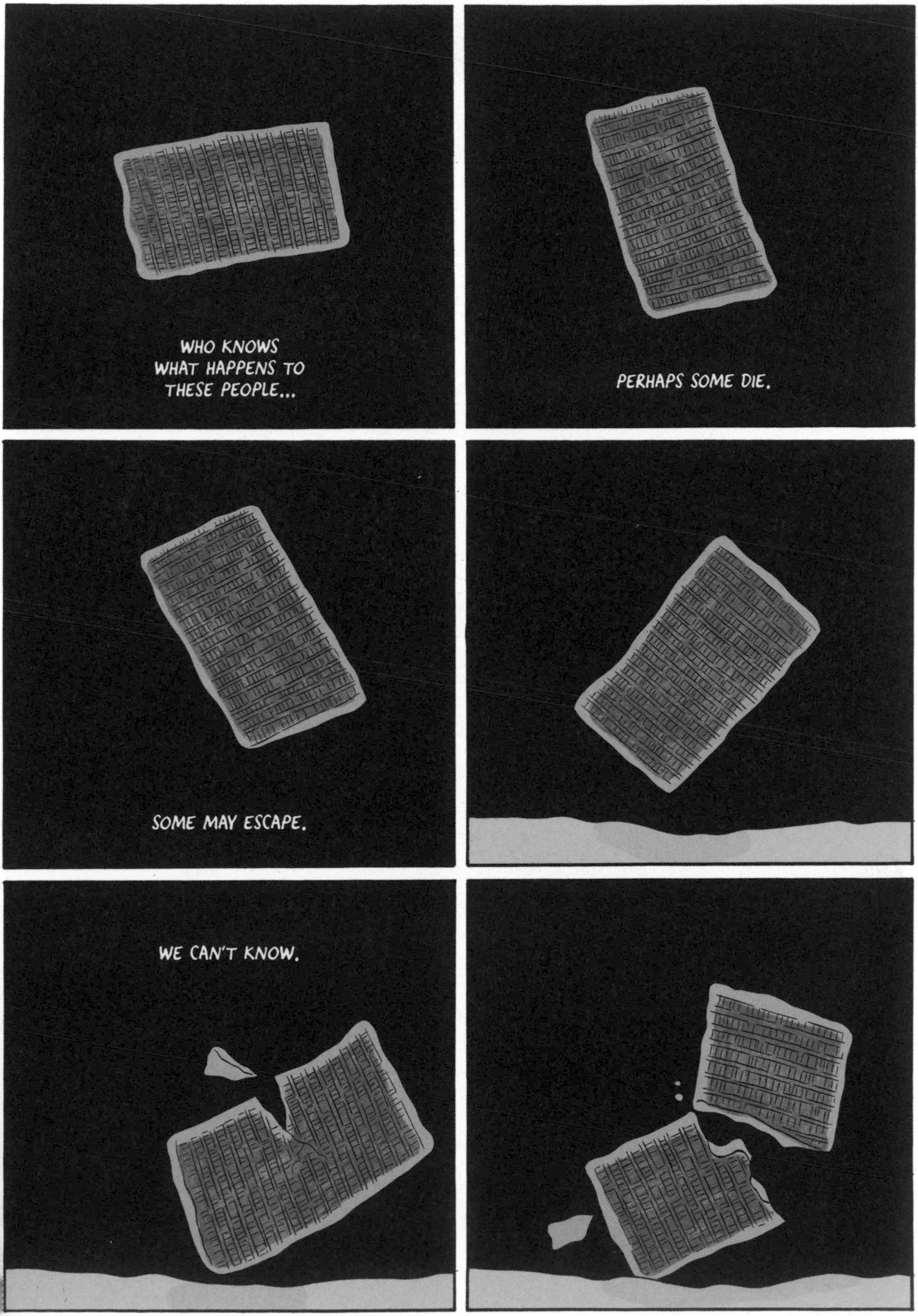
WHO KNOWS WHAT HAPPENS TO THESE PEOPLE...
PERHAPS SOME DIE.
SOME MAY ESCAPE.
WE CAN'T KNOW.

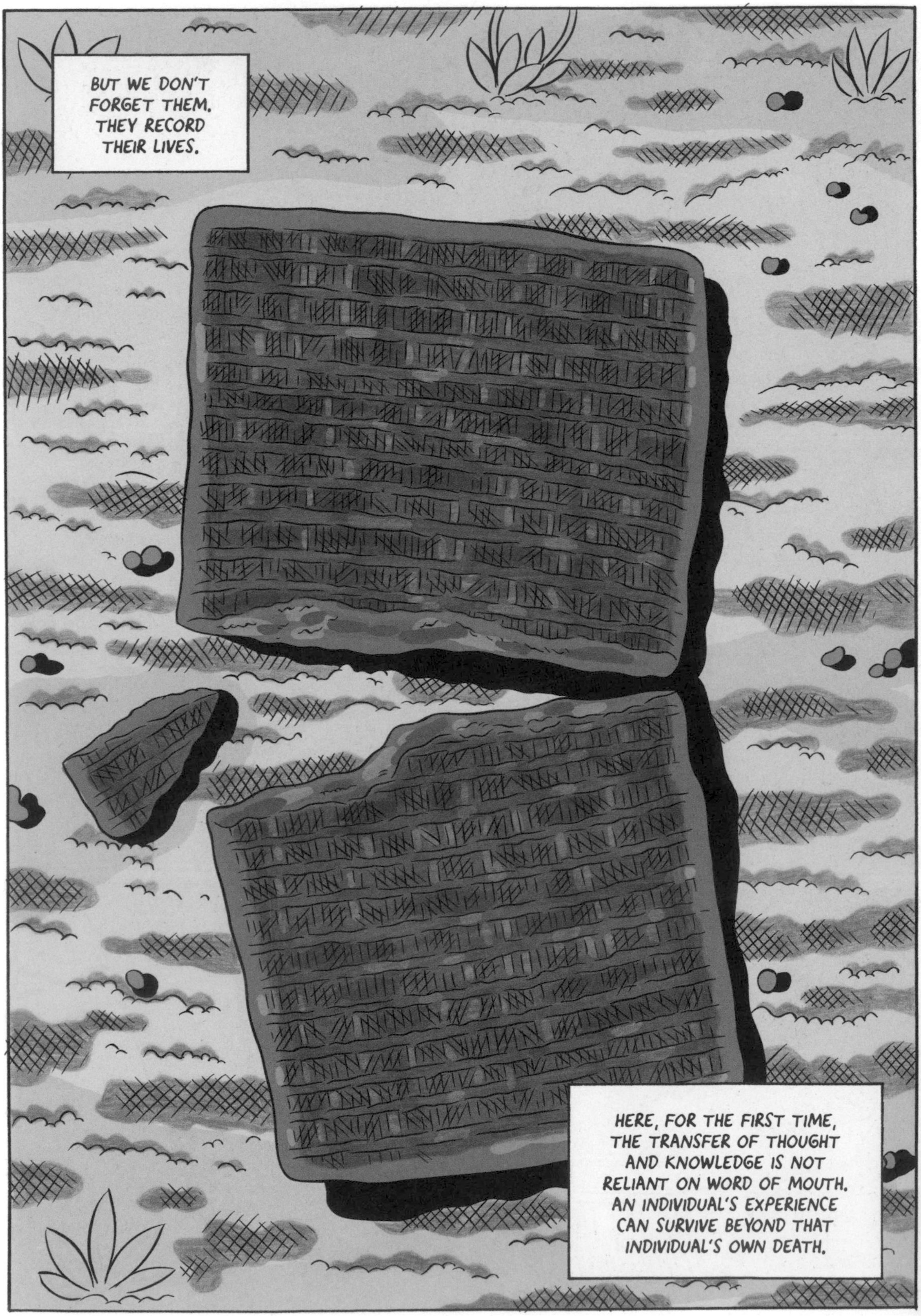
BUT WE DON'T FORGET THEM. THEY RECORD THEIR LIVES.
HERE, FOR THE FIRST TIME, THE TRANSFER OF THOUGHT AND KNOWLEDGE IS NOT RELIANT ON WORD OF MOUTH. AN INDIVIDUAL'S EXPERIENCE CAN SURVIVE BEYOND THAT INDIVIDUAL'S OWN DEATH.

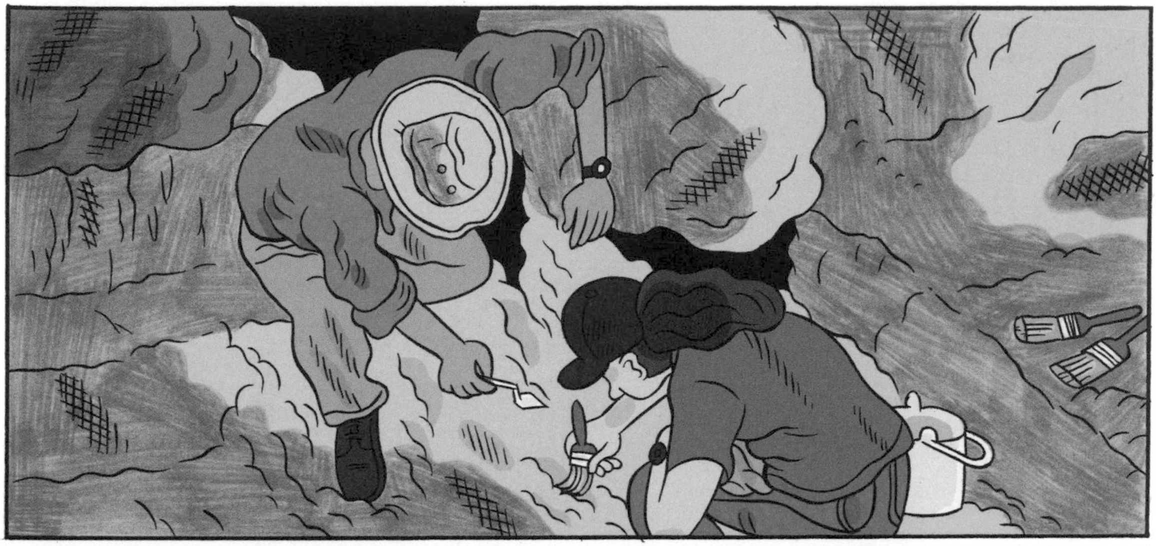

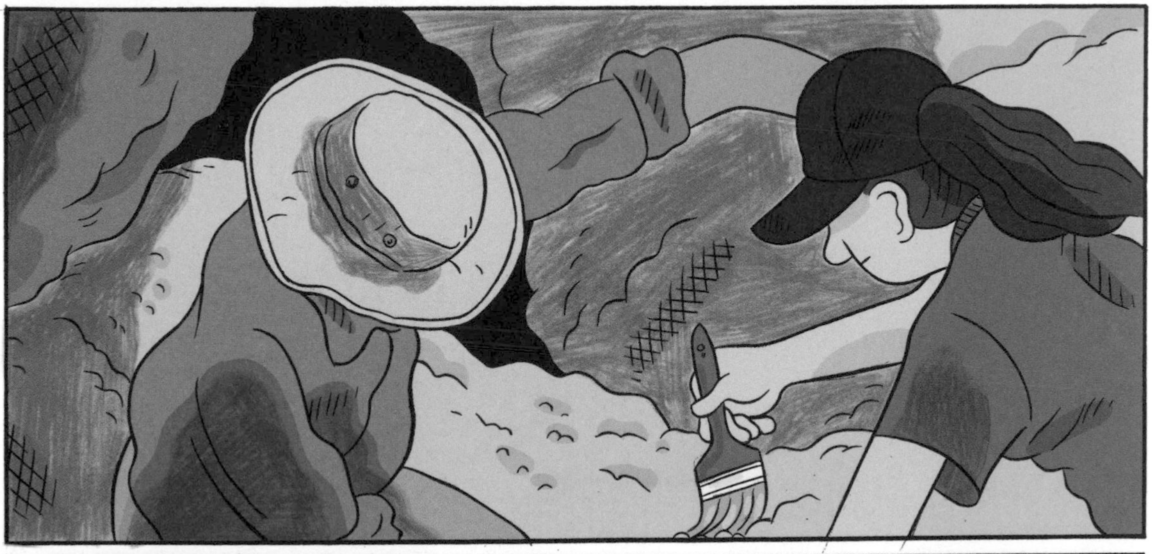

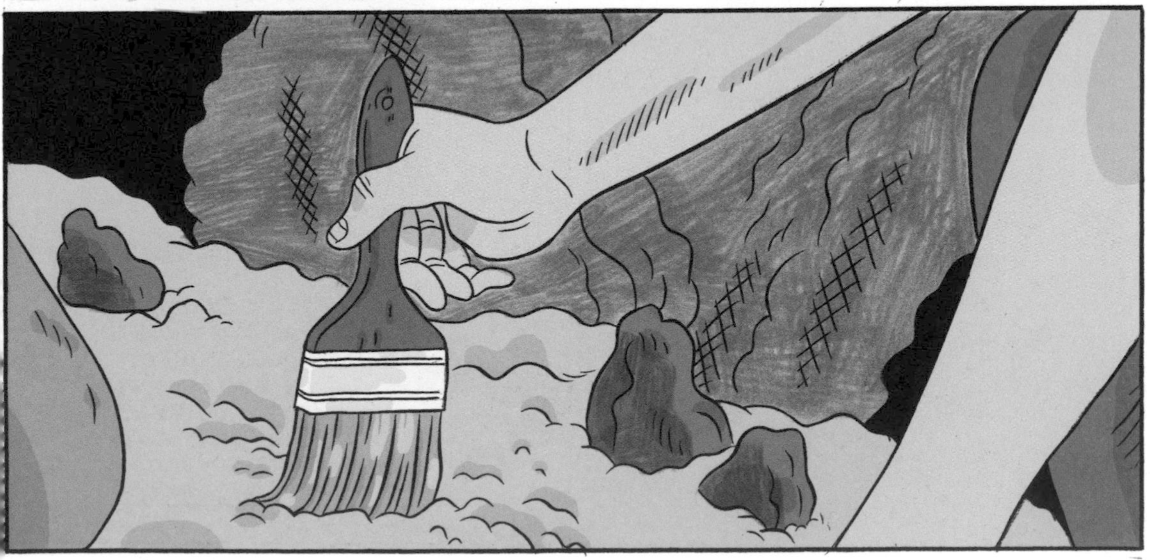

BRUSH

CHIP

SCRAPE

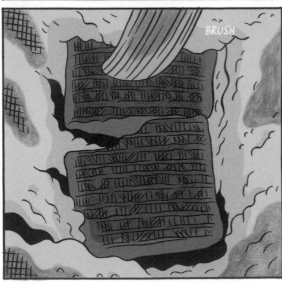

BRUSH

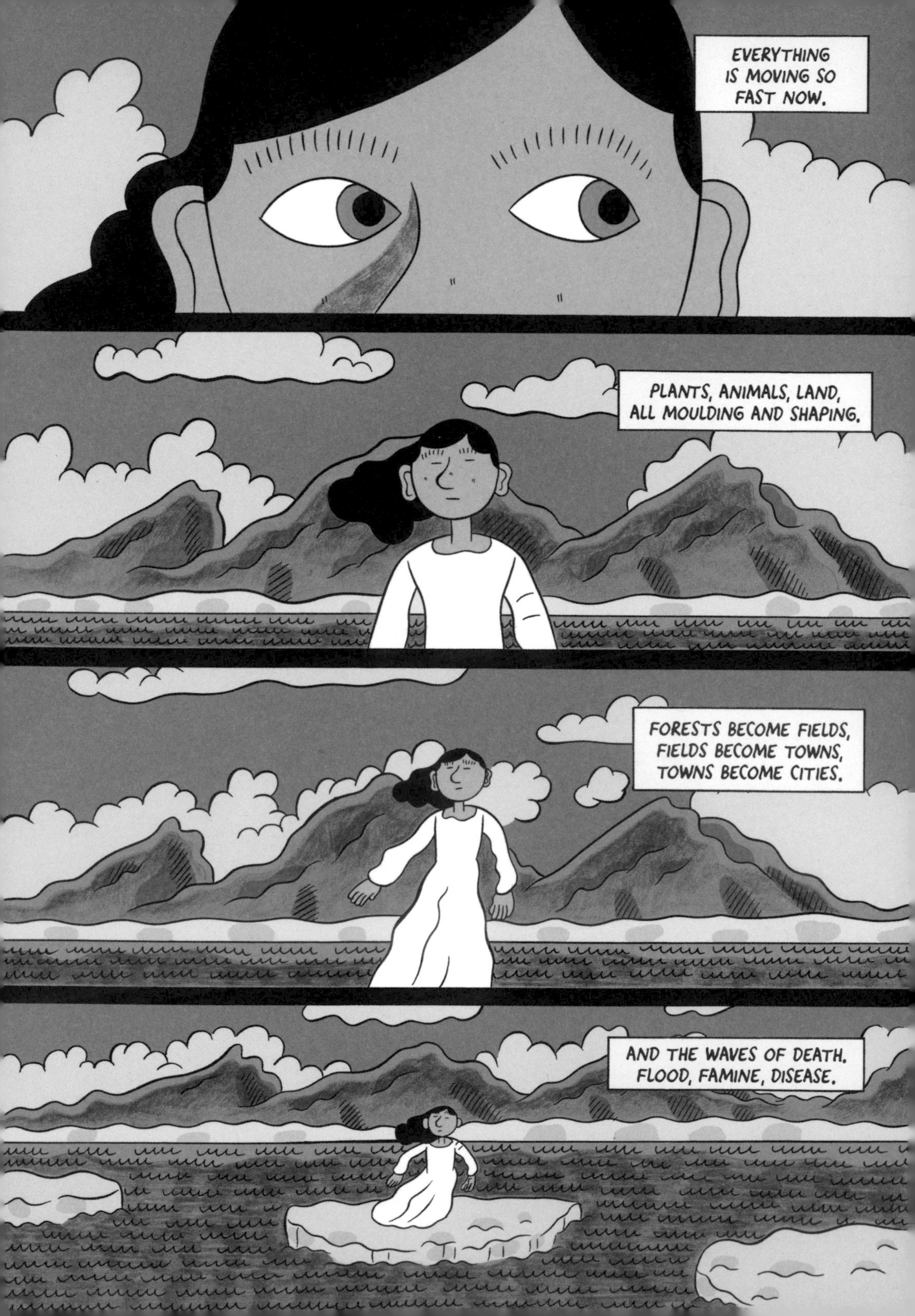
EVERYTHING IS MOVING SO FAST NOW.
PLANTS, ANIMALS, LAND, ALL MOULDING AND SHAPING.
FORESTS BECOME FIELDS, FIELDS BECOME TOWNS, TOWNS BECOME CITIES.
AND THE WAVES OF DEATH. FLOOD, FAMINE, DISEASE.

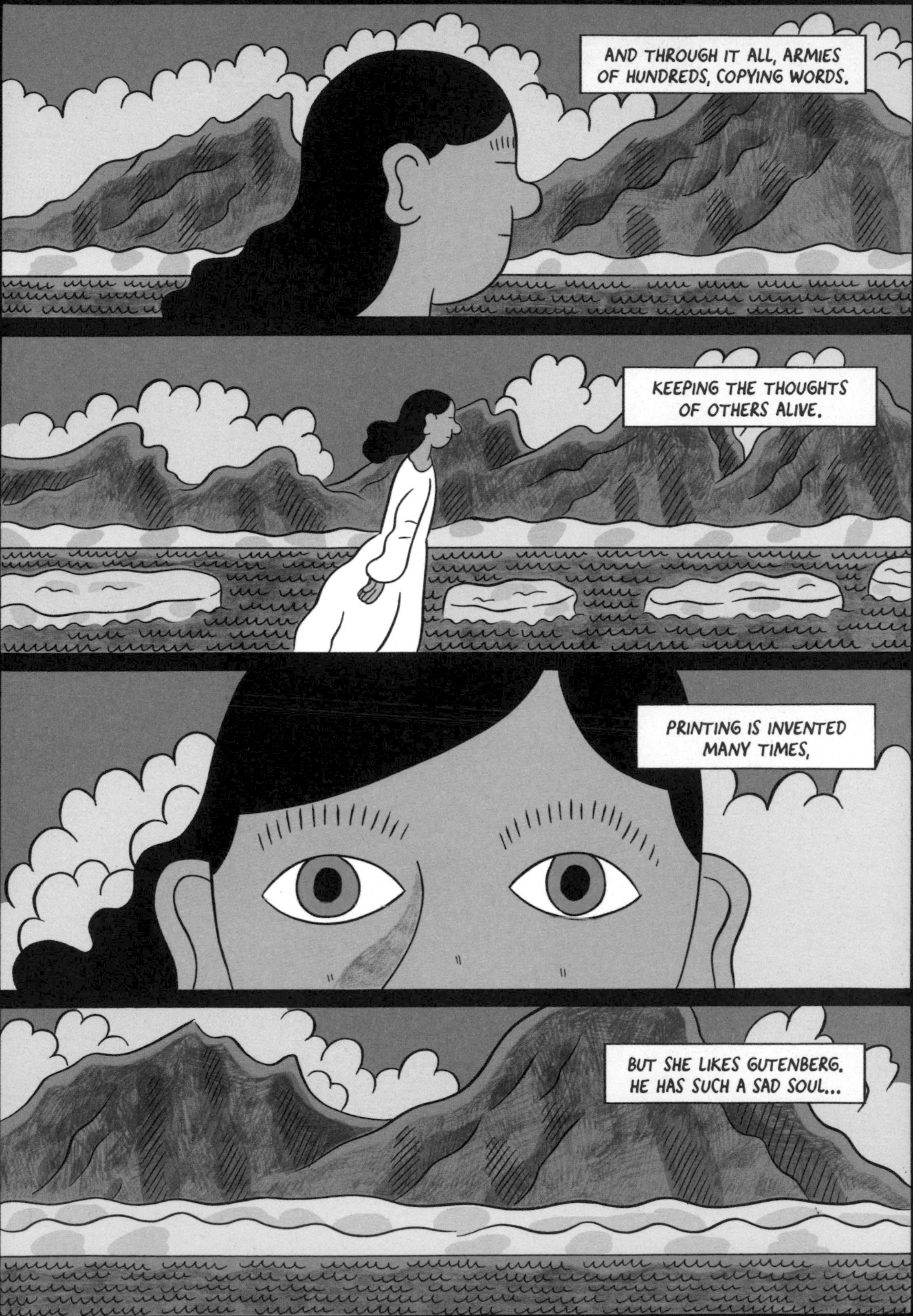
AND THROUGH IT ALL, ARMIES OF HUNDREDS, COPYING WORDS.
KEEPING THE THOUGHTS OF OTHERS ALIVE.
PRINTING IS INVENTED MANY TIMES,
BUT SHE LIKES GUTENBERG. HE HAS SUCH A SAD SOUL...

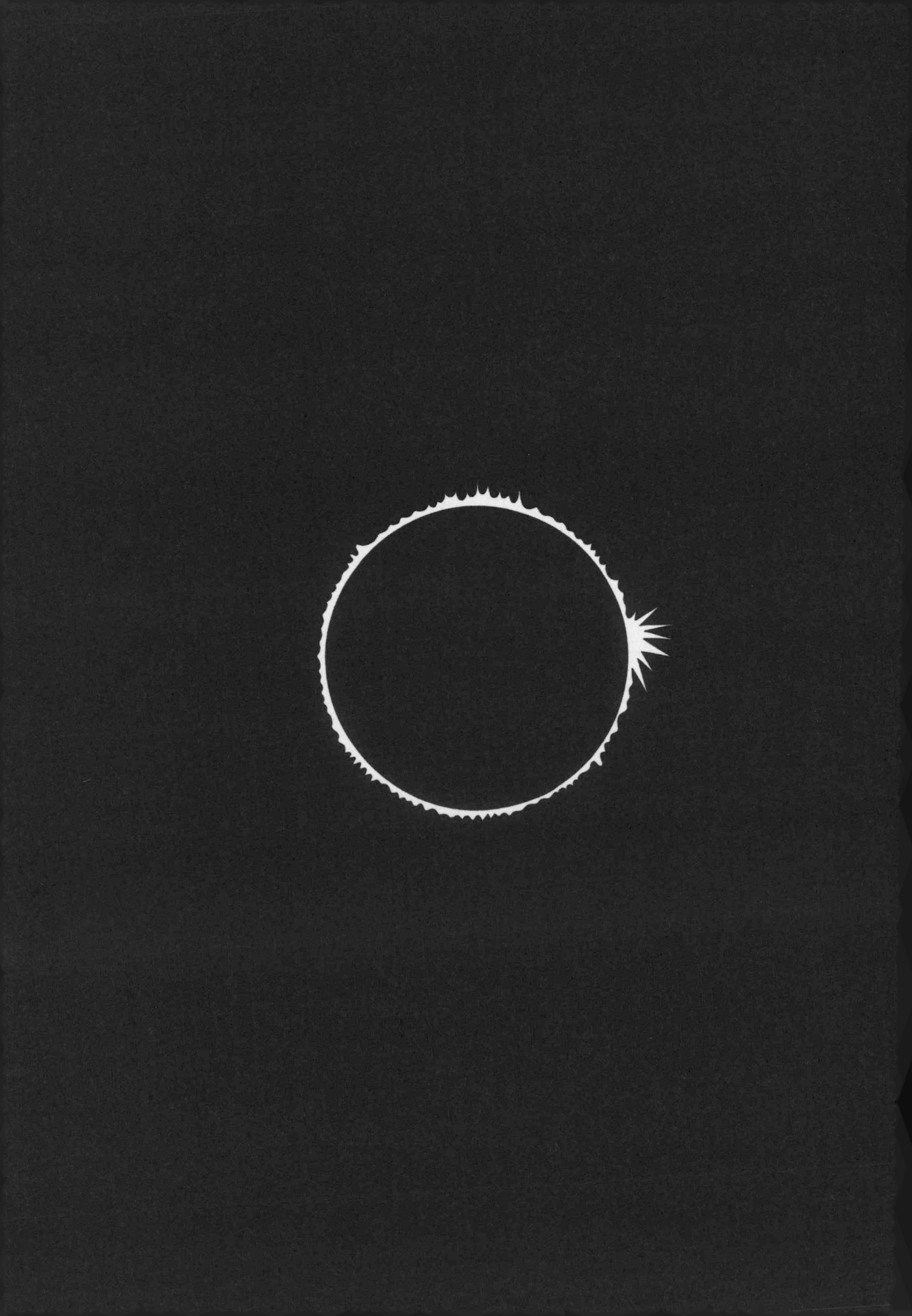

4622
years later...

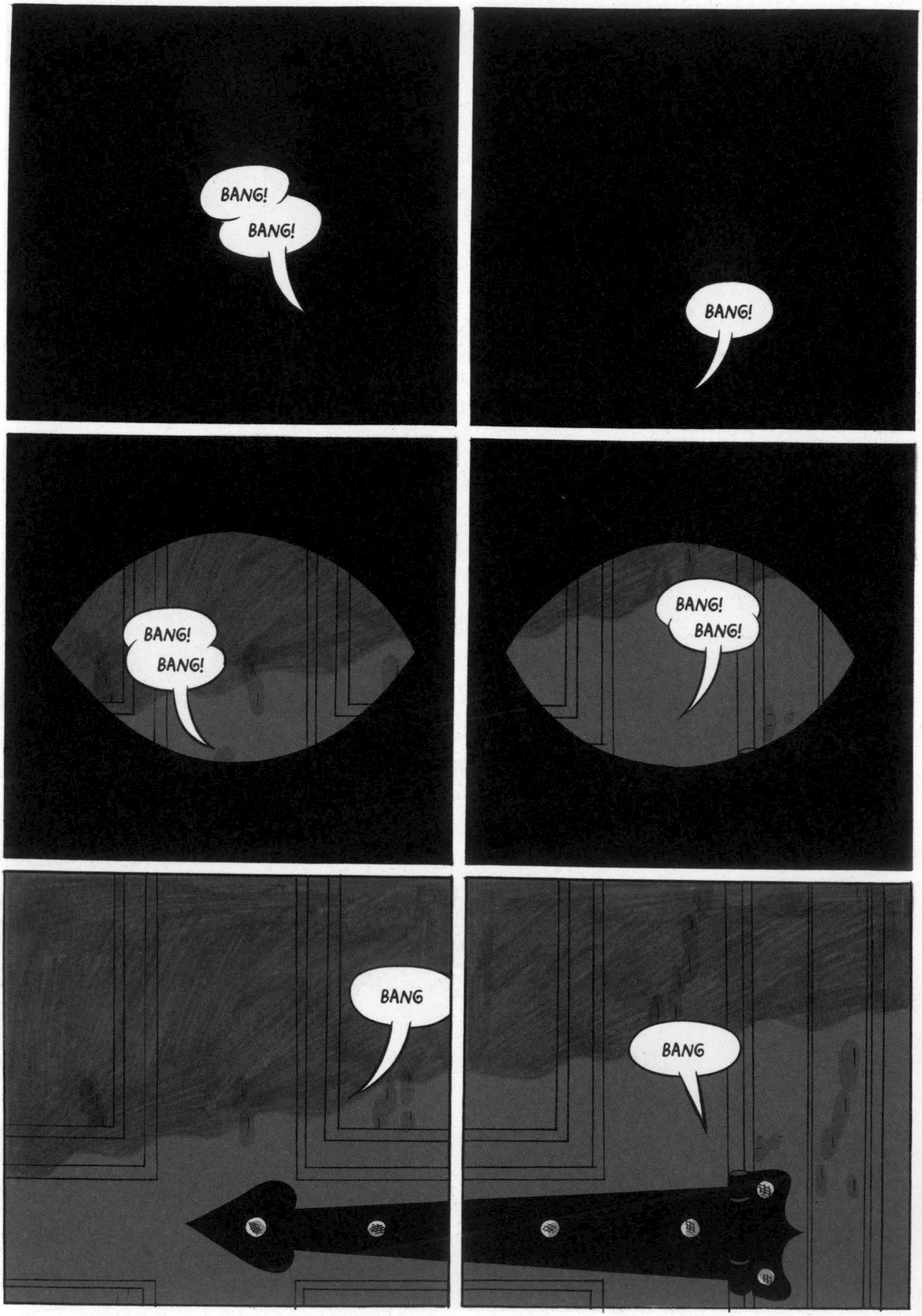
BANG!
BANG!
BANG!
BANG!
BANG!
BANG!
BANG!
BANG
BANG

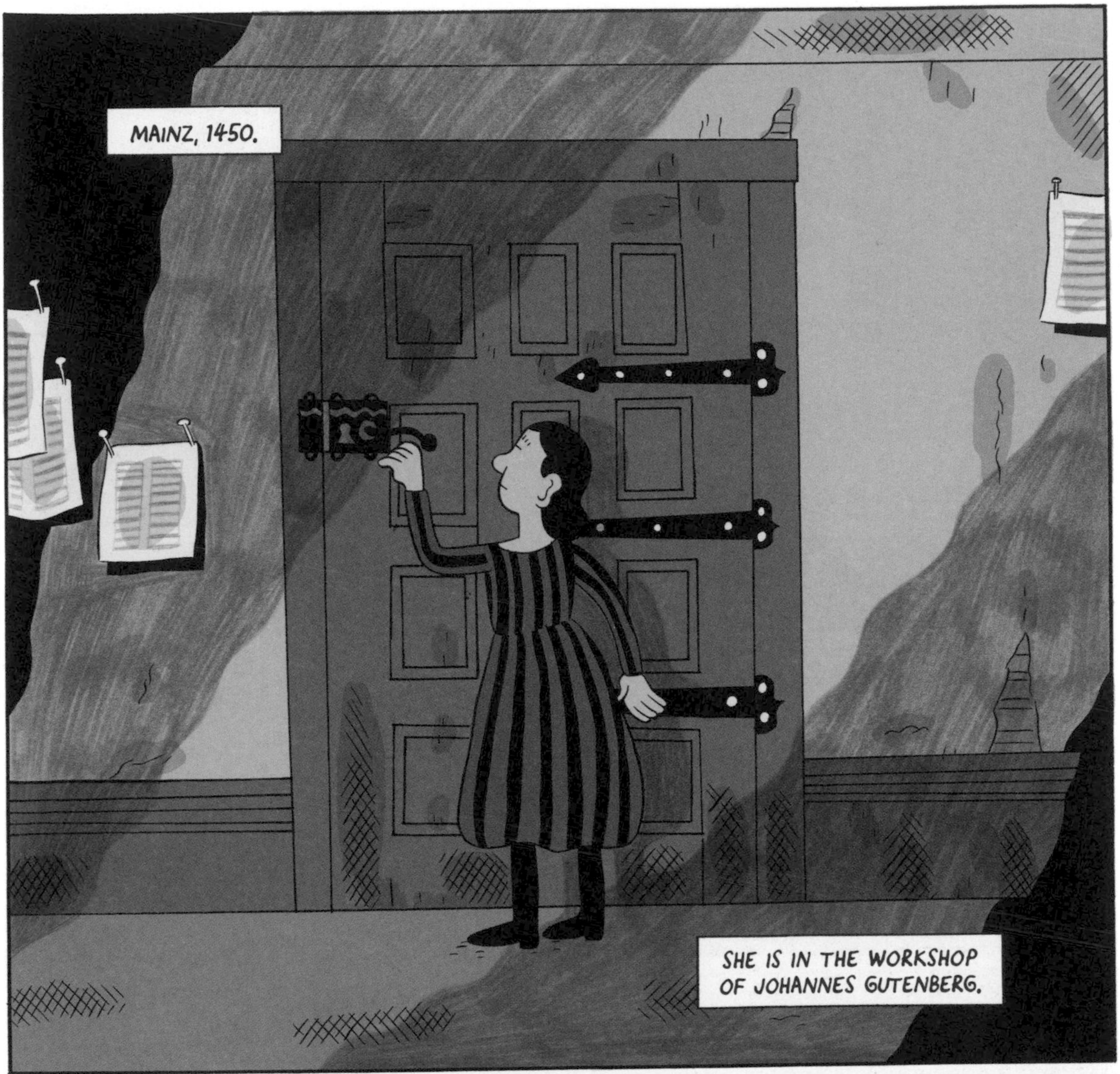
MAINZ, 1450.
SHE IS IN THE WORKSHOP OF JOHANNES GUTENBERG.

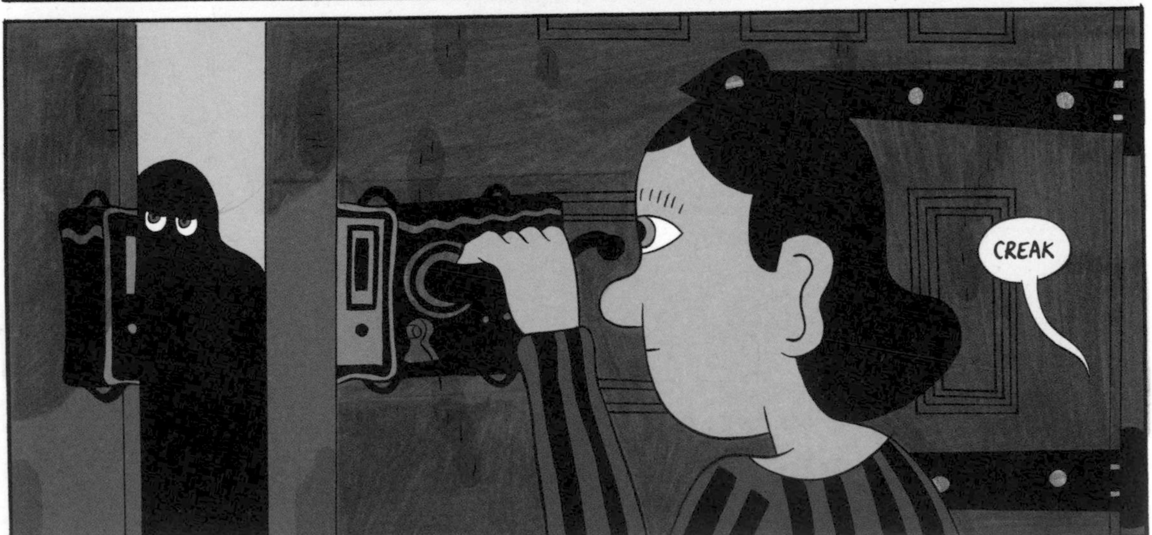
CREAK

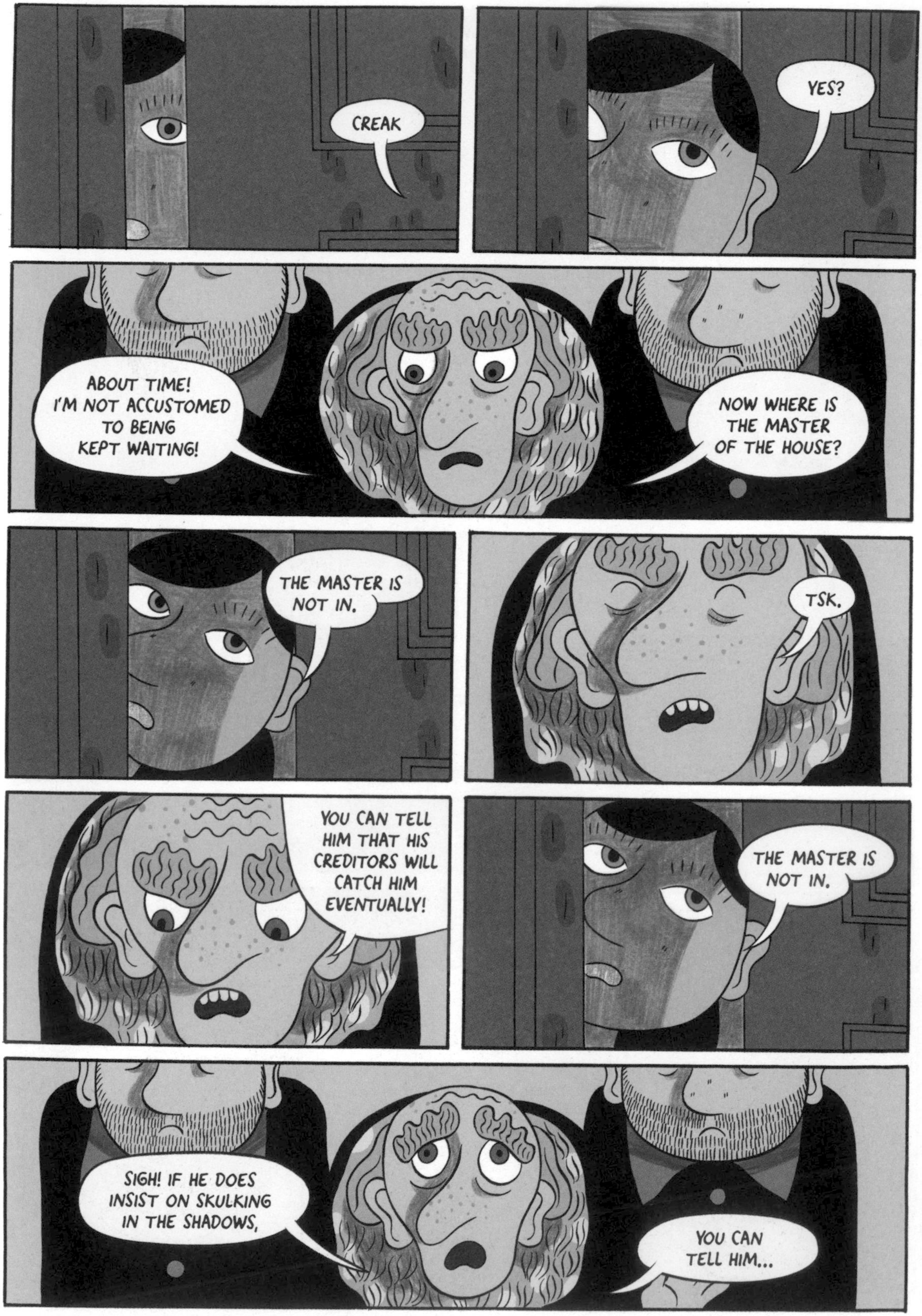
CREAK
YES?
ABOUT TIME! I'M NOT ACCUSTOMED TO BEING KEPT WAITING!
NOW WHERE IS THE MASTER OF THE HOUSE?
THE MASTER IS NOT IN.
TSK.
YOU CAN TELL HIM THAT HIS CREDITORS WILL CATCH HIM EVENTUALLY!
THE MASTER IS NOT IN.
SIGH! IF HE DOES INSIST ON SKULKING IN THE SHADOWS,
YOU CAN TELL HIM...

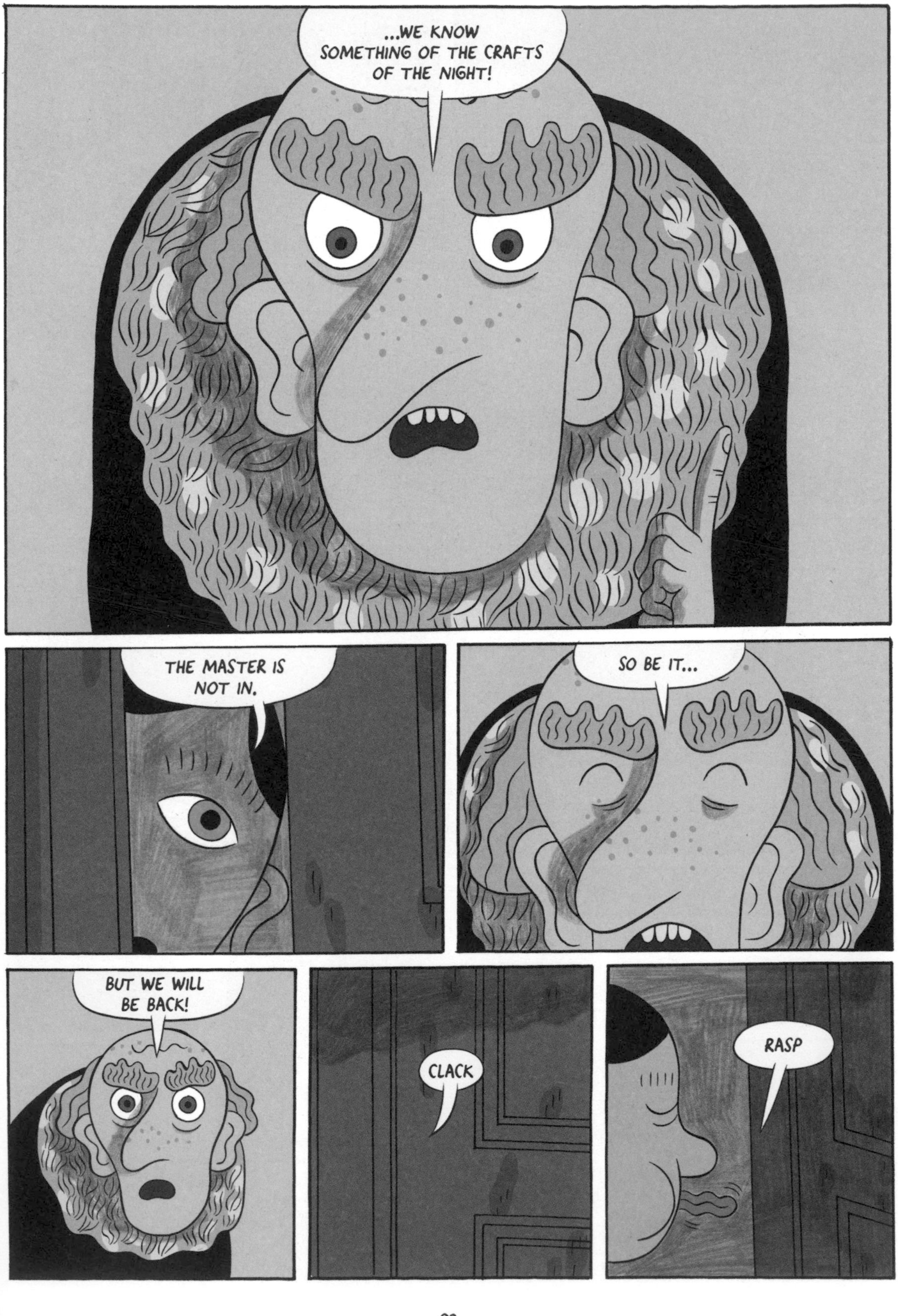
...WE KNOW SOMETHING OF THE CRAFTS OF THE NIGHT!
THE MASTER IS NOT IN.
SO BE IT...
BUT WE WILL BE BACK!
CLACK
RASP

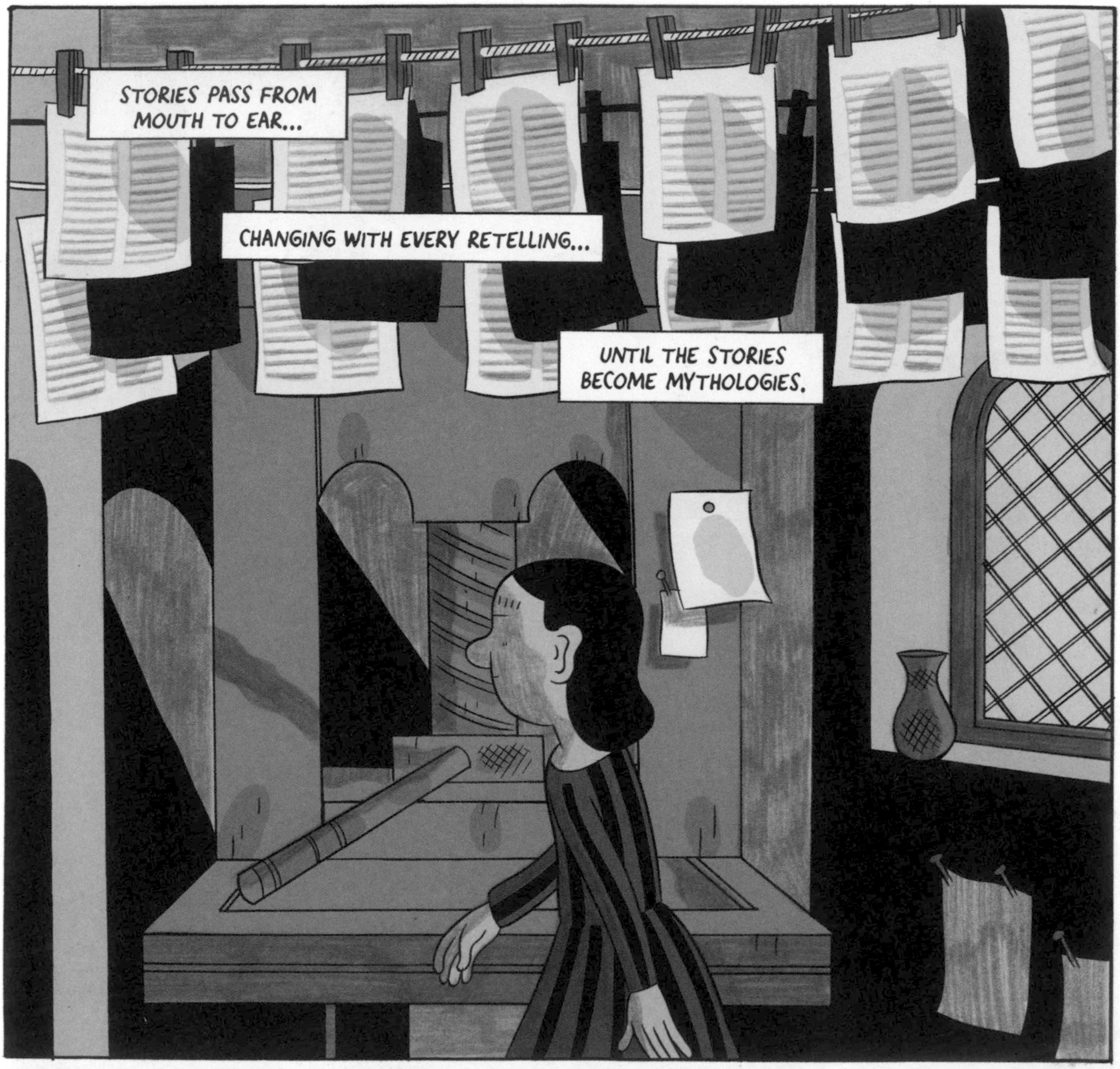
STORIES PASS FROM MOUTH TO EAR...
CHANGING WITH EVERY RETELLING...
UNTIL THE STORIES BECOME MYTHOLOGIES.

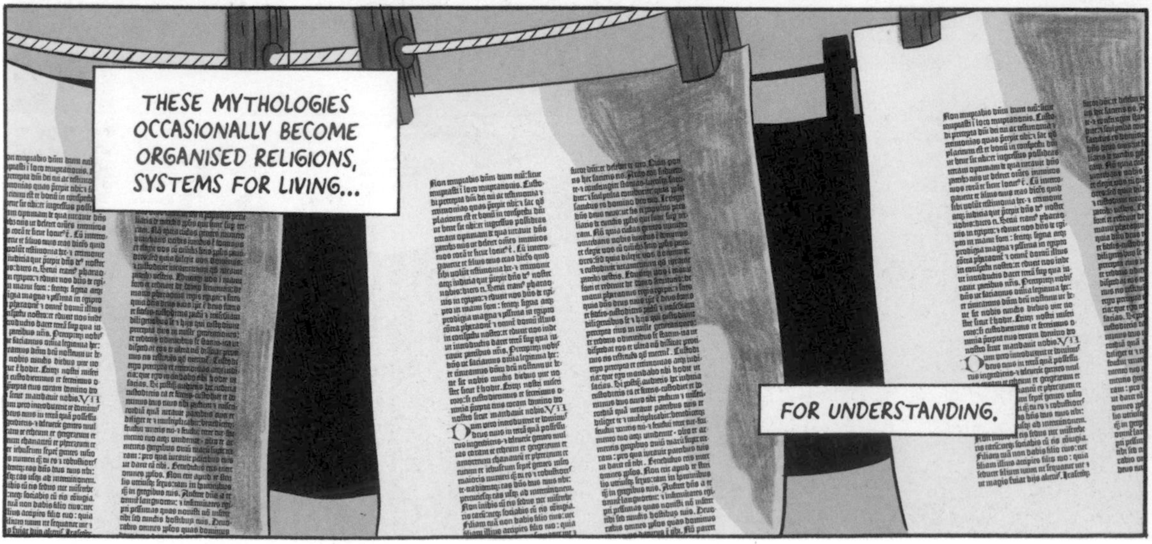
THESE MYTHOLOGIES OCCASIONALLY BECOME ORGANISED RELIGIONS, SYSTEMS FOR LIVING...
FOR UNDERSTANDING.

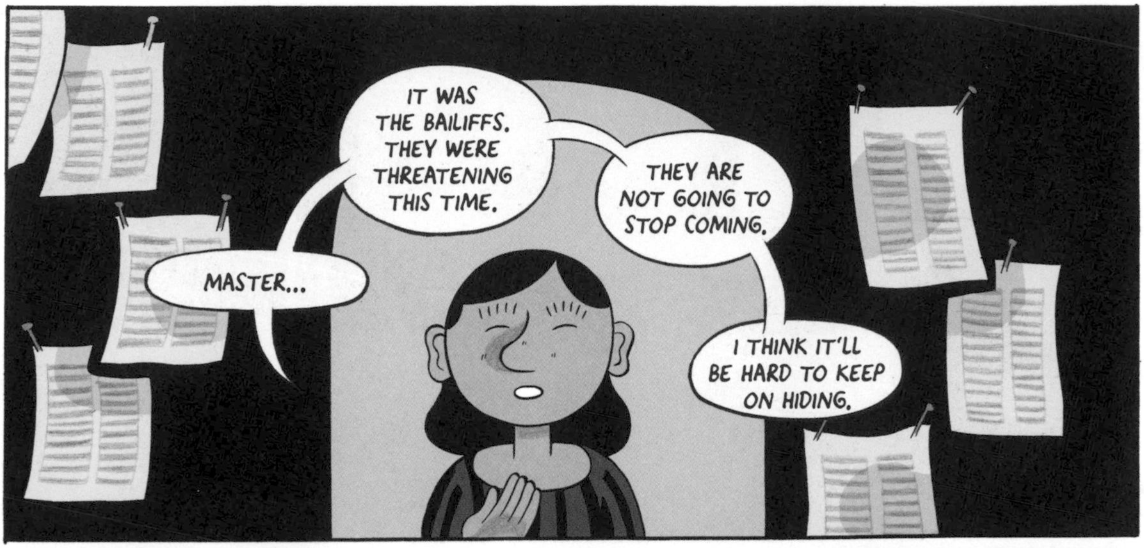
MASTER...
IT WAS THE BAILIFFS. THEY WERE THREATENING THIS TIME.
THEY ARE NOT GOING TO STOP COMING.
I THINK IT'LL BE HARD TO KEEP ON HIDING.

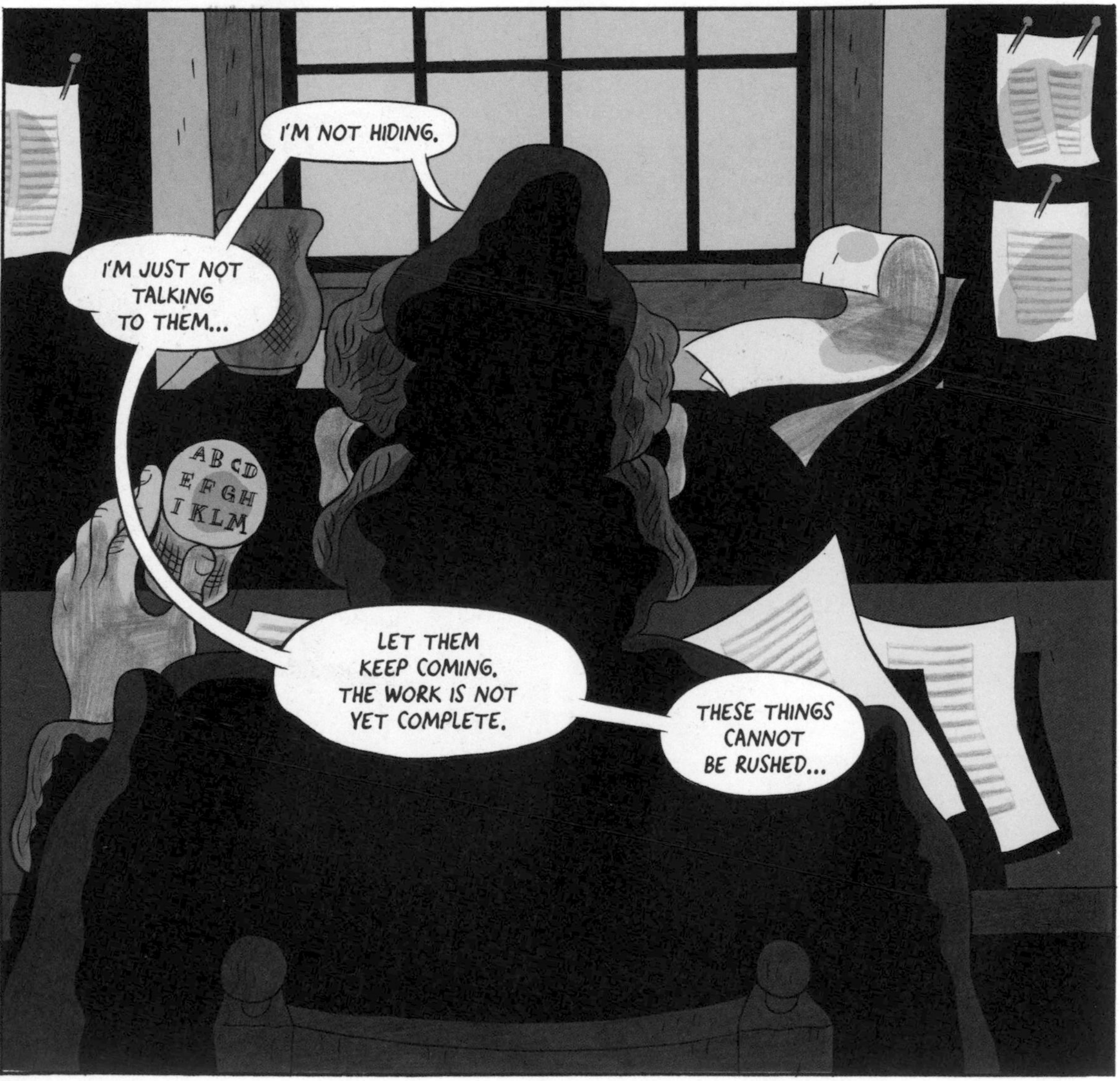
I'M NOT HIDING.
I'M JUST NOT TALKING TO THEM...
ABCD EFGH IKLM
LET THEM KEEP COMING. THE WORK IS NOT YET COMPLETE.
THESE THINGS CANNOT BE RUSHED...

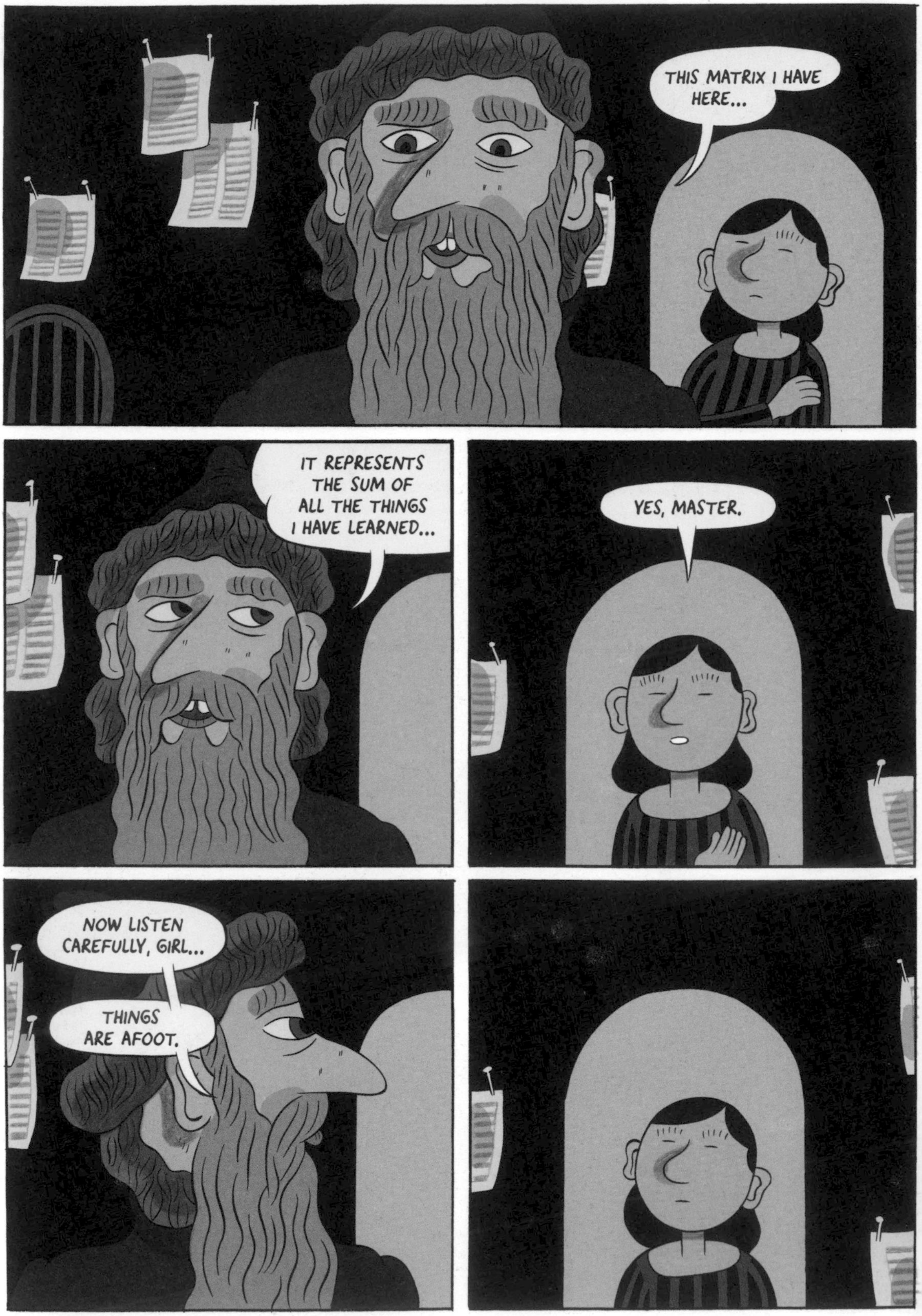
THIS MATRIX I HAVE HERE...
IT REPRESENTS THE SUM OF ALL THE THINGS I HAVE LEARNED...
YES, MASTER.
NOW LISTEN CAREFULLY, GIRL...
THINGS ARE AFOOT.

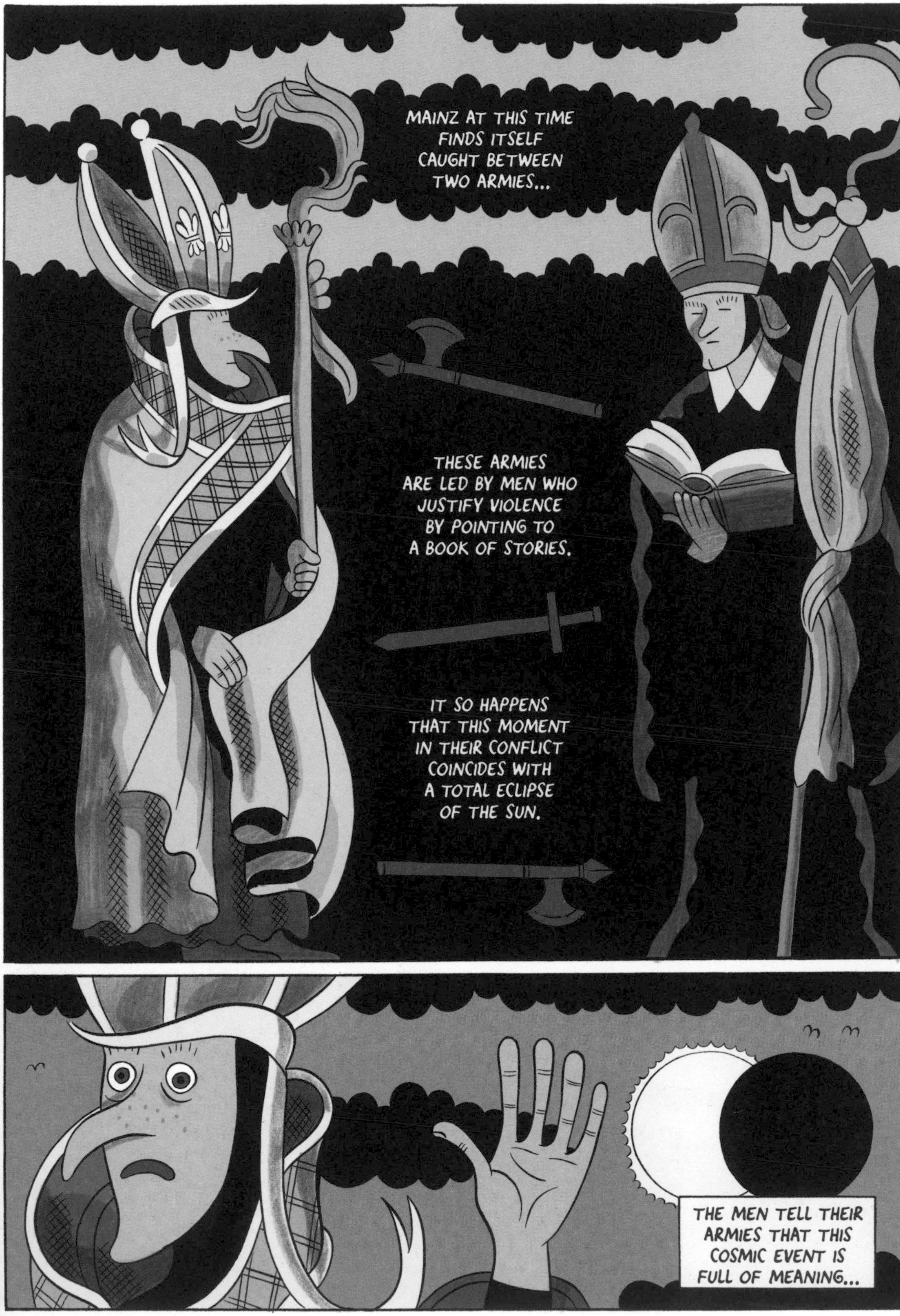
MAINZ AT THIS TIME FINDS ITSELF CAUGHT BETWEEN TWO ARMIES...
THESE ARMIES ARE LED BY MEN WHO JUSTIFY VIOLENCE BY POINTING TO A BOOK OF STORIES.
IT SO HAPPENS THAT THIS MOMENT IN THEIR CONFLICT COINCIDES WITH A TOTAL ECLIPSE OF THE SUN.
THE MEN TELL THEIR ARMIES THAT THIS COSMIC EVENT IS FULL OF MEANING...

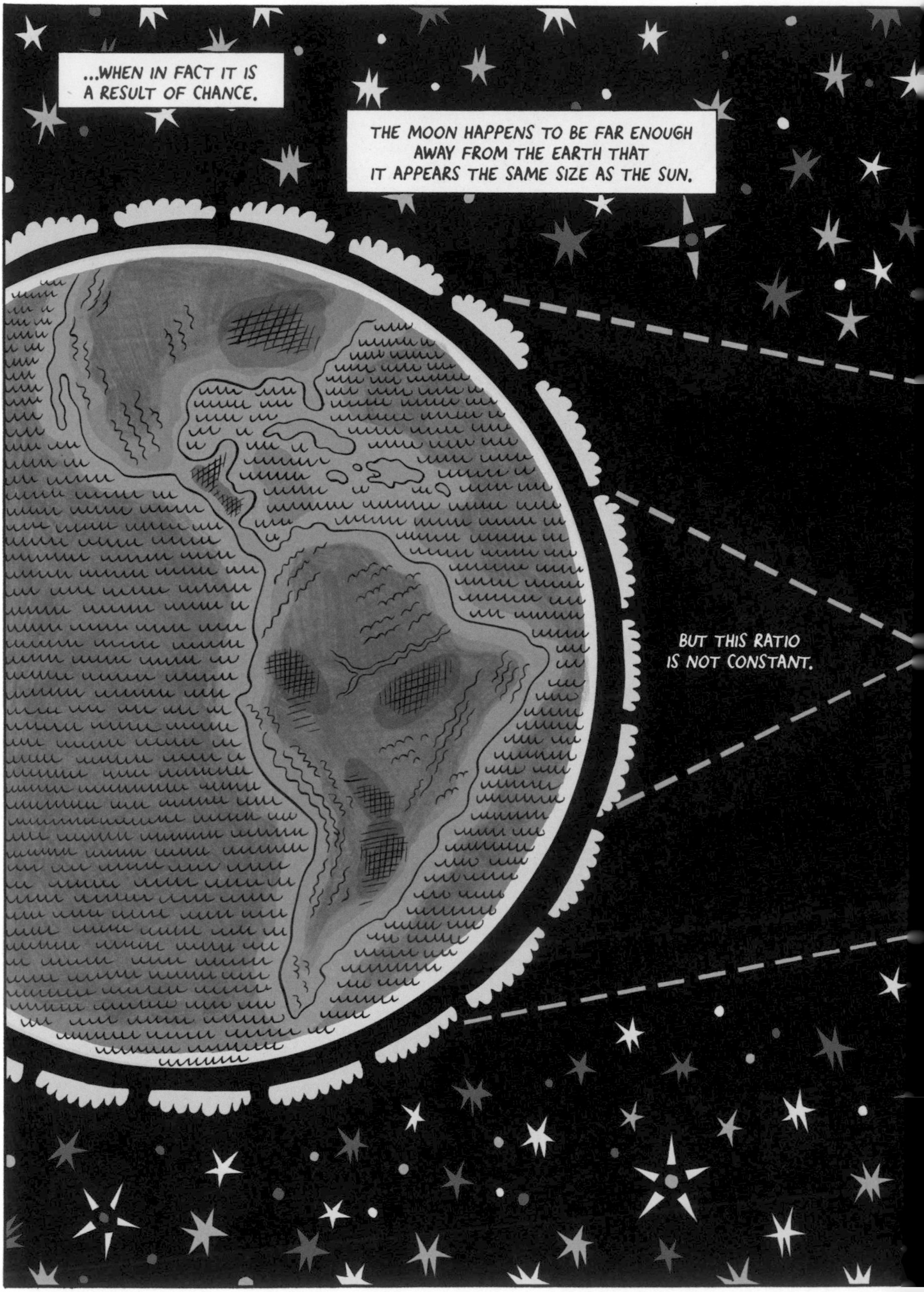
...WHEN IN FACT IT IS A RESULT OF CHANCE.
THE MOON HAPPENS TO BE FAR ENOUGH AWAY FROM THE EARTH THAT IT APPEARS THE SAME SIZE AS THE SUN.
BUT THIS RATIO IS NOT CONSTANT.

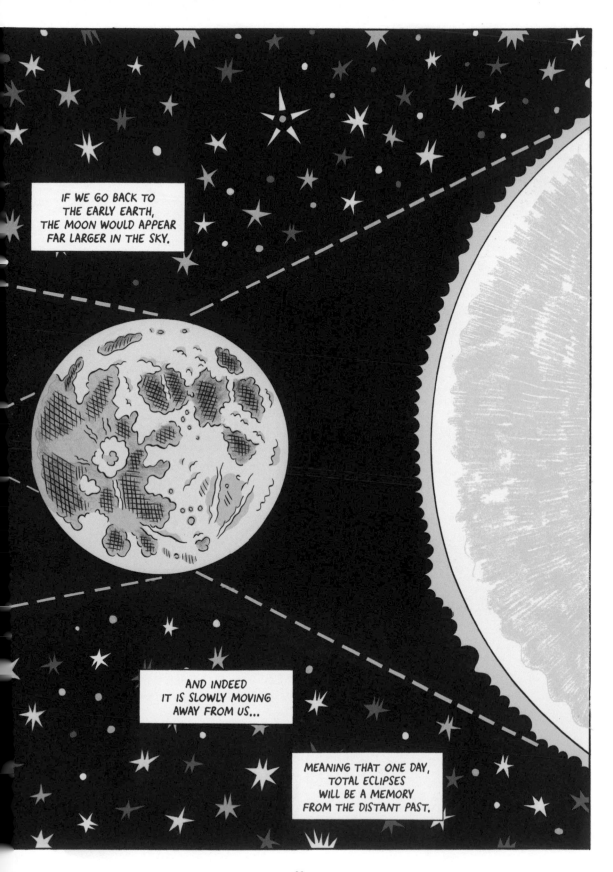

IF WE GO BACK TO THE EARLY EARTH, THE MOON WOULD APPEAR FAR LARGER IN THE SKY.

AND INDEED IT IS SLOWLY MOVING AWAY FROM US...

MEANING THAT ONE DAY, TOTAL ECLIPSES WILL BE A MEMORY FROM THE DISTANT PAST.

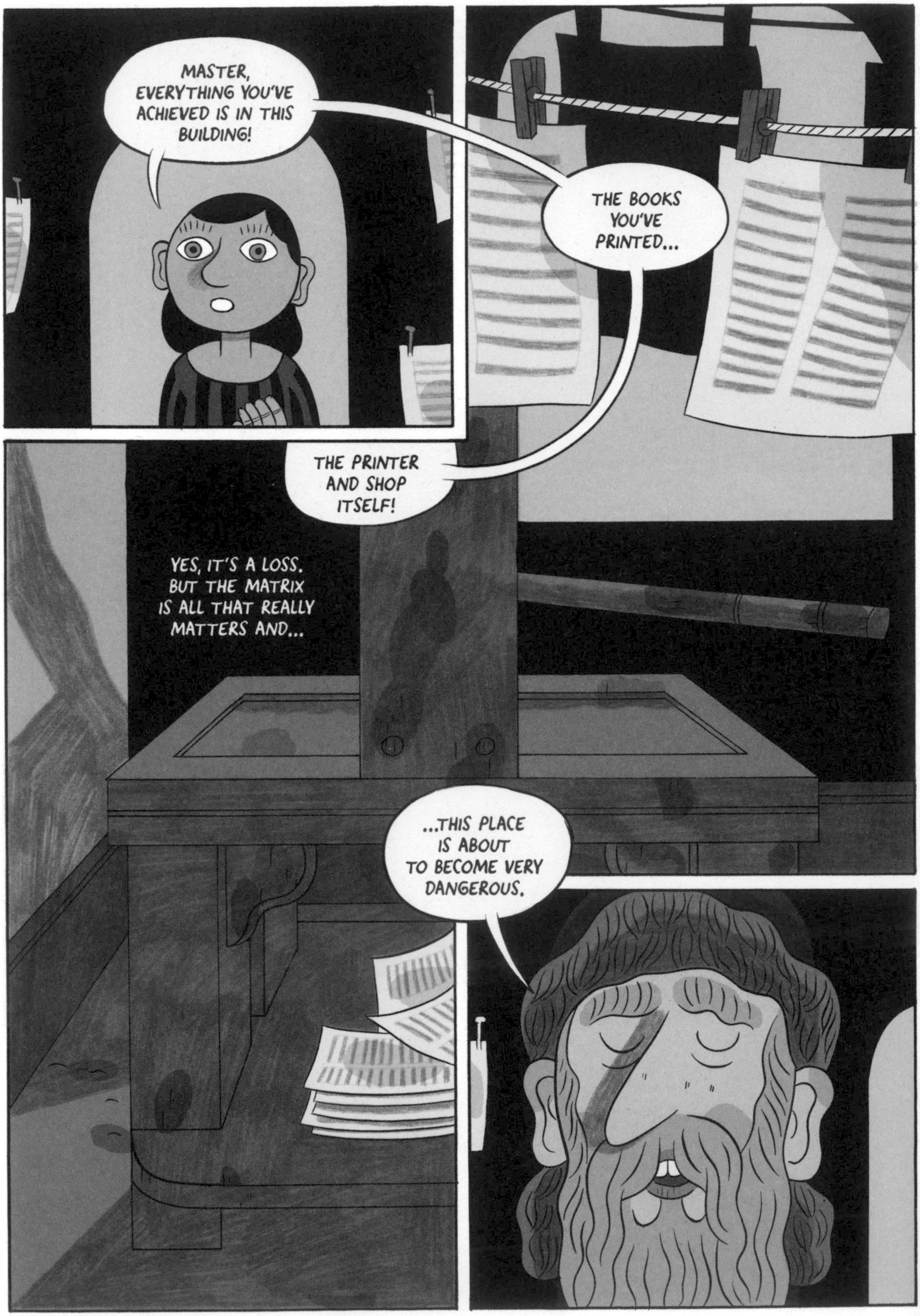
MASTER, EVERYTHING YOU'VE ACHIEVED IS IN THIS BUILDING!
THE BOOKS YOU'VE PRINTED...
THE PRINTER AND SHOP ITSELF!
YES, IT'S A LOSS. BUT THE MATRIX IS ALL THAT REALLY MATTERS AND...
...THIS PLACE IS ABOUT TO BECOME VERY DANGEROUS.

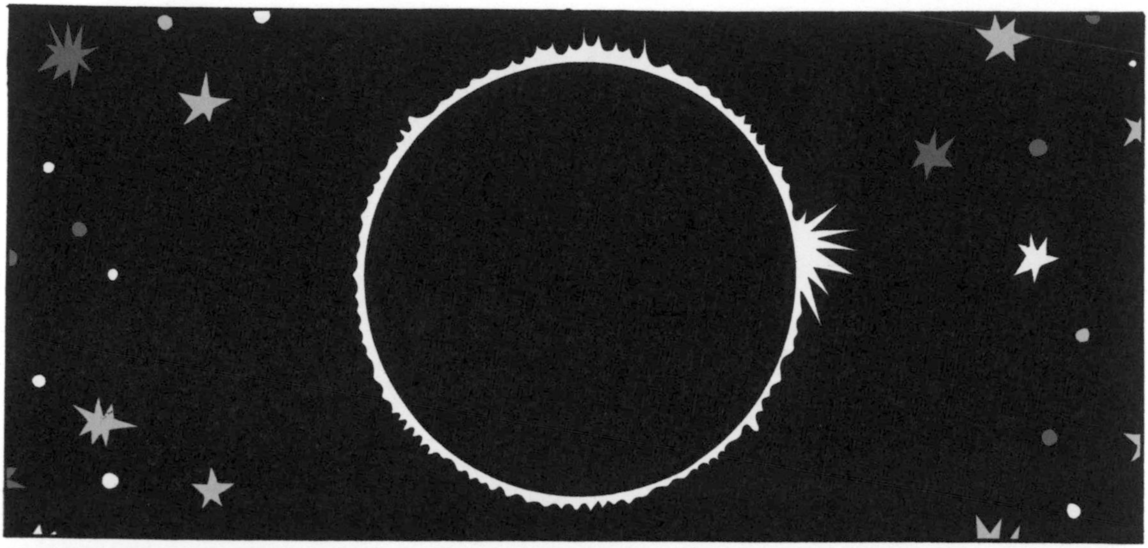

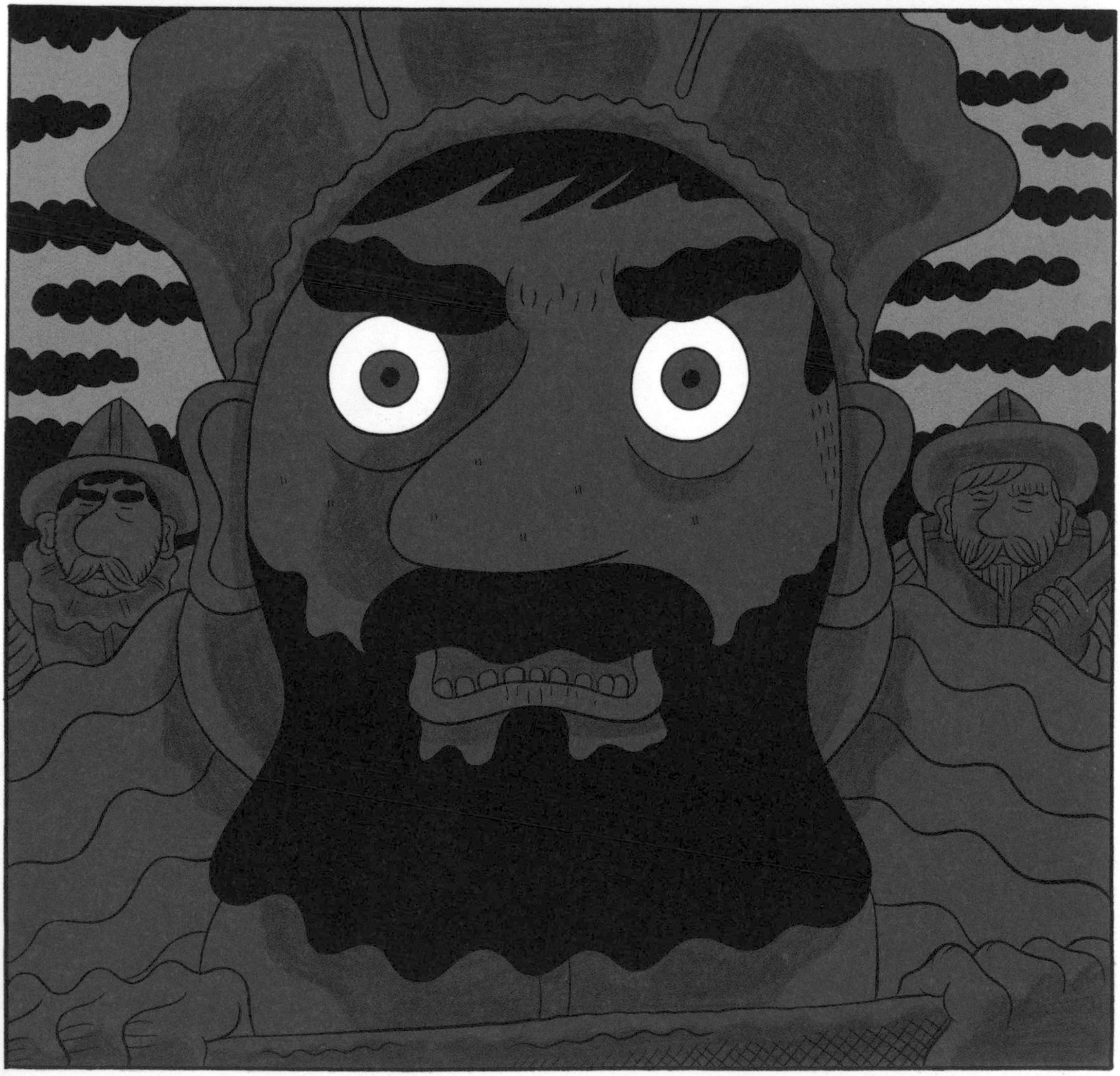

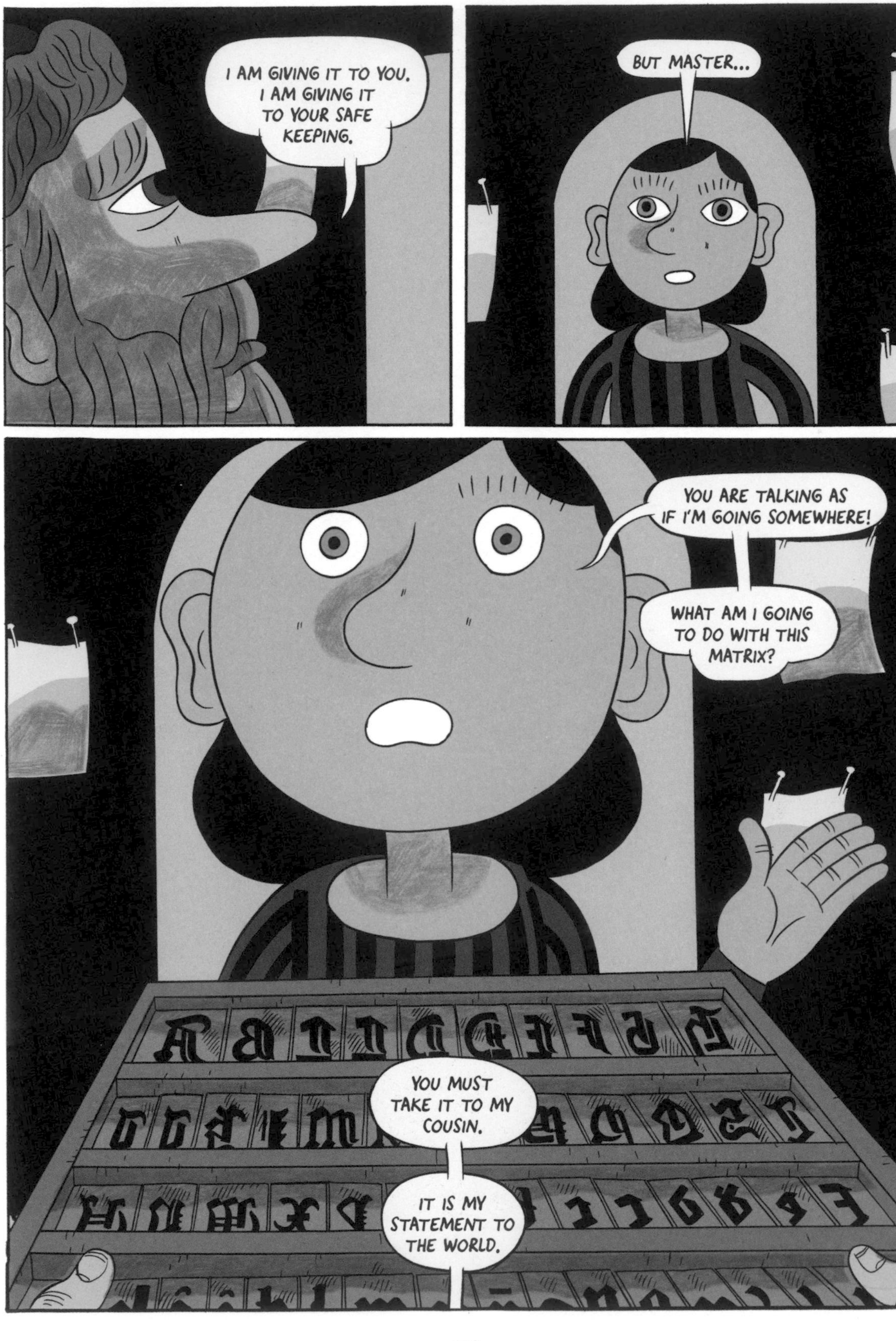
I AM GIVING IT TO YOU. I AM GIVING IT TO YOUR SAFE KEEPING.
BUT MASTER...
YOU ARE TALKING AS IF I'M GOING SOMEWHERE!
WHAT AM I GOING TO DO WITH THIS MATRIX?
YOU MUST TAKE IT TO MY COUSIN.
IT IS MY STATEMENT TO THE WORLD.

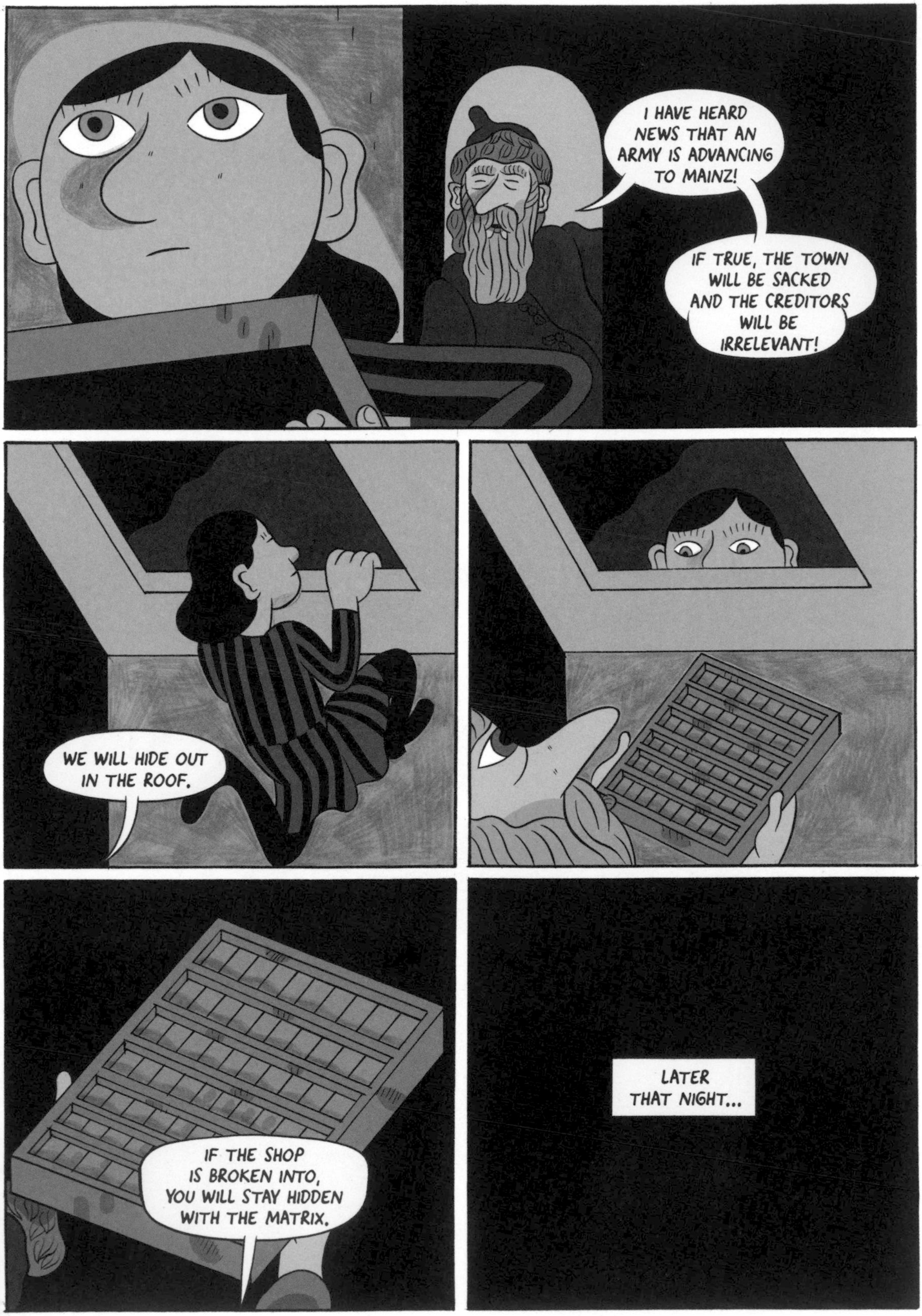
I HAVE HEARD NEWS THAT AN ARMY IS ADVANCING TO MAINZ!
IF TRUE, THE TOWN WILL BE SACKED AND THE CREDITORS WILL BE IRRELEVANT!
WE WILL HIDE OUT IN THE ROOF.
IF THE SHOP IS BROKEN INTO, YOU WILL STAY HIDDEN WITH THE MATRIX.
LATER THAT NIGHT...

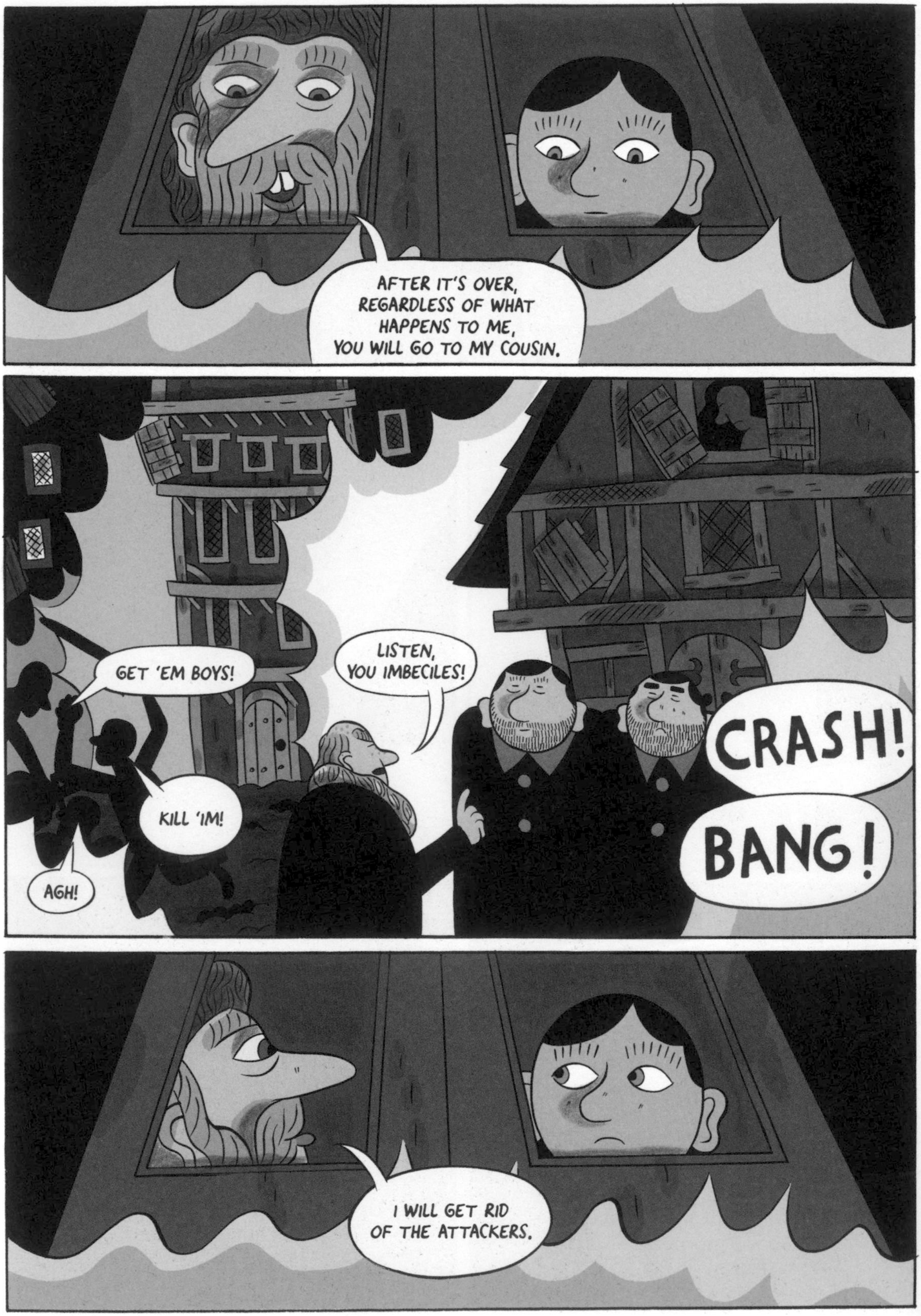
AFTER IT'S OVER, REGARDLESS OF WHAT HAPPENS TO ME, YOU WILL GO TO MY COUSIN.
GET 'EM BOYS!
KILL 'IM!
AGH!
LISTEN, YOU IMBECILES!
CRASH!
BANG!
I WILL GET RID OF THE ATTACKERS.

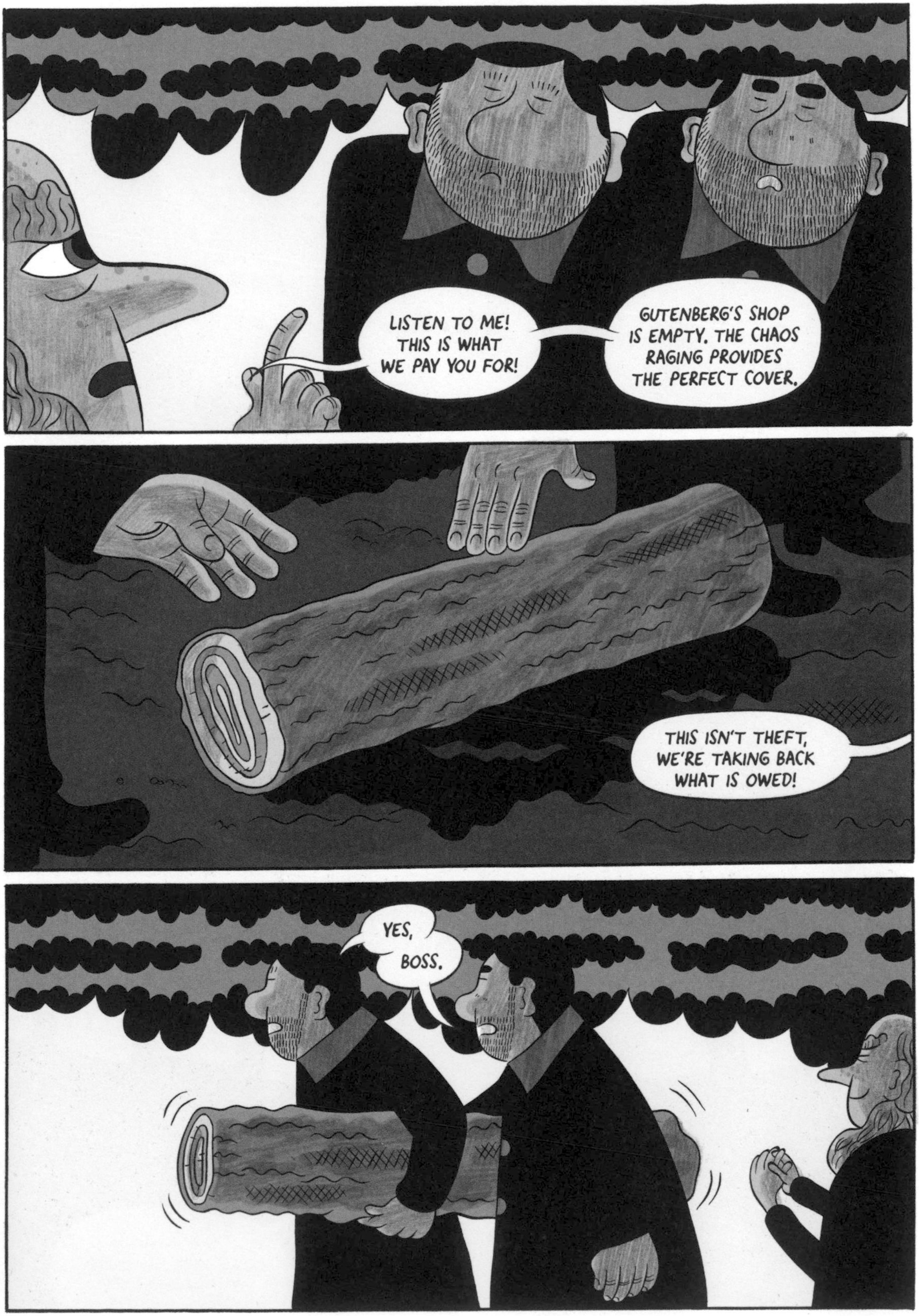
LISTEN TO ME! THIS IS WHAT WE PAY YOU FOR!
GUTENBERG'S SHOP IS EMPTY. THE CHAOS RAGING PROVIDES THE PERFECT COVER.
THIS ISN'T THEFT, WE'RE TAKING BACK WHAT IS OWED!
YES, BOSS.

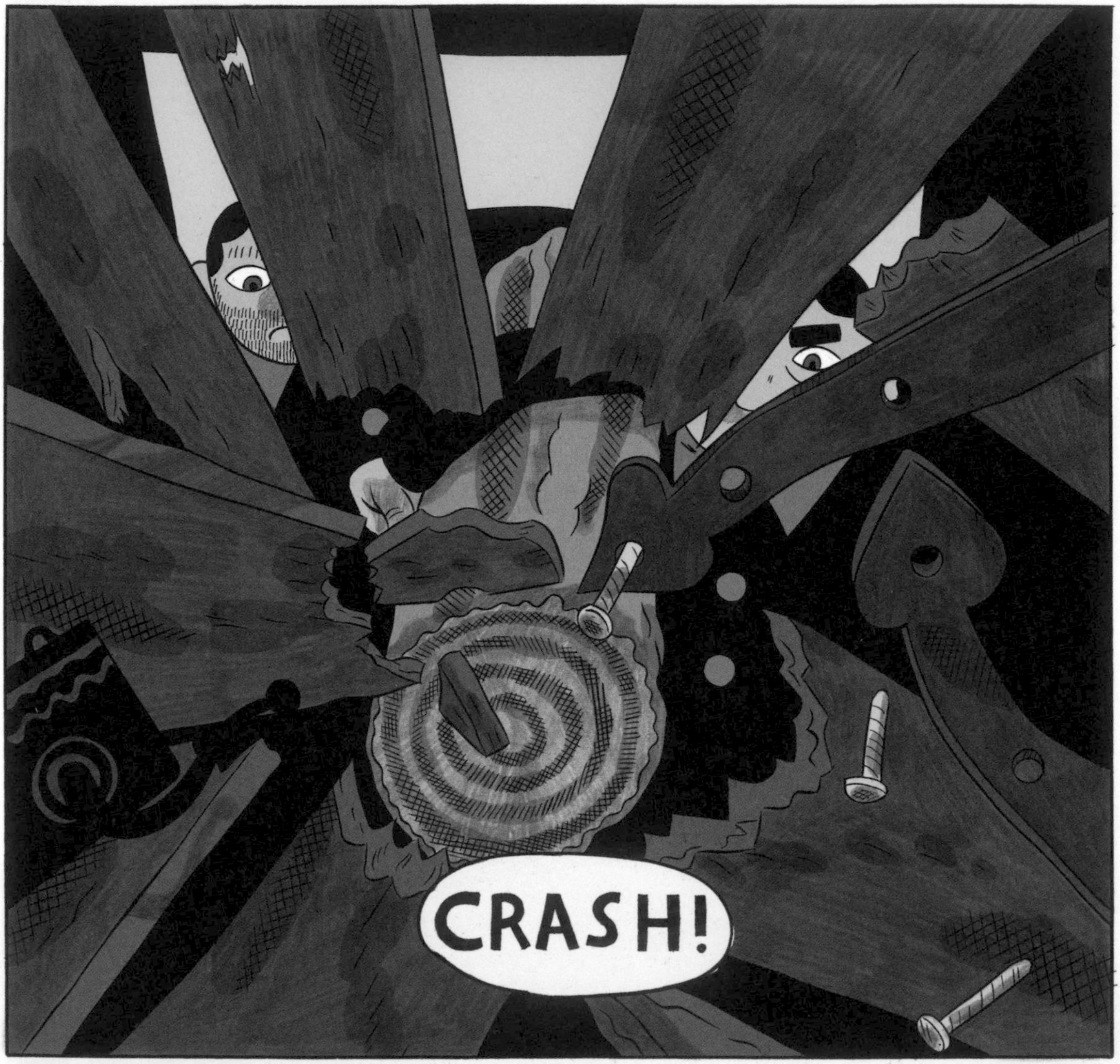
CRASH!

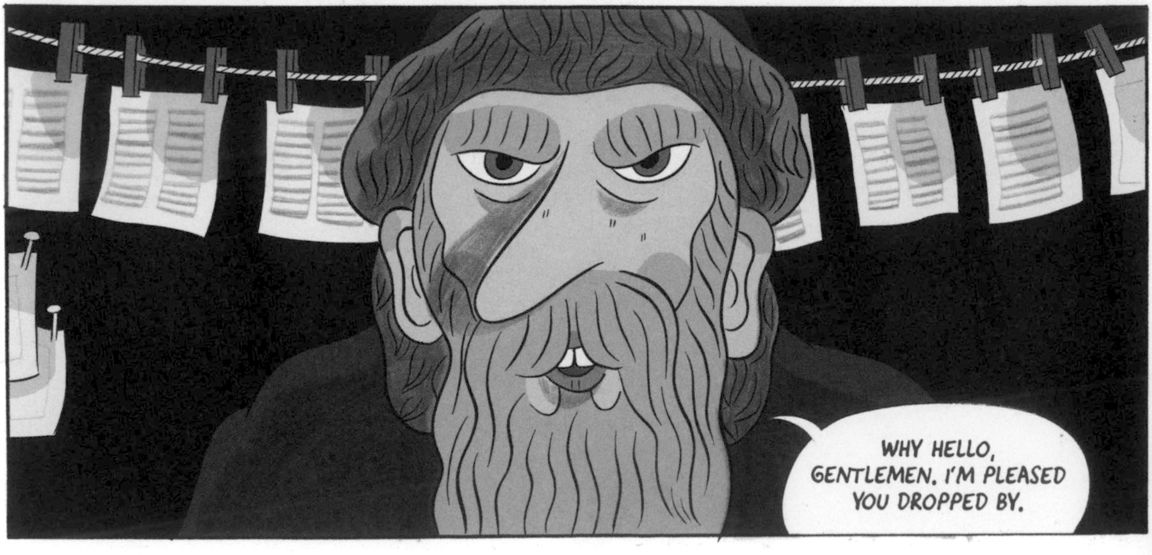
WHY HELLO, GENTLEMEN. I'M PLEASED YOU DROPPED BY.

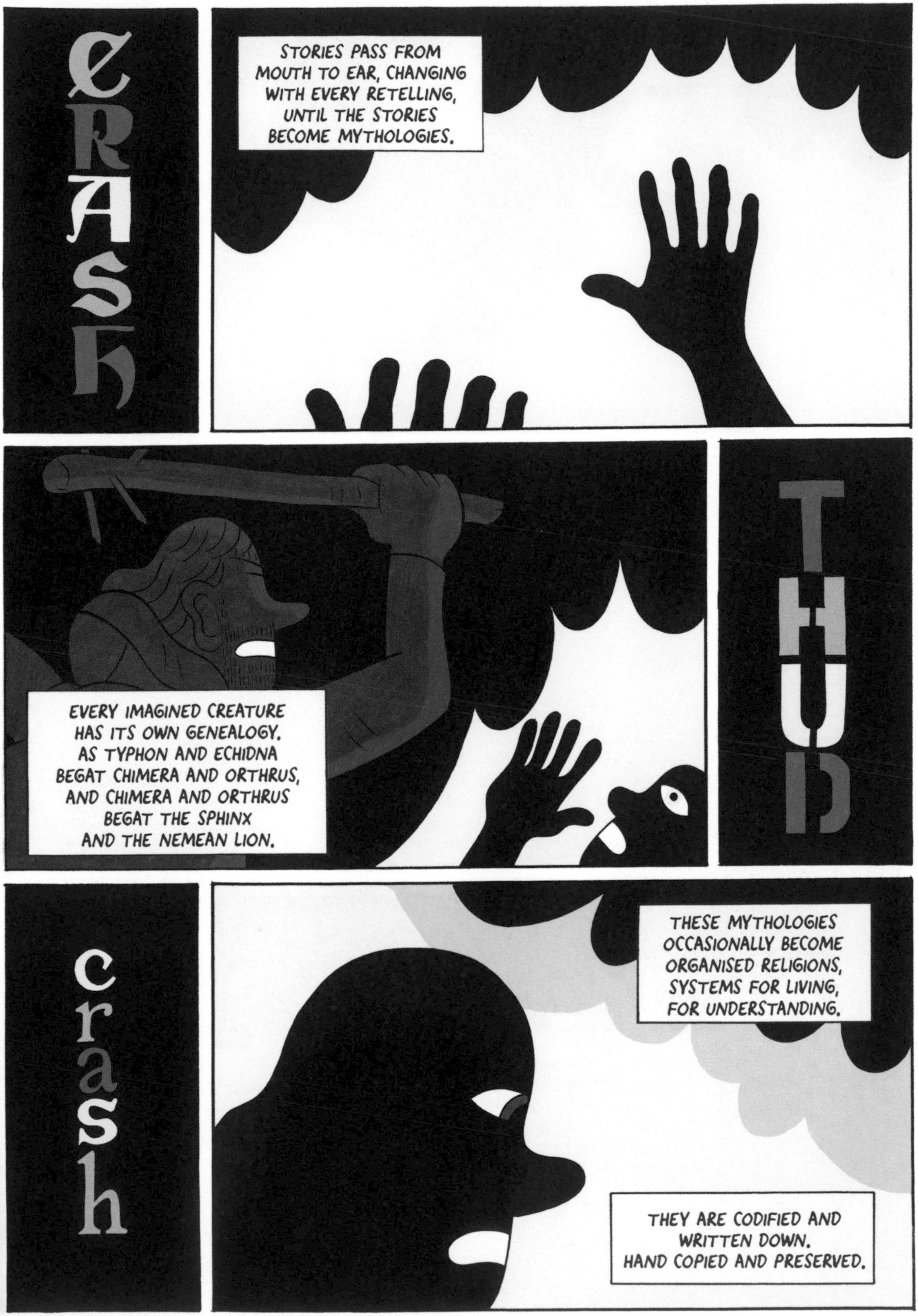
CRASH
STORIES PASS FROM MOUTH TO EAR, CHANGING WITH EVERY RETELLING, UNTIL THE STORIES BECOME MYTHOLOGIES.
EVERY IMAGINED CREATURE HAS ITS OWN GENEALOGY. AS TYPHON AND ECHIDNA BEGAT CHIMERA AND ORTHRUS, AND CHIMERA AND ORTHRUS BEGAT THE SPHINX AND THE NEMEAN LION.
THUD
crash
THESE MYTHOLOGIES OCCASIONALLY BECOME ORGANISED RELIGIONS, SYSTEMS FOR LIVING, FOR UNDERSTANDING.
THEY ARE CODIFIED AND WRITTEN DOWN. HAND COPIED AND PRESERVED.

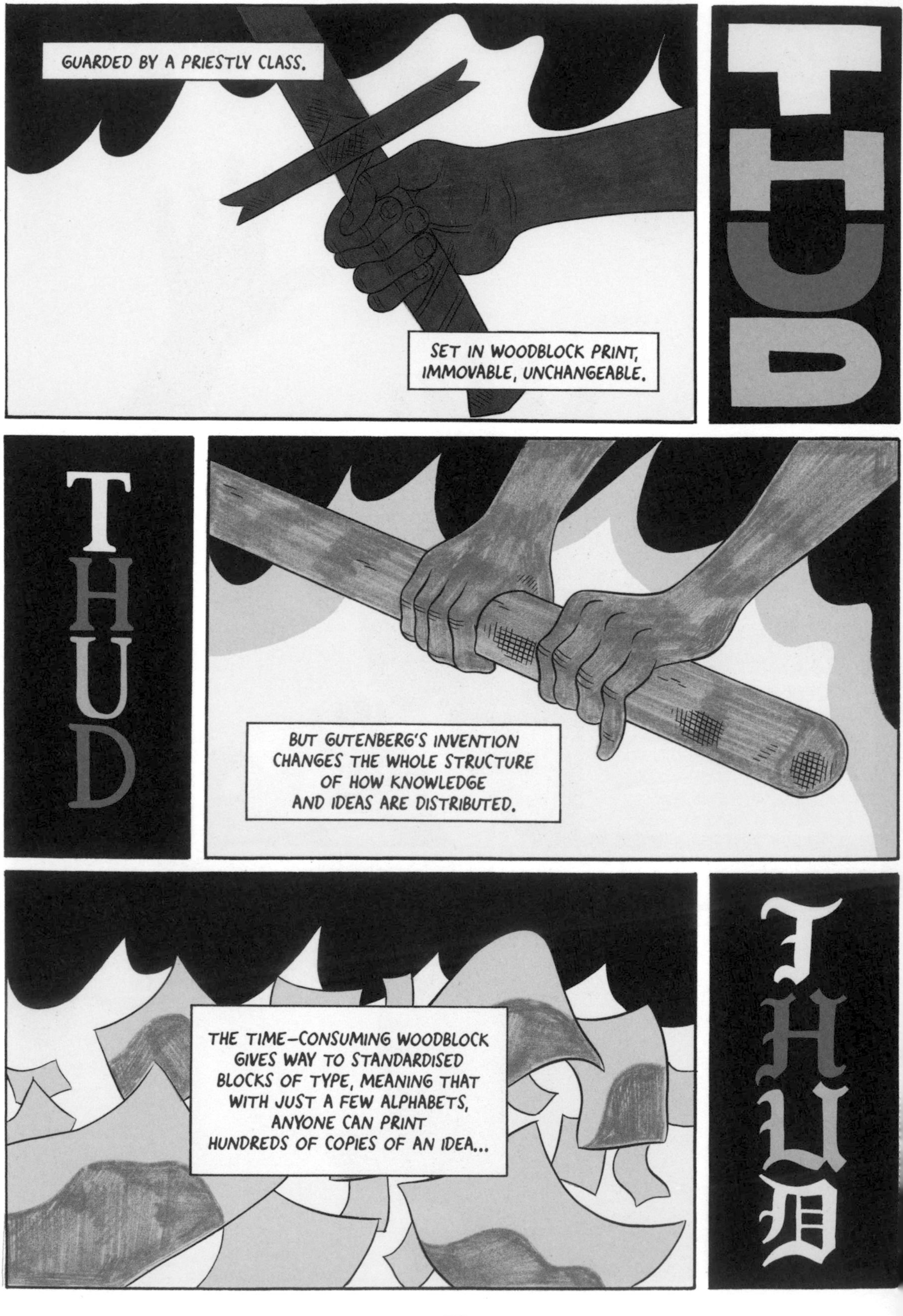
GUARDED BY A PRIESTLY CLASS.
SET IN WOODBLOCK PRINT, IMMOVABLE, UNCHANGEABLE.
THUD
THUD
BUT GUTENBERG'S INVENTION CHANGES THE WHOLE STRUCTURE OF HOW KNOWLEDGE AND IDEAS ARE DISTRIBUTED.
THE TIME-CONSUMING WOODBLOCK GIVES WAY TO STANDARDISED BLOCKS OF TYPE, MEANING THAT WITH JUST A FEW ALPHABETS, ANYONE CAN PRINT HUNDREDS OF COPIES OF AN IDEA...
THUD

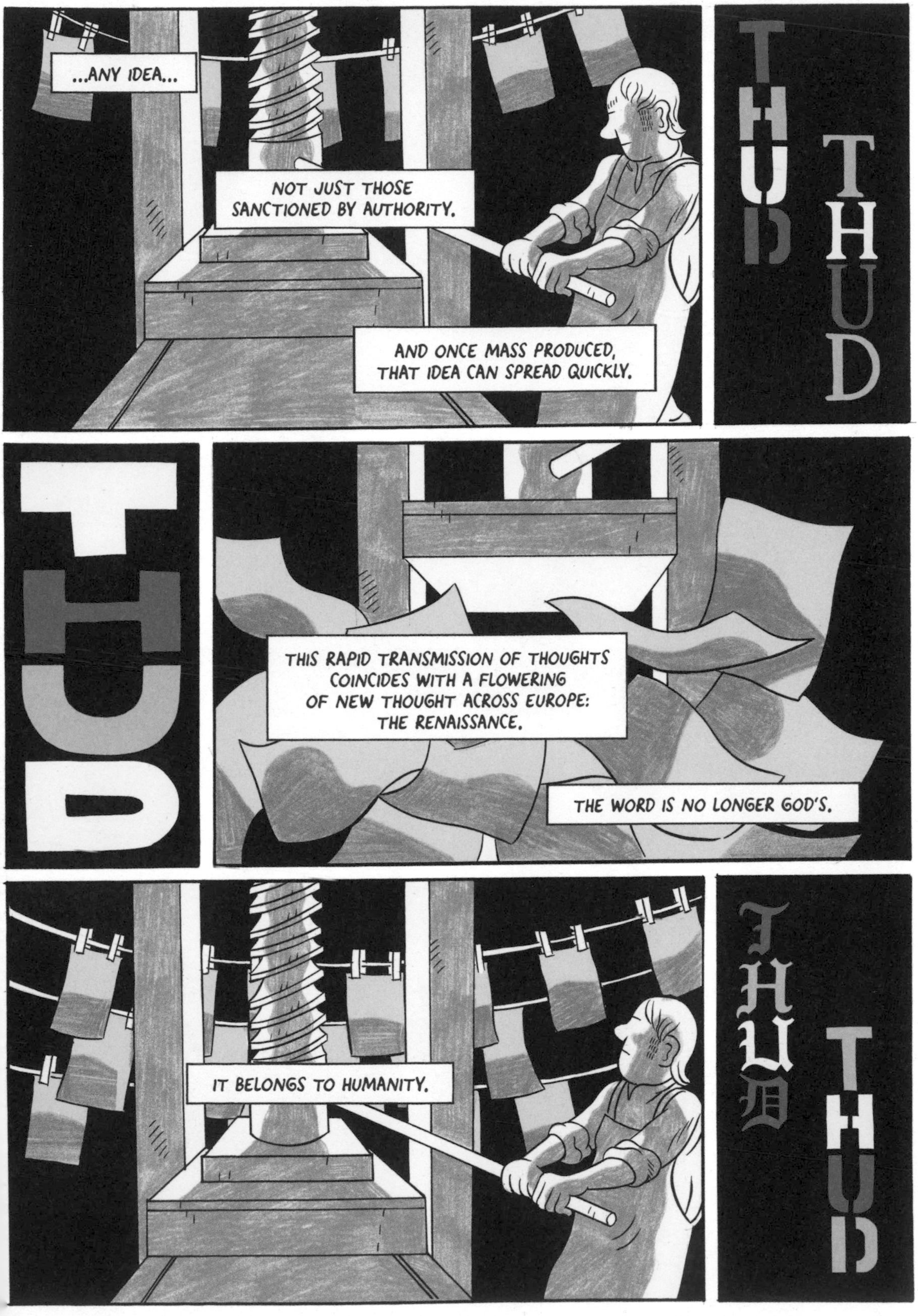
...ANY IDEA...
NOT JUST THOSE SANCTIONED BY AUTHORITY.
AND ONCE MASS PRODUCED, THAT IDEA CAN SPREAD QUICKLY.
THUD
THUD
THUD
THIS RAPID TRANSMISSION OF THOUGHTS COINCIDES WITH A FLOWERING OF NEW THOUGHT ACROSS EUROPE: THE RENAISSANCE.
THE WORD IS NO LONGER GOD'S.
IT BELONGS TO HUMANITY.
THUD
THUD

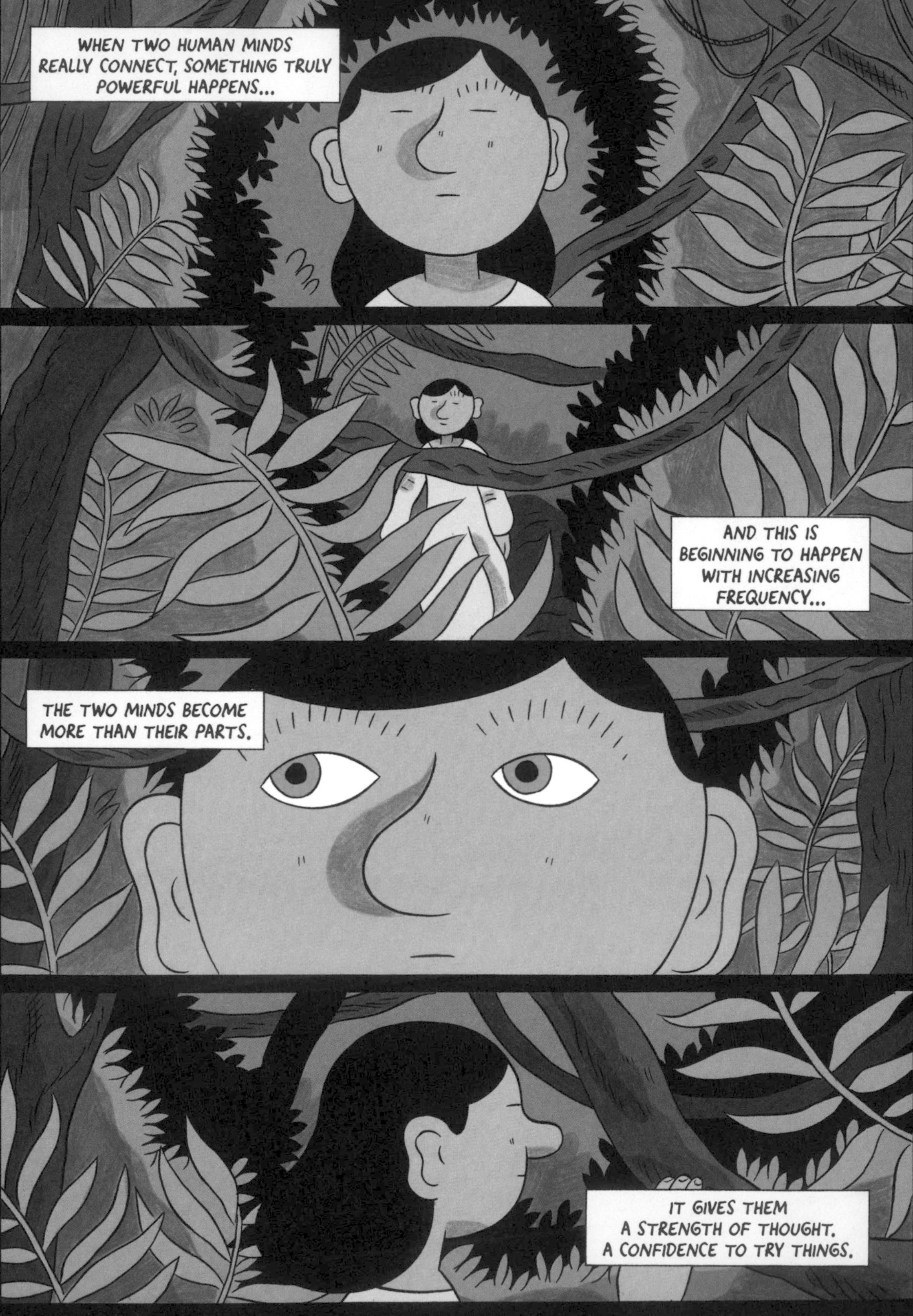
WHEN TWO HUMAN MINDS REALLY CONNECT, SOMETHING TRULY POWERFUL HAPPENS...
AND THIS IS BEGINNING TO HAPPEN WITH INCREASING FREQUENCY...
THE TWO MINDS BECOME MORE THAN THEIR PARTS.
IT GIVES THEM A STRENGTH OF THOUGHT. A CONFIDENCE TO TRY THINGS.

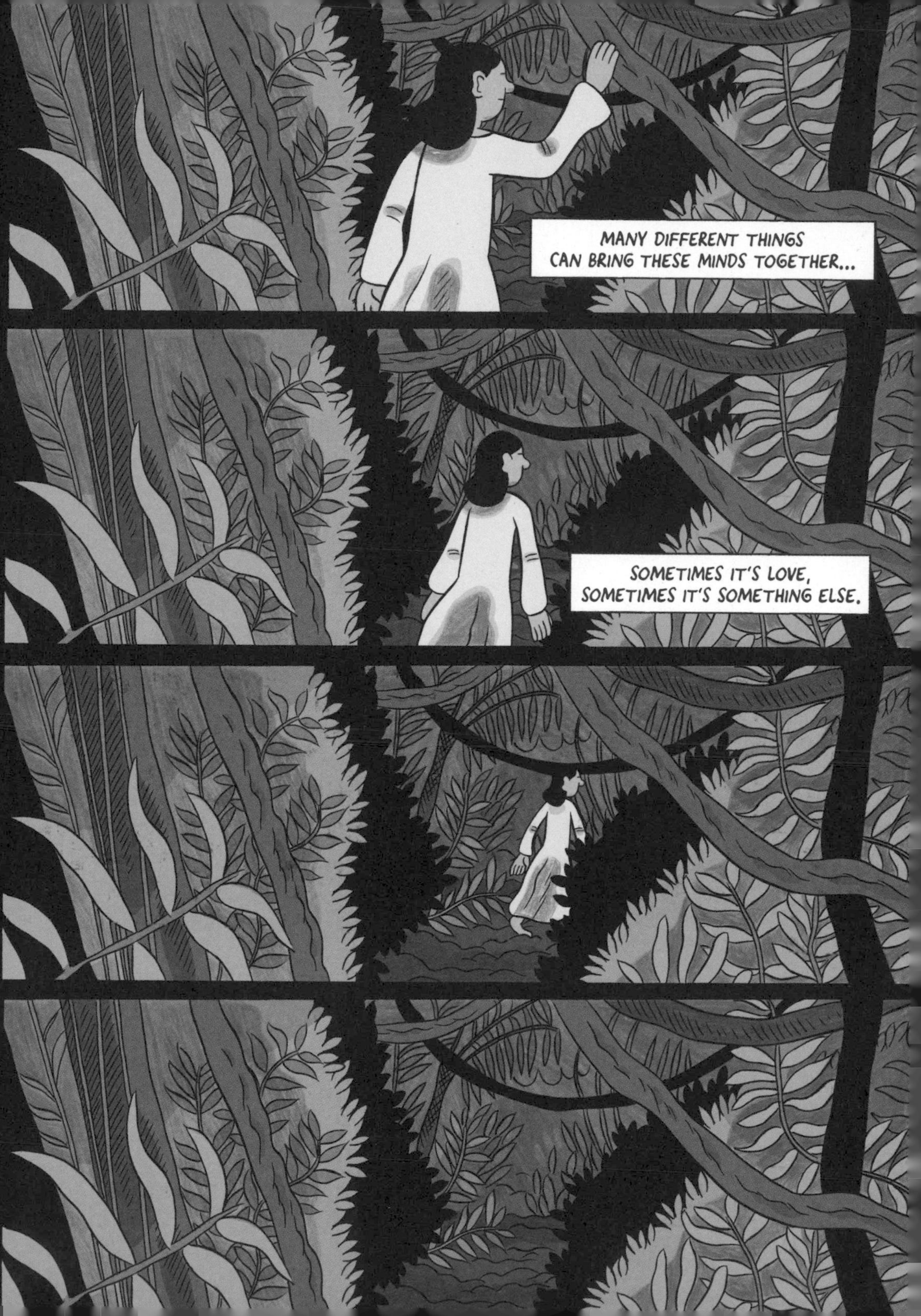
MANY DIFFERENT THINGS CAN BRING THESE MINDS TOGETHER...
SOMETIMES IT'S LOVE, SOMETIMES IT'S SOMETHING ELSE.

457

years later...

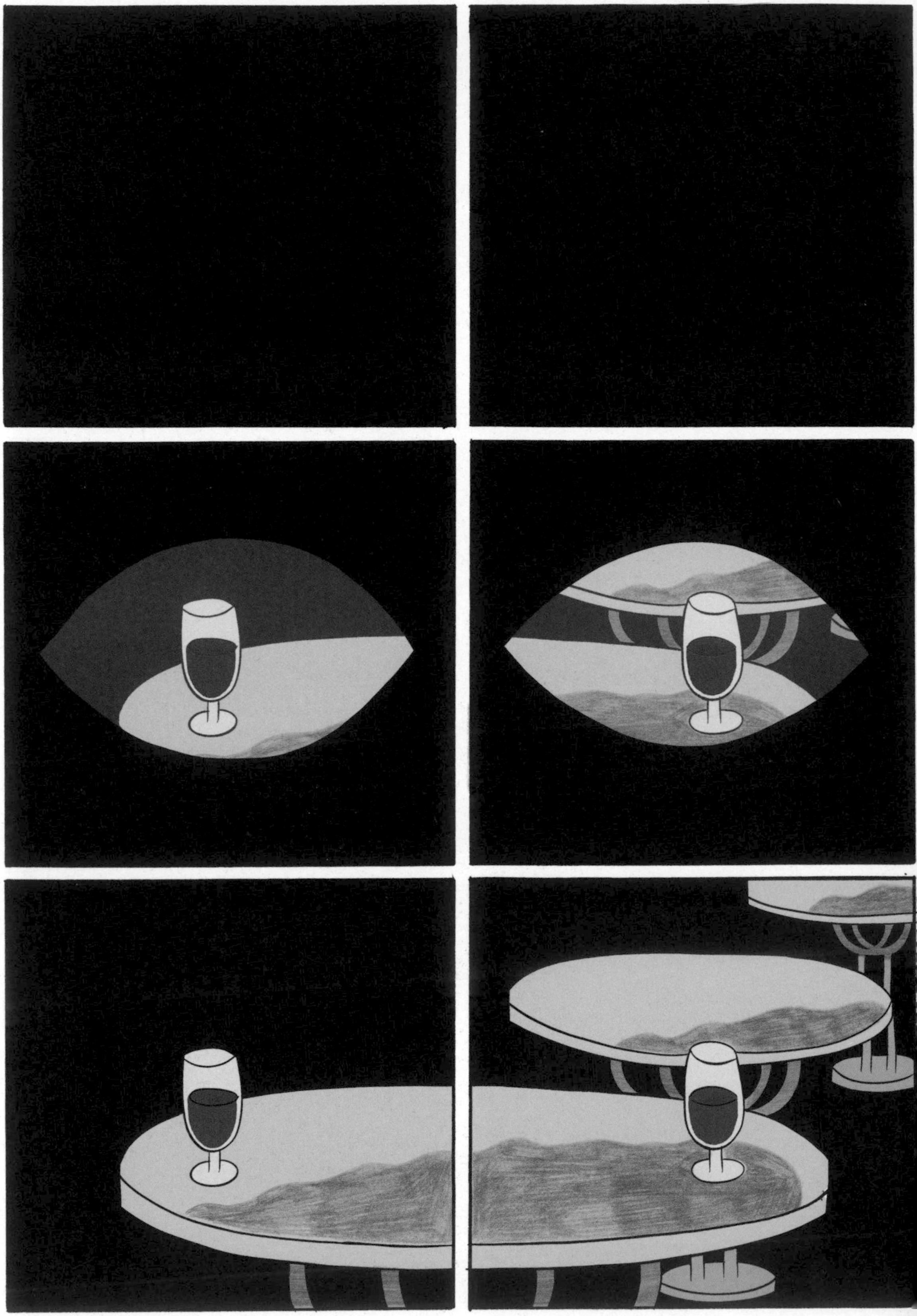

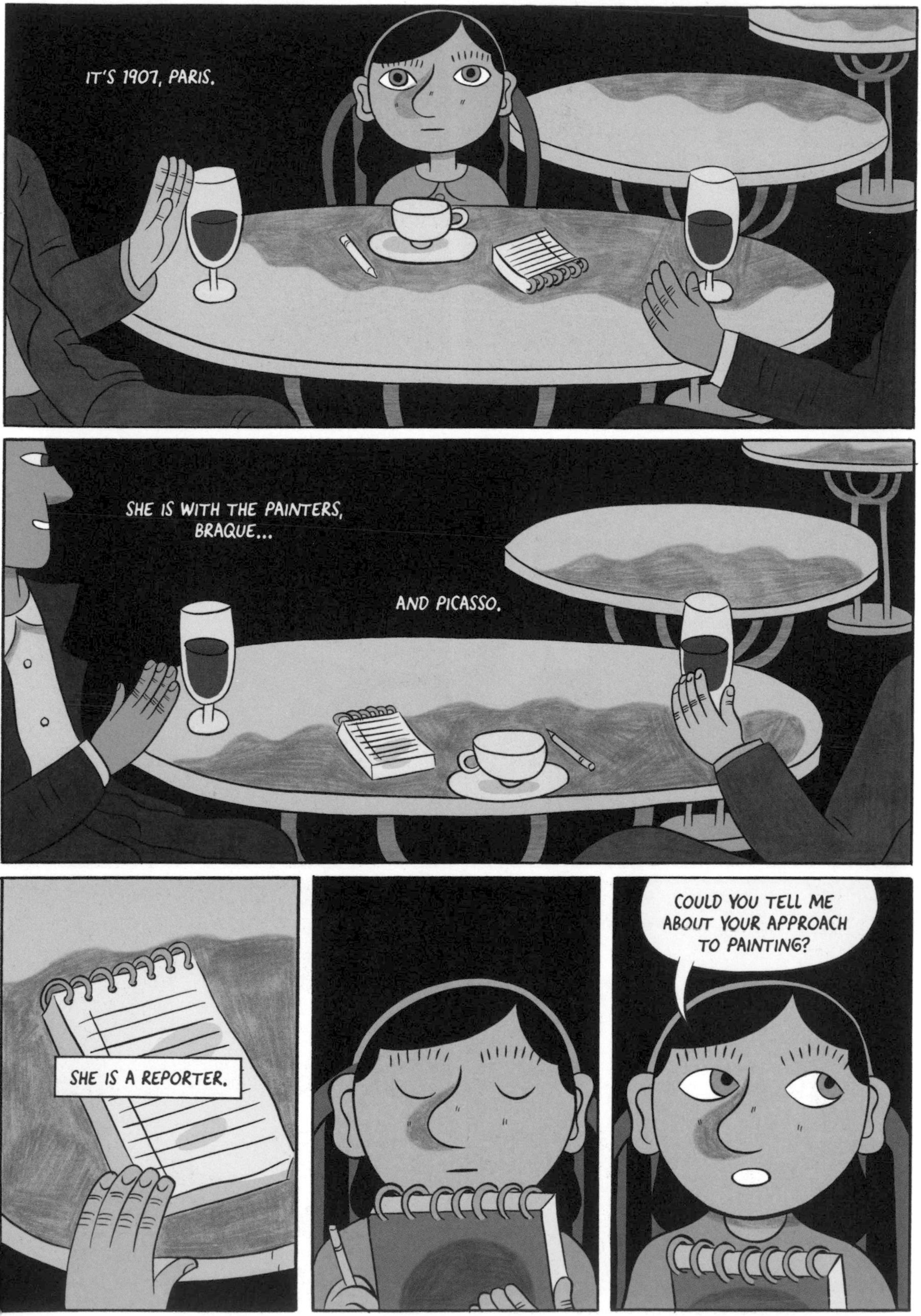
IT'S 1907, PARIS.
SHE IS WITH THE PAINTERS, BRAQUE...
AND PICASSO.
SHE IS A REPORTER.
COULD YOU TELL ME ABOUT YOUR APPROACH TO PAINTING?

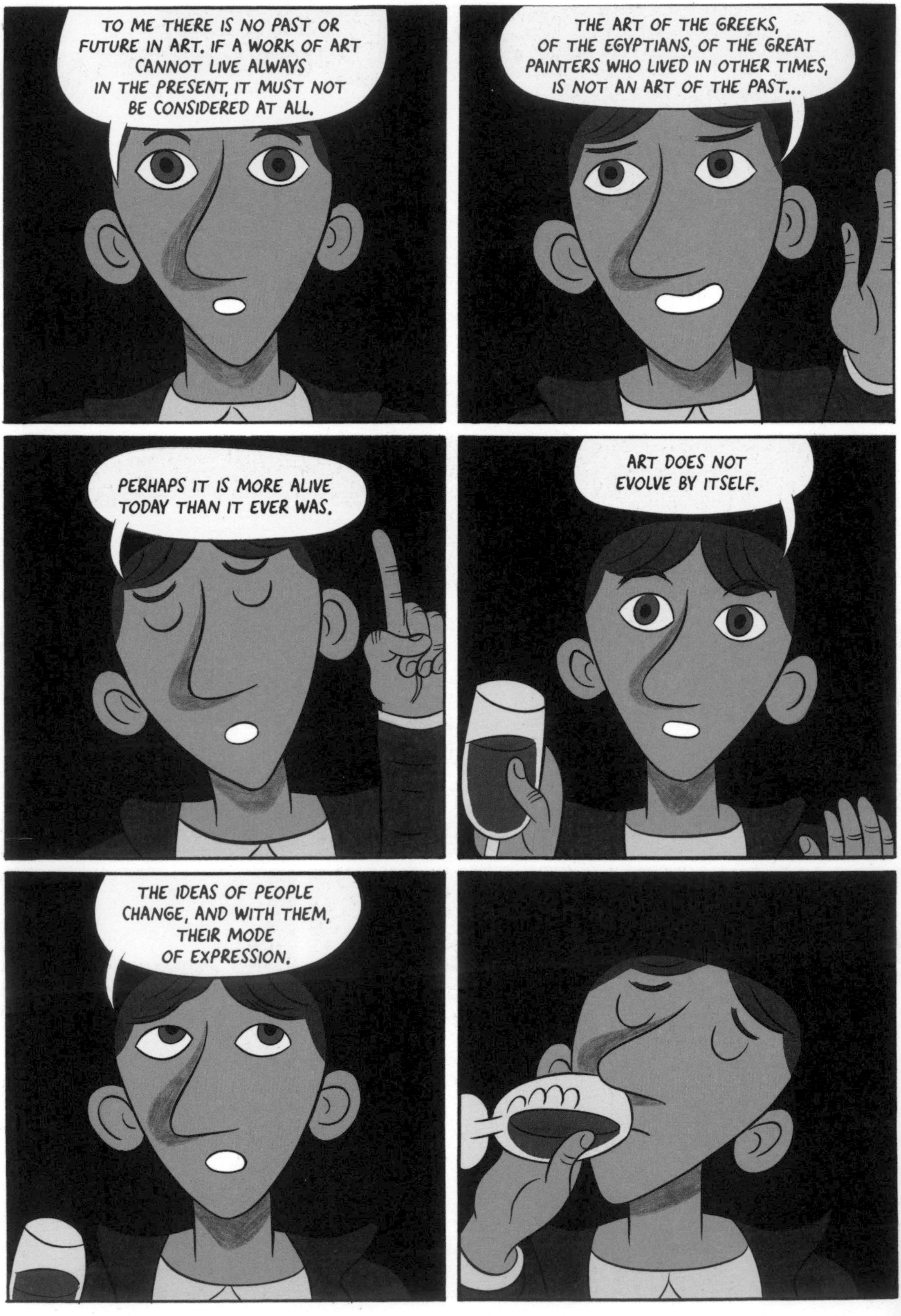
TO ME THERE IS NO PAST OR FUTURE IN ART. IF A WORK OF ART CANNOT LIVE ALWAYS IN THE PRESENT, IT MUST NOT BE CONSIDERED AT ALL.
THE ART OF THE GREEKS, OF THE EGYPTIANS, OF THE GREAT PAINTERS WHO LIVED IN OTHER TIMES, IS NOT AN ART OF THE PAST...
PERHAPS IT IS MORE ALIVE TODAY THAN IT EVER WAS.
ART DOES NOT EVOLVE BY ITSELF.
THE IDEAS OF PEOPLE CHANGE, AND WITH THEM, THEIR MODE OF EXPRESSION.

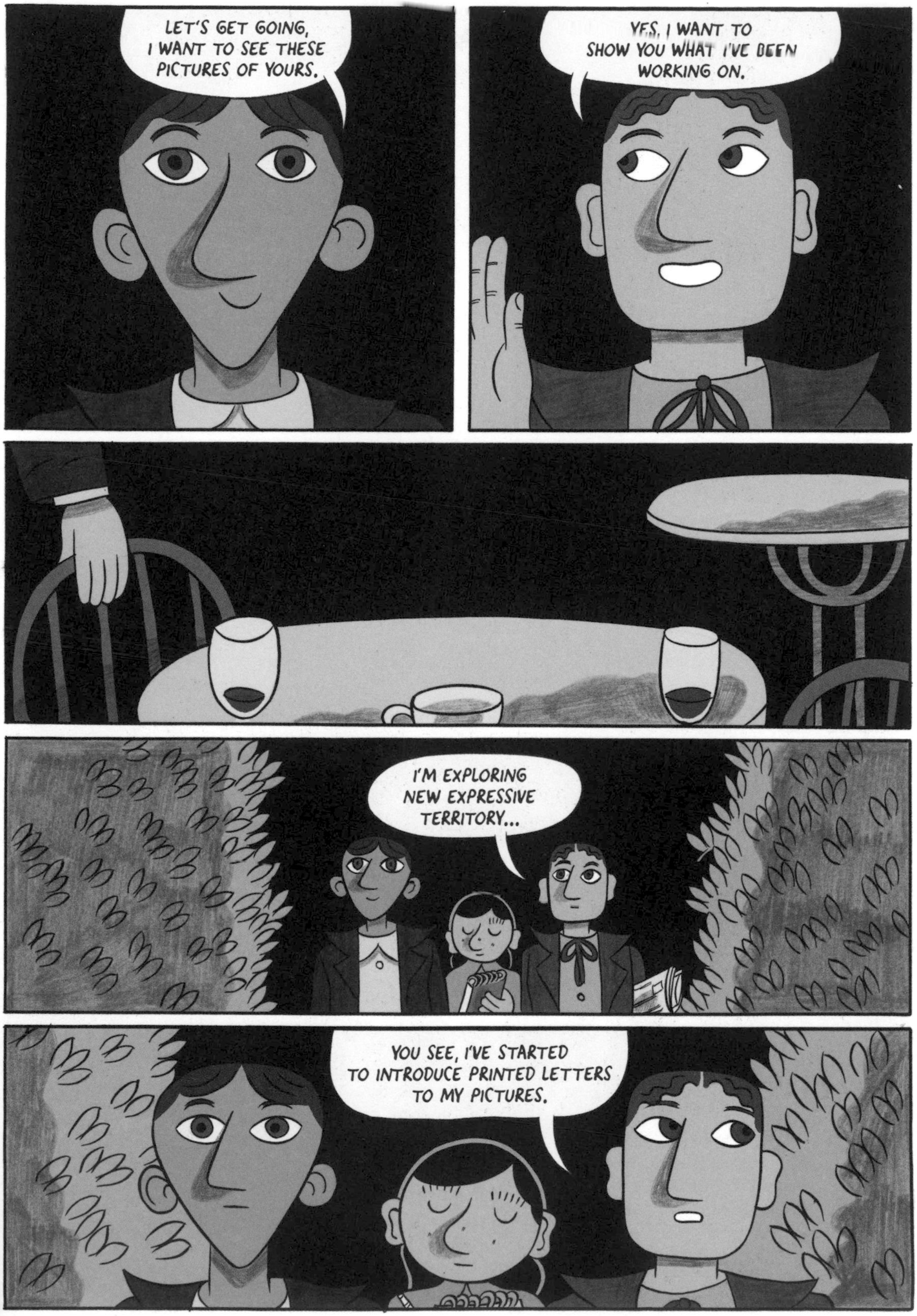
LET'S GET GOING, I WANT TO SEE THESE PICTURES OF YOURS.
YES, I WANT TO SHOW YOU WHAT I'VE BEEN WORKING ON.
I'M EXPLORING NEW EXPRESSIVE TERRITORY...
YOU SEE, I'VE STARTED TO INTRODUCE PRINTED LETTERS TO MY PICTURES.

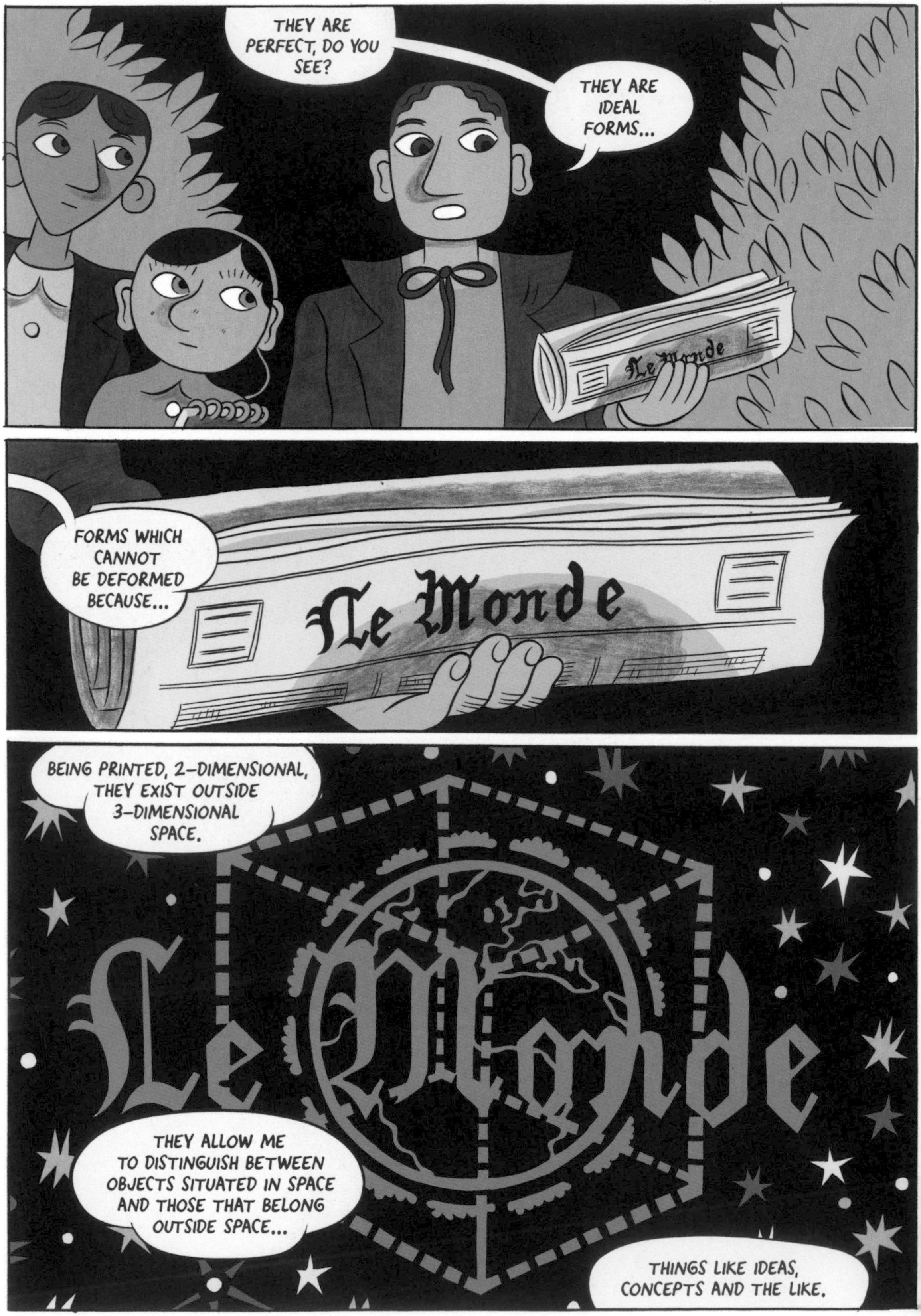
THEY ARE PERFECT, DO YOU SEE?
THEY ARE IDEAL FORMS...
Le Monde
FORMS WHICH CANNOT BE DEFORMED BECAUSE...
Le Monde
BEING PRINTED, 2-DIMENSIONAL, THEY EXIST OUTSIDE 3-DIMENSIONAL SPACE.
Le Monde
THEY ALLOW ME TO DISTINGUISH BETWEEN OBJECTS SITUATED IN SPACE AND THOSE THAT BELONG OUTSIDE SPACE...
THINGS LIKE IDEAS, CONCEPTS AND THE LIKE.

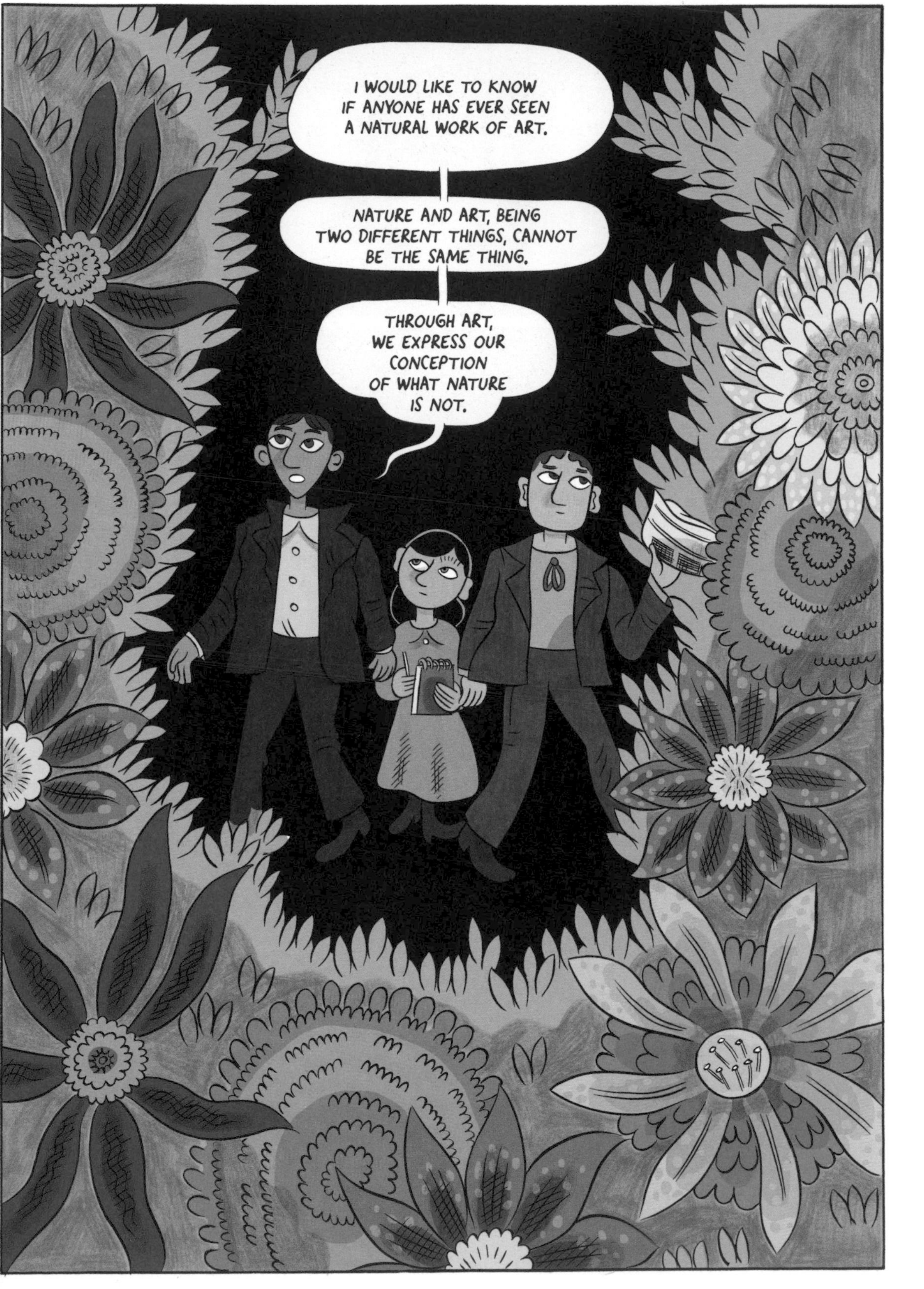
I WOULD LIKE TO KNOW IF ANYONE HAS EVER SEEN A NATURAL WORK OF ART.
NATURE AND ART, BEING TWO DIFFERENT THINGS, CANNOT BE THE SAME THING.
THROUGH ART, WE EXPRESS OUR CONCEPTION OF WHAT NATURE IS NOT.

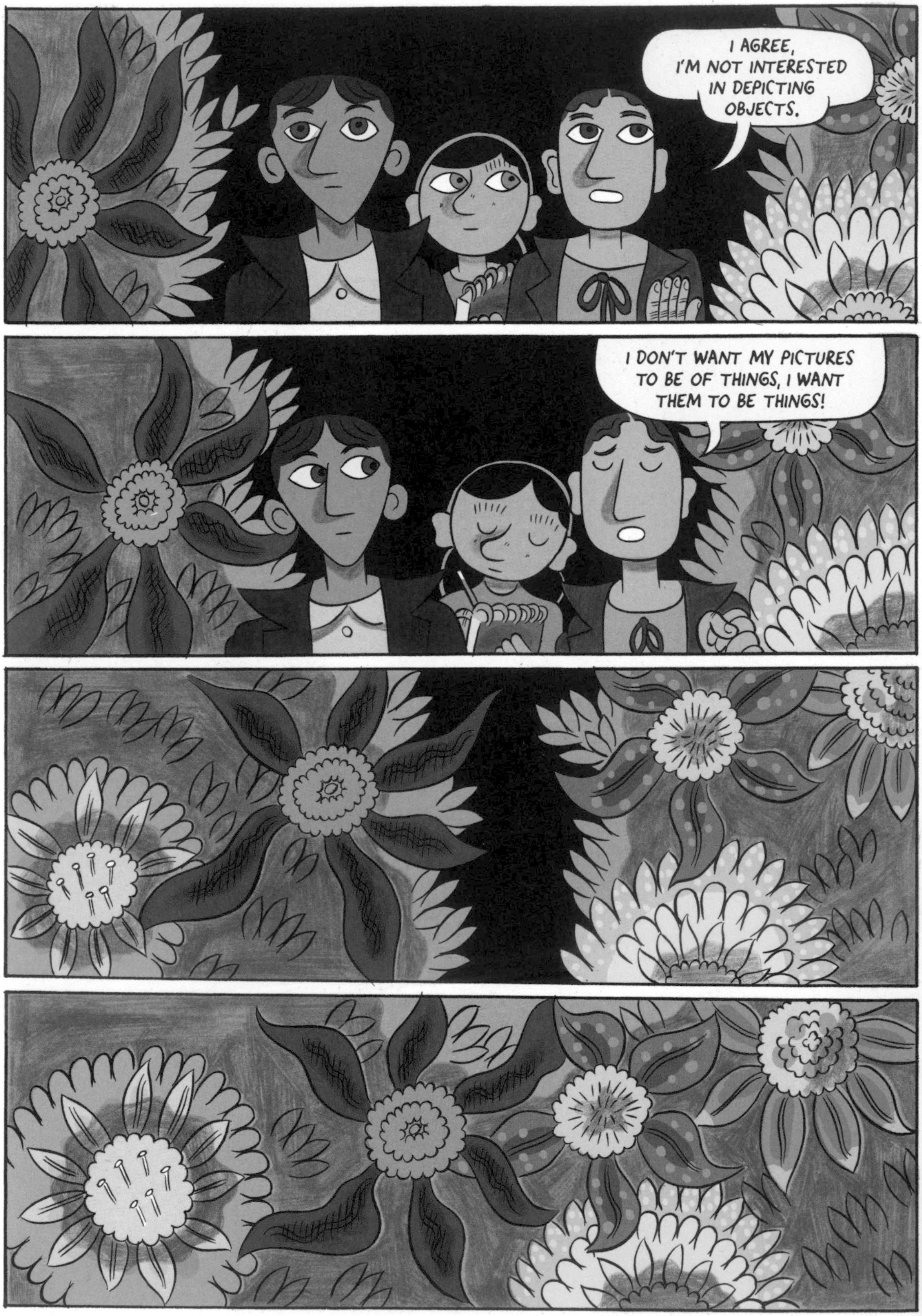
I AGREE, I'M NOT INTERESTED IN DEPICTING OBJECTS.
I DON'T WANT MY PICTURES TO BE OF THINGS, I WANT THEM TO BE THINGS!

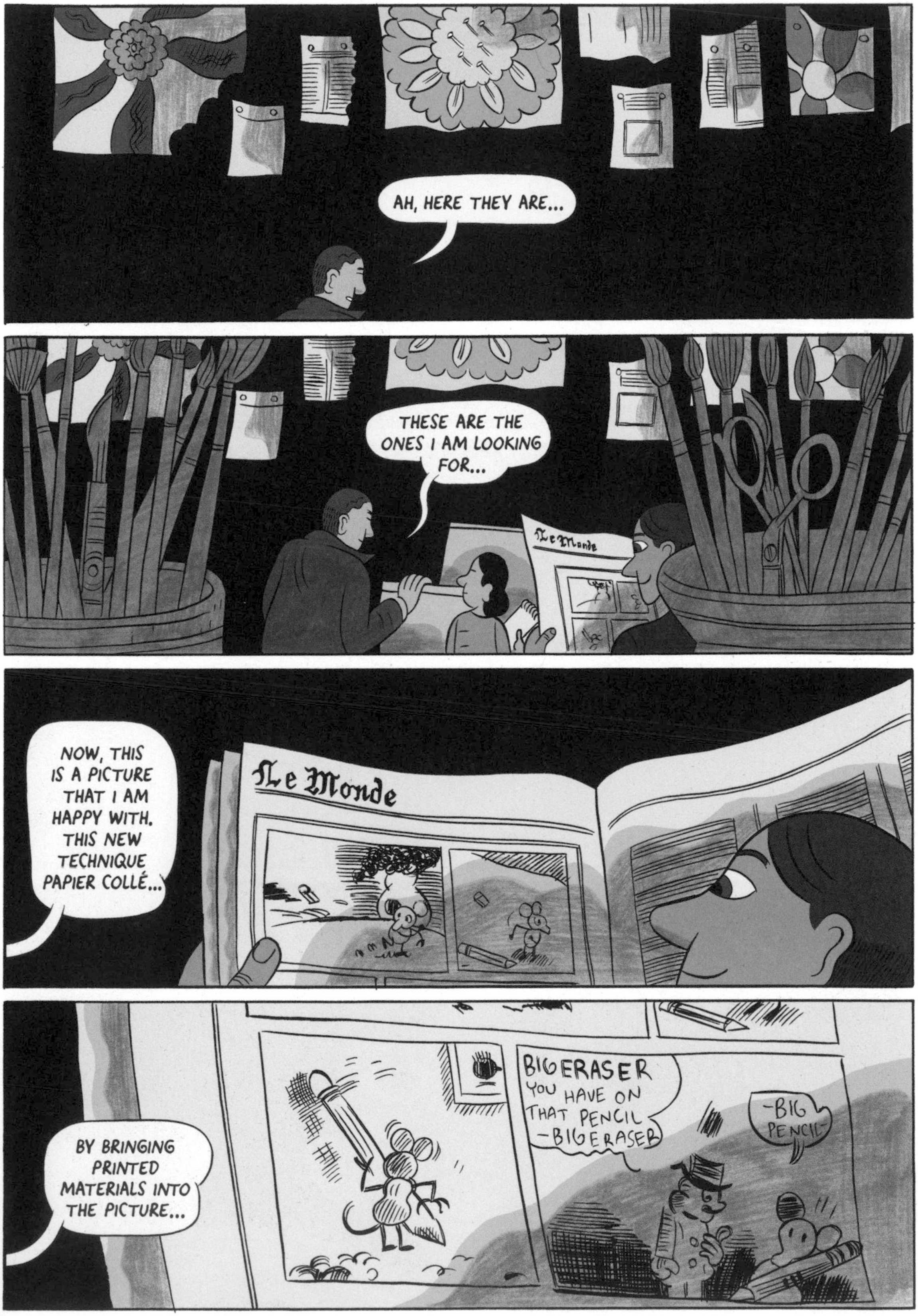
AH, HERE THEY ARE...
THESE ARE THE ONES I AM LOOKING FOR...
Le Monde
NOW, THIS IS A PICTURE THAT I AM HAPPY WITH. THIS NEW TECHNIQUE PAPIER COLLÉ...
Le Monde
BY BRINGING PRINTED MATERIALS INTO THE PICTURE...
BIGERASER YOU HAVE ON THAT PENCIL -BIGERASER
-BIG PENCIL-

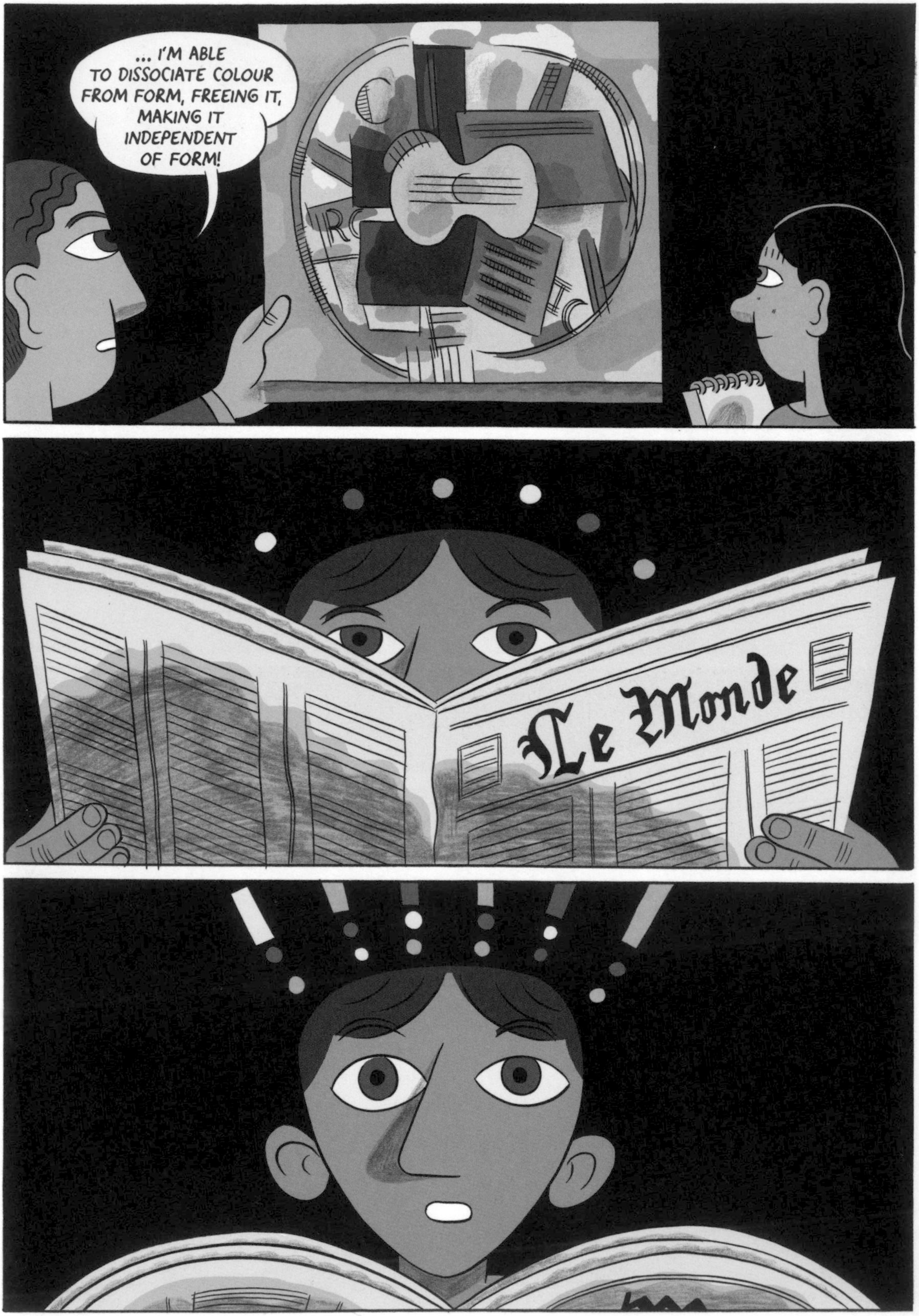
... I'M ABLE TO DISSOCIATE COLOUR FROM FORM, FREEING IT, MAKING IT INDEPENDENT OF FORM!
Le Monde

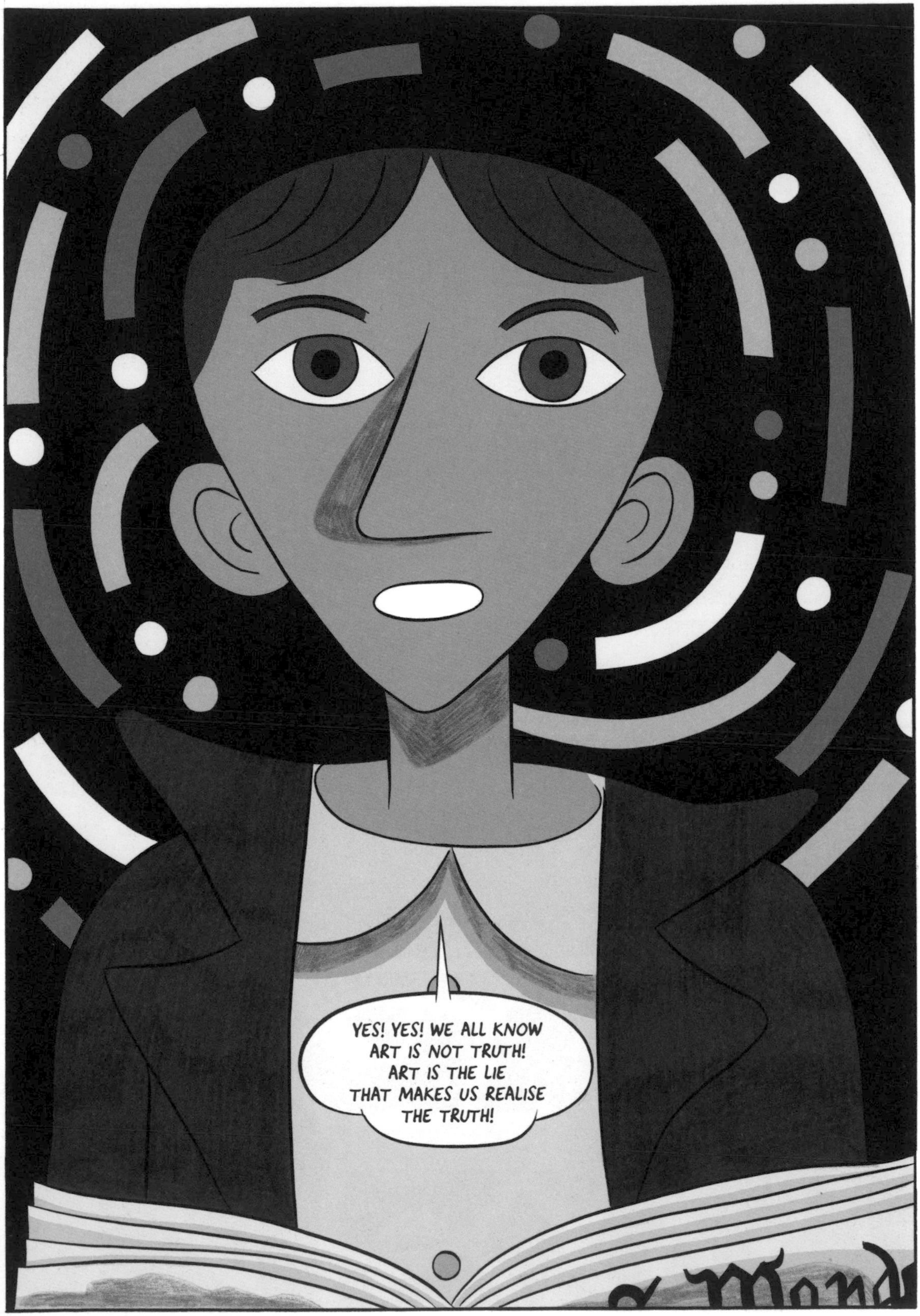
YES! YES! WE ALL KNOW ART IS NOT TRUTH! ART IS THE LIE THAT MAKES US REALISE THE TRUTH!

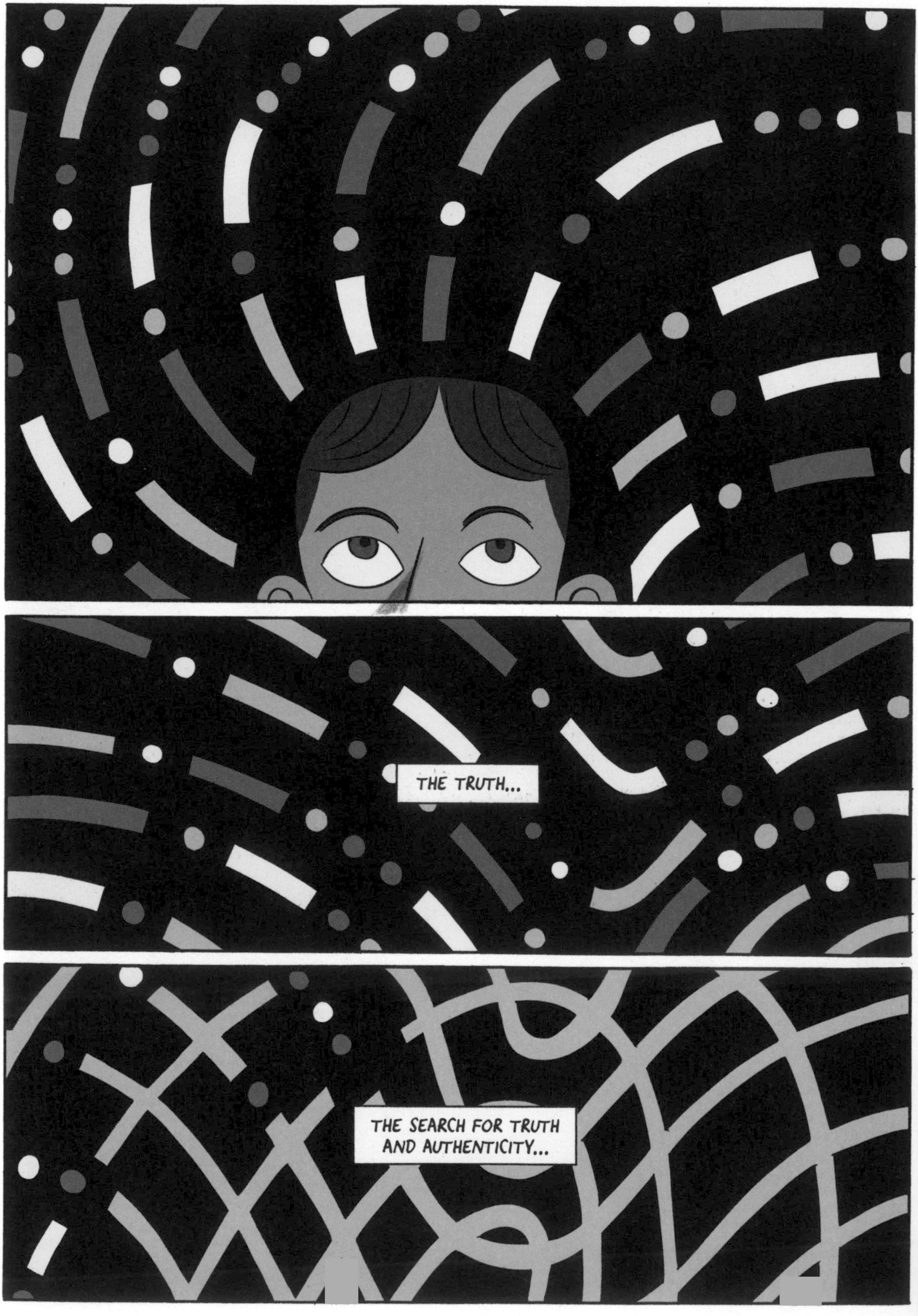
THE TRUTH...
THE SEARCH FOR TRUTH AND AUTHENTICITY...

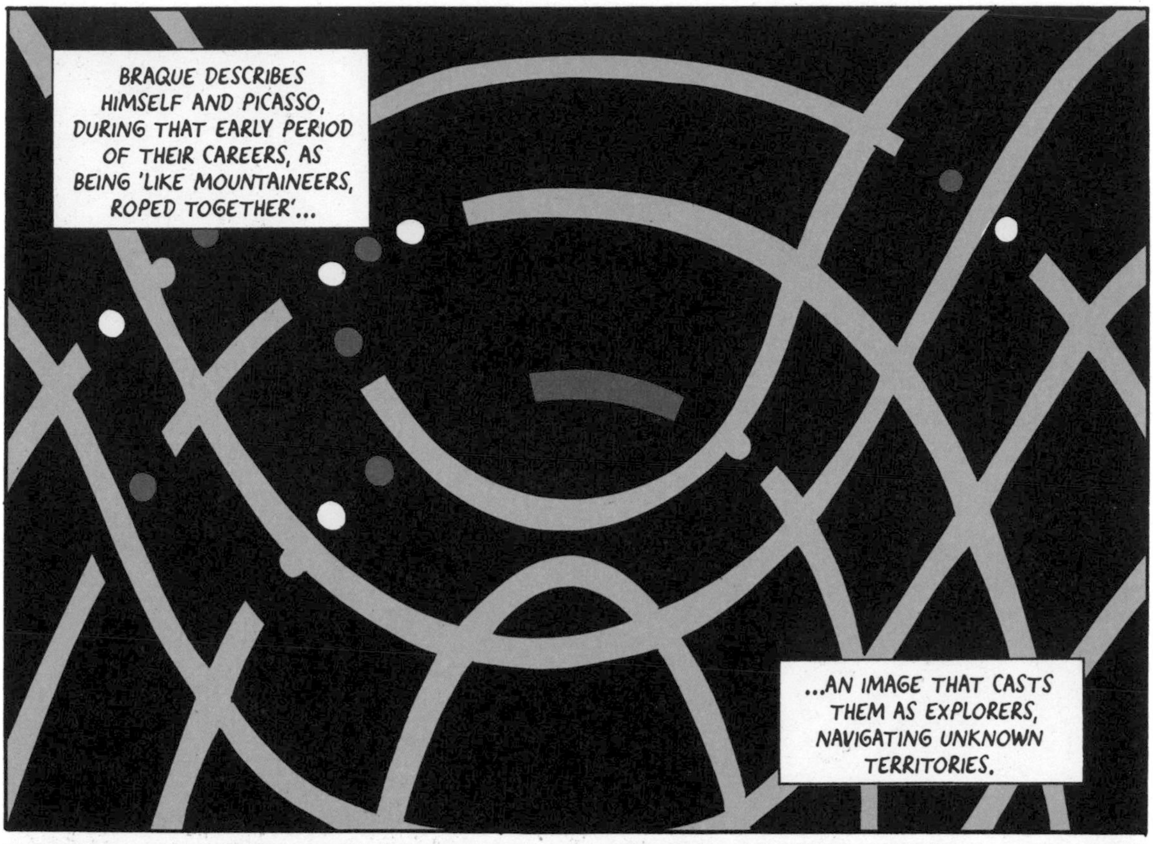
BRAQUE DESCRIBES HIMSELF AND PICASSO, DURING THAT EARLY PERIOD OF THEIR CAREERS, AS BEING 'LIKE MOUNTAINEERS, ROPED TOGETHER'...
...AN IMAGE THAT CASTS THEM AS EXPLORERS, NAVIGATING UNKNOWN TERRITORIES.

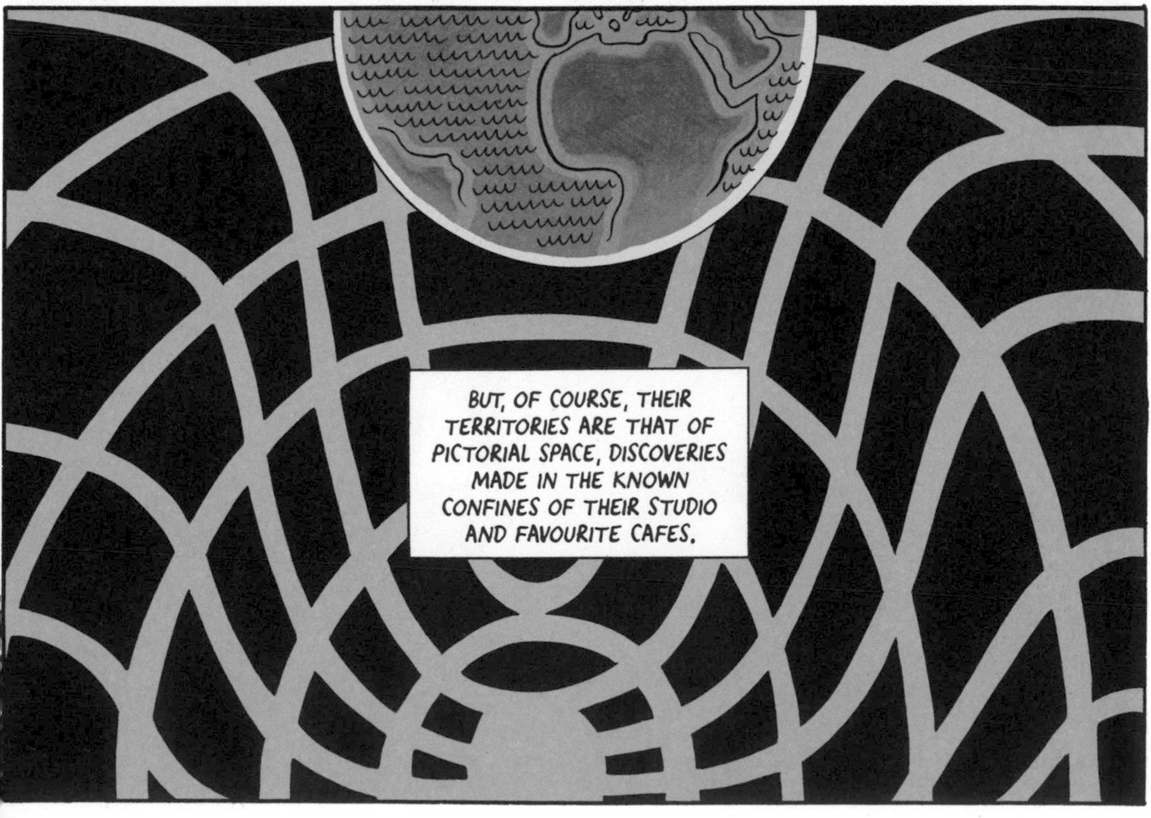
BUT, OF COURSE, THEIR TERRITORIES ARE THAT OF PICTORIAL SPACE, DISCOVERIES MADE IN THE KNOWN CONFINES OF THEIR STUDIO AND FAVOURITE CAFES.

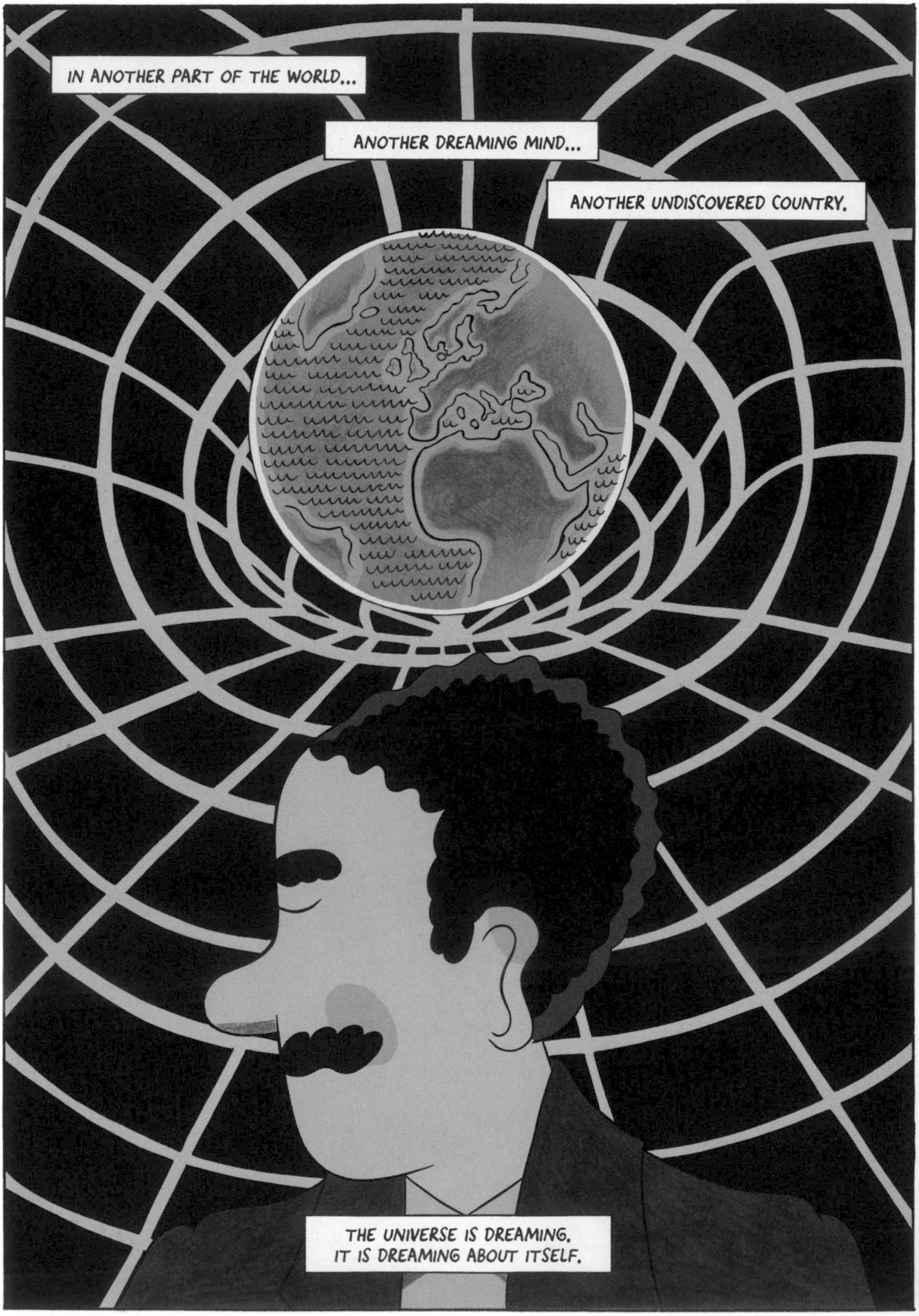
IN ANOTHER PART OF THE WORLD...
ANOTHER DREAMING MIND...
ANOTHER UNDISCOVERED COUNTRY.
THE UNIVERSE IS DREAMING.
IT IS DREAMING ABOUT ITSELF.

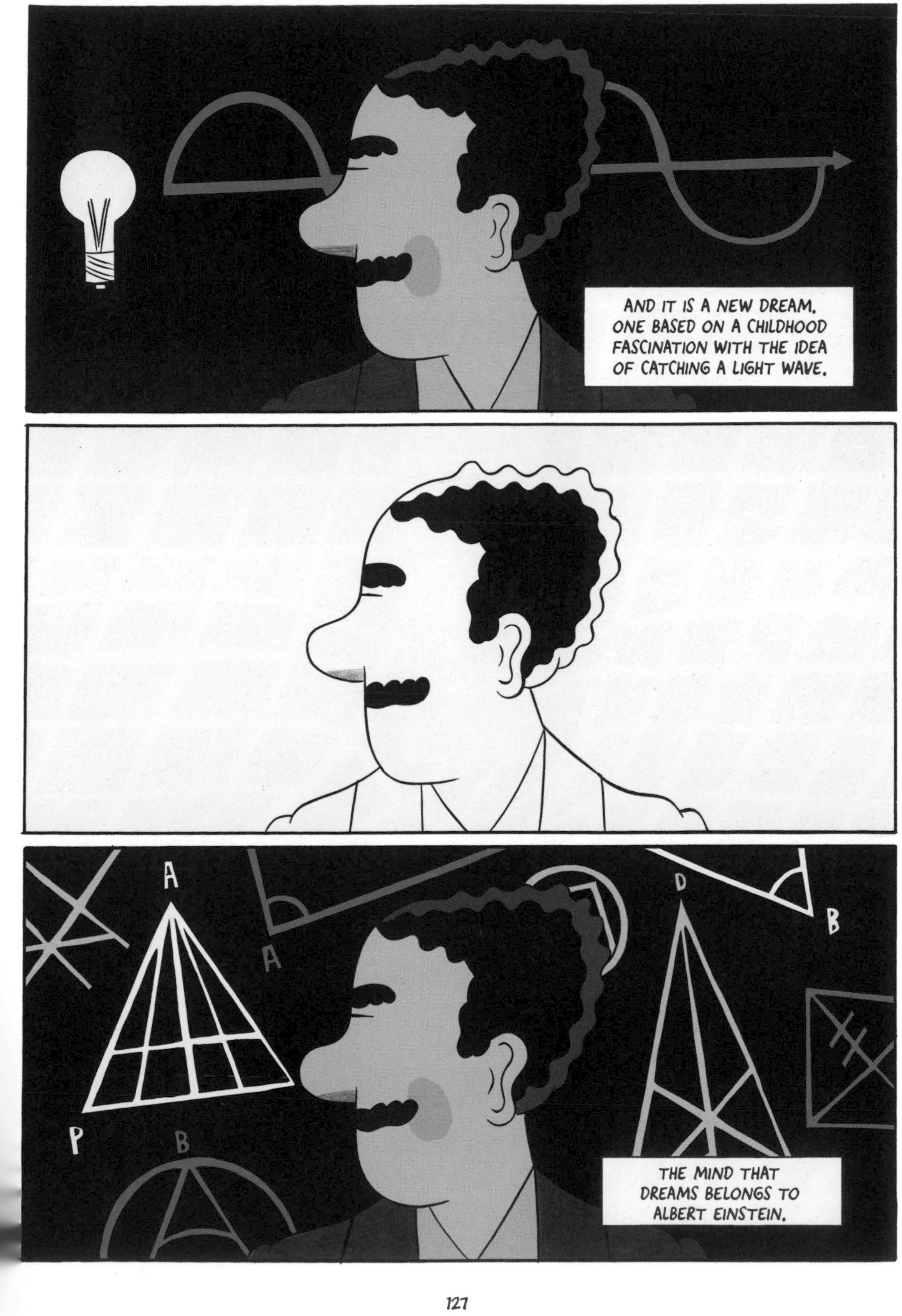
AND IT IS A NEW DREAM. ONE BASED ON A CHILDHOOD FASCINATION WITH THE IDEA OF CATCHING A LIGHT WAVE.
THE MIND THAT DREAMS BELONGS TO ALBERT EINSTEIN.

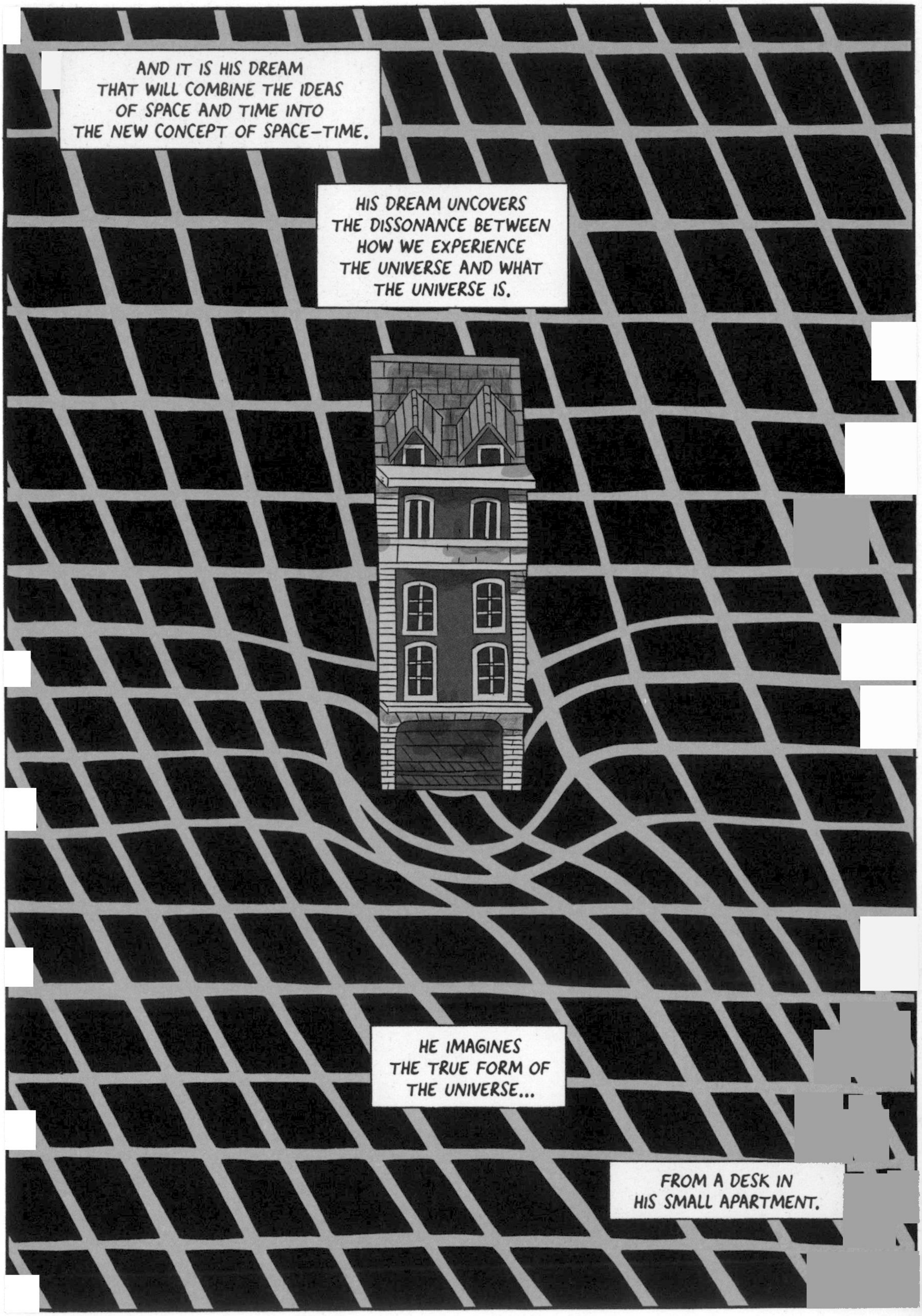
AND IT IS HIS DREAM THAT WILL COMBINE THE IDEAS OF SPACE AND TIME INTO THE NEW CONCEPT OF SPACE-TIME.
HIS DREAM UNCOVERS THE DISSONANCE BETWEEN HOW WE EXPERIENCE THE UNIVERSE AND WHAT THE UNIVERSE IS.
HE IMAGINES THE TRUE FORM OF THE UNIVERSE...
FROM A DESK IN HIS SMALL APARTMENT.

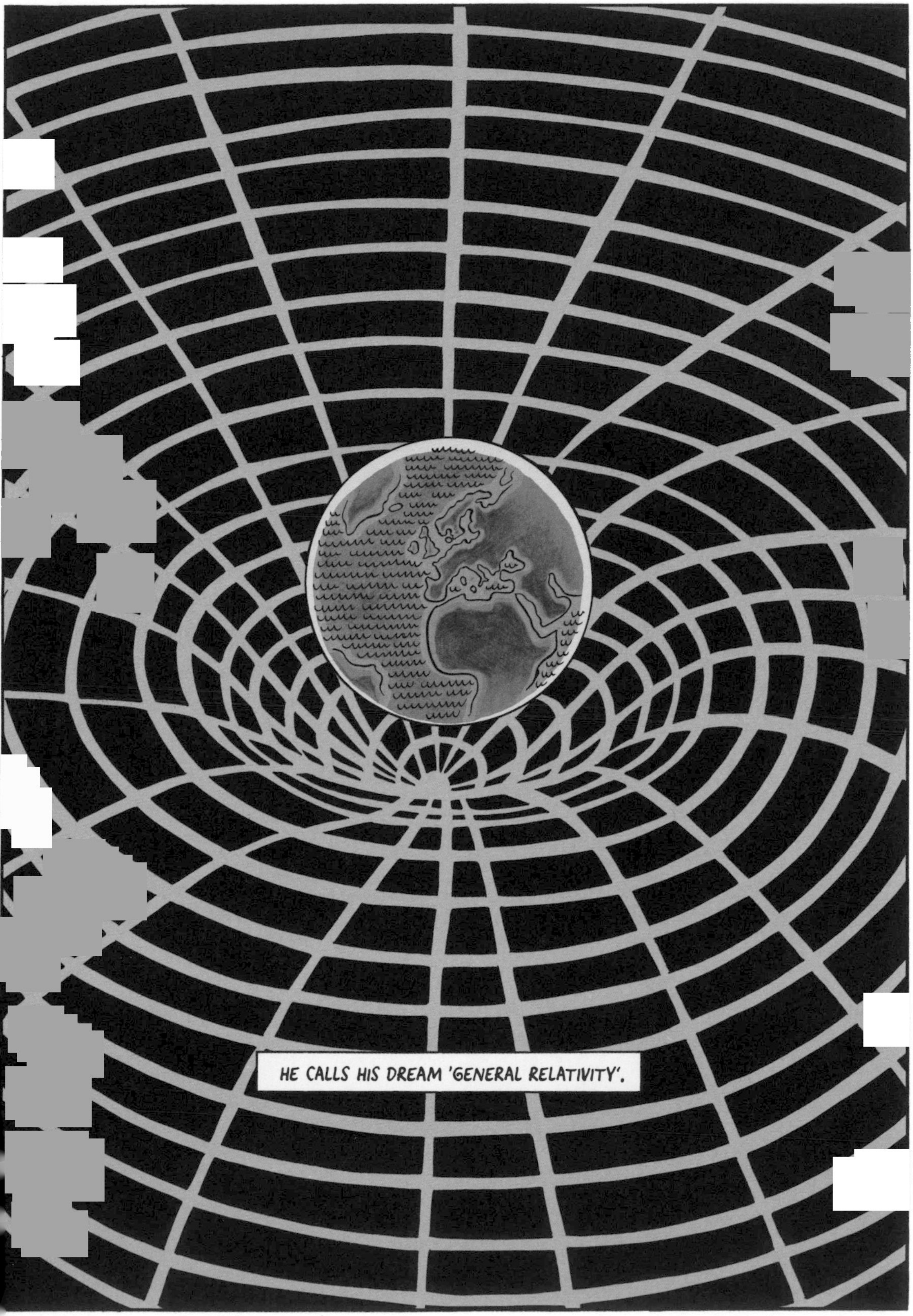
HE CALLS HIS DREAM 'GENERAL RELATIVITY'.

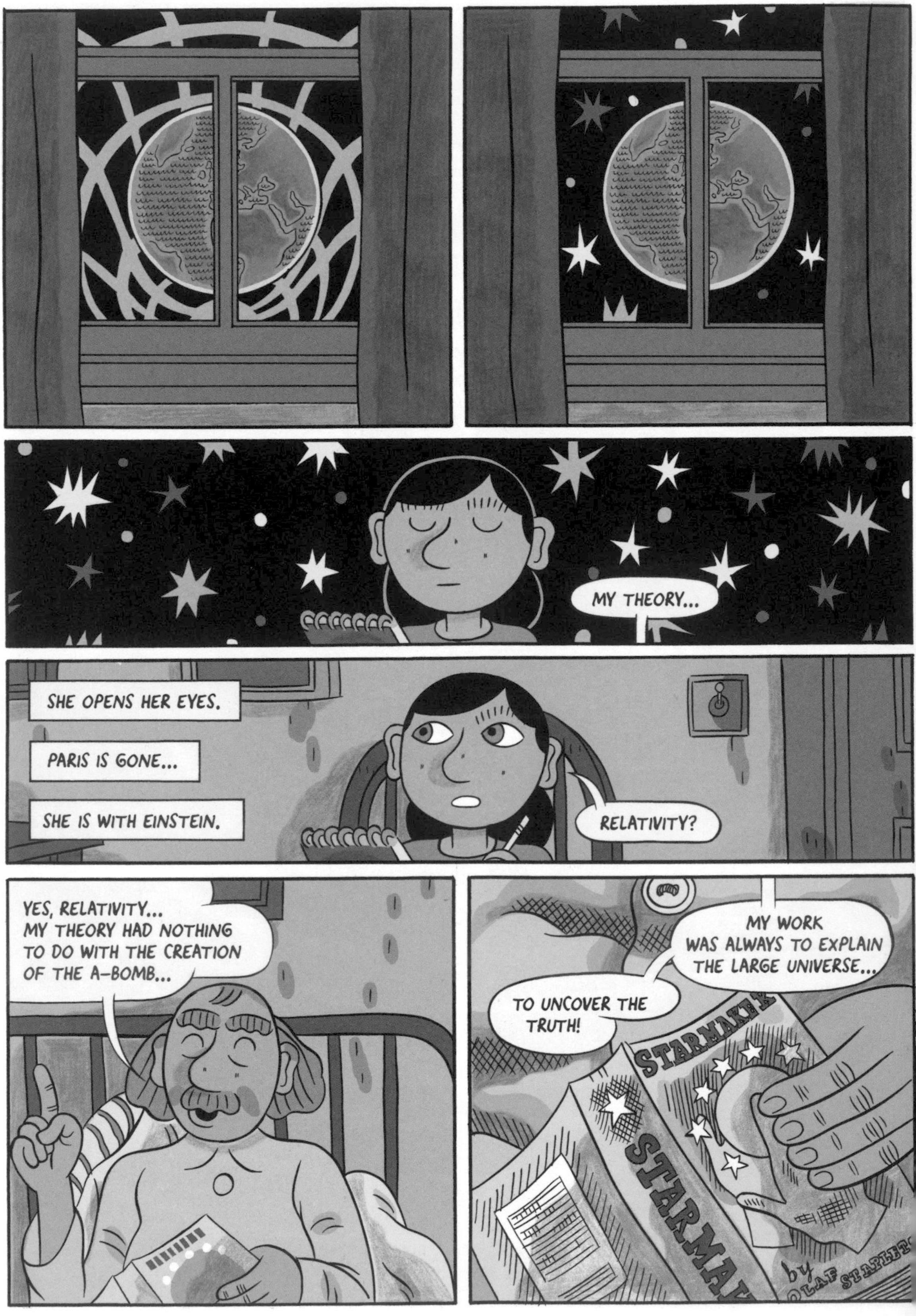
MY THEORY...
SHE OPENS HER EYES.
PARIS IS GONE...
SHE IS WITH EINSTEIN.
RELATIVITY?
YES, RELATIVITY... MY THEORY HAD NOTHING TO DO WITH THE CREATION OF THE A-BOMB...
MY WORK WAS ALWAYS TO EXPLAIN THE LARGE UNIVERSE...
TO UNCOVER THE TRUTH!
STARMAKER
STARMAK
by
OLAF STAPLET

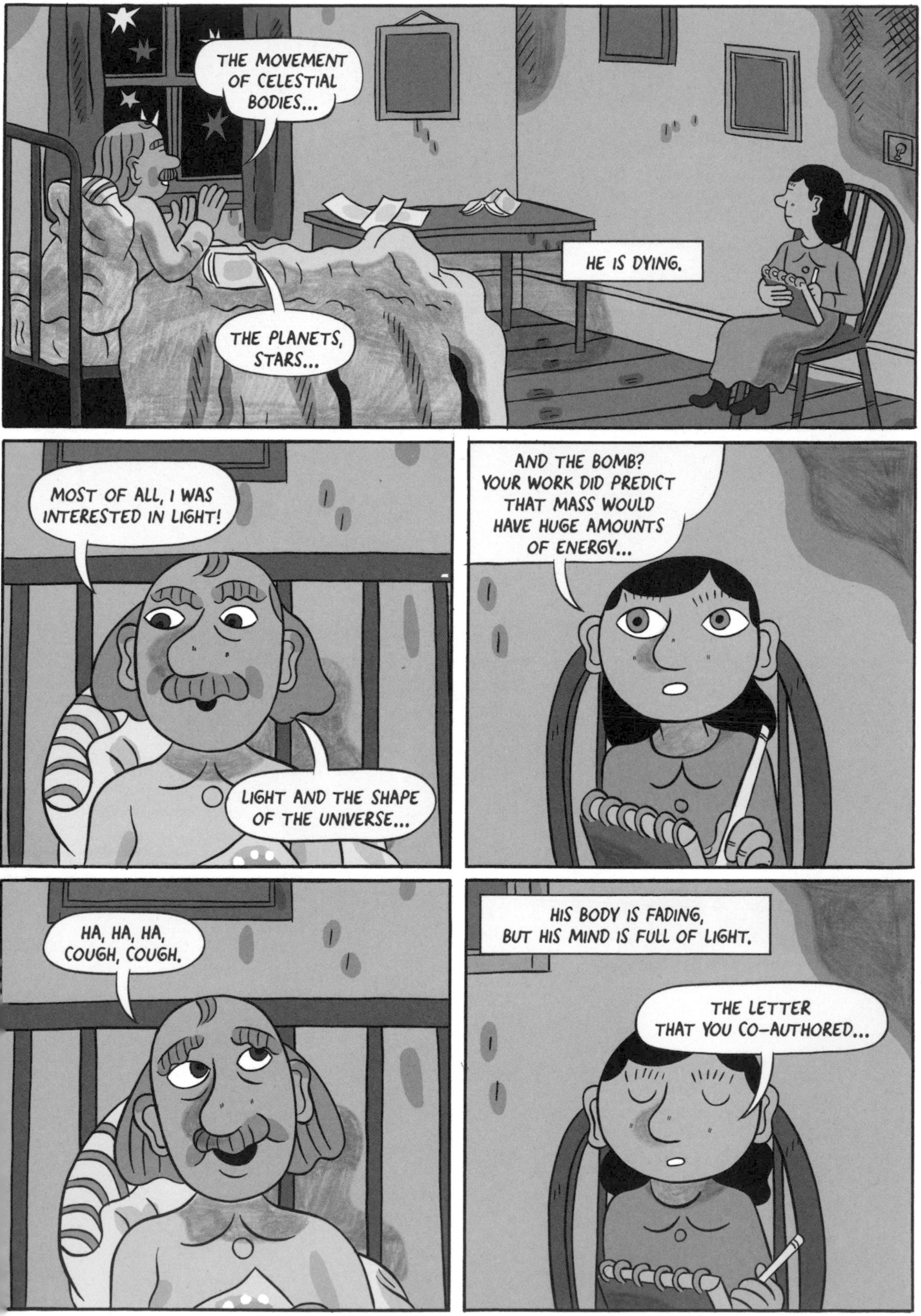
THE MOVEMENT OF CELESTIAL BODIES...
HE IS DYING.
THE PLANETS, STARS...
MOST OF ALL, I WAS INTERESTED IN LIGHT!
LIGHT AND THE SHAPE OF THE UNIVERSE...
AND THE BOMB? YOUR WORK DID PREDICT THAT MASS WOULD HAVE HUGE AMOUNTS OF ENERGY...
HA, HA, HA, COUGH, COUGH.
HIS BODY IS FADING, BUT HIS MIND IS FULL OF LIGHT.
THE LETTER THAT YOU CO-AUTHORED...

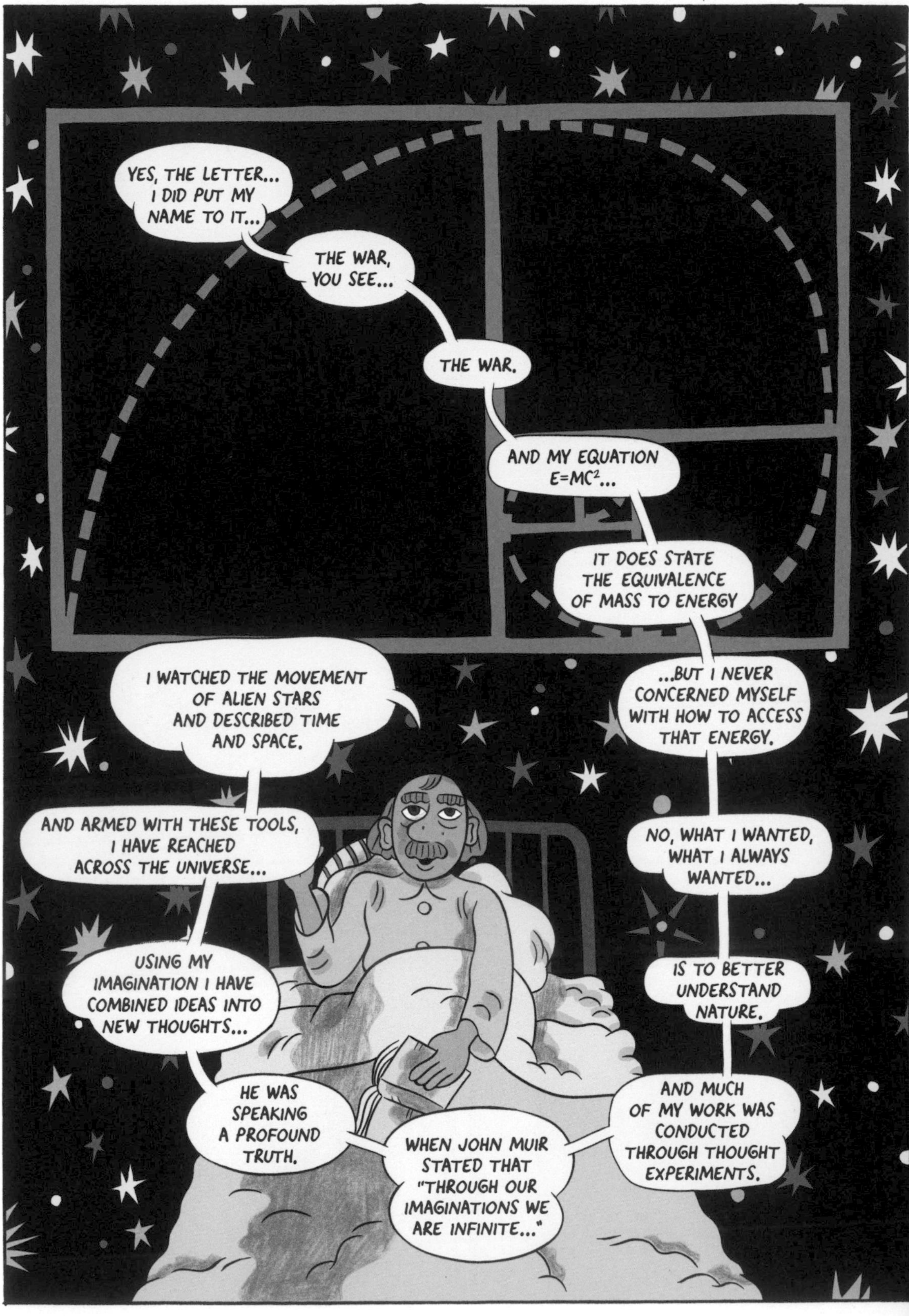
YES, THE LETTER... I DID PUT MY NAME TO IT...
THE WAR, YOU SEE...
THE WAR.
AND MY EQUATION E=MC²...
IT DOES STATE THE EQUIVALENCE OF MASS TO ENERGY
...BUT I NEVER CONCERNED MYSELF WITH HOW TO ACCESS THAT ENERGY.
I WATCHED THE MOVEMENT OF ALIEN STARS AND DESCRIBED TIME AND SPACE.
AND ARMED WITH THESE TOOLS, I HAVE REACHED ACROSS THE UNIVERSE...
USING MY IMAGINATION I HAVE COMBINED IDEAS INTO NEW THOUGHTS...
NO, WHAT I WANTED, WHAT I ALWAYS WANTED...
IS TO BETTER UNDERSTAND NATURE.
AND MUCH OF MY WORK WAS CONDUCTED THROUGH THOUGHT EXPERIMENTS.
WHEN JOHN MUIR STATED THAT "THROUGH OUR IMAGINATIONS WE ARE INFINITE..."
HE WAS SPEAKING A PROFOUND TRUTH.

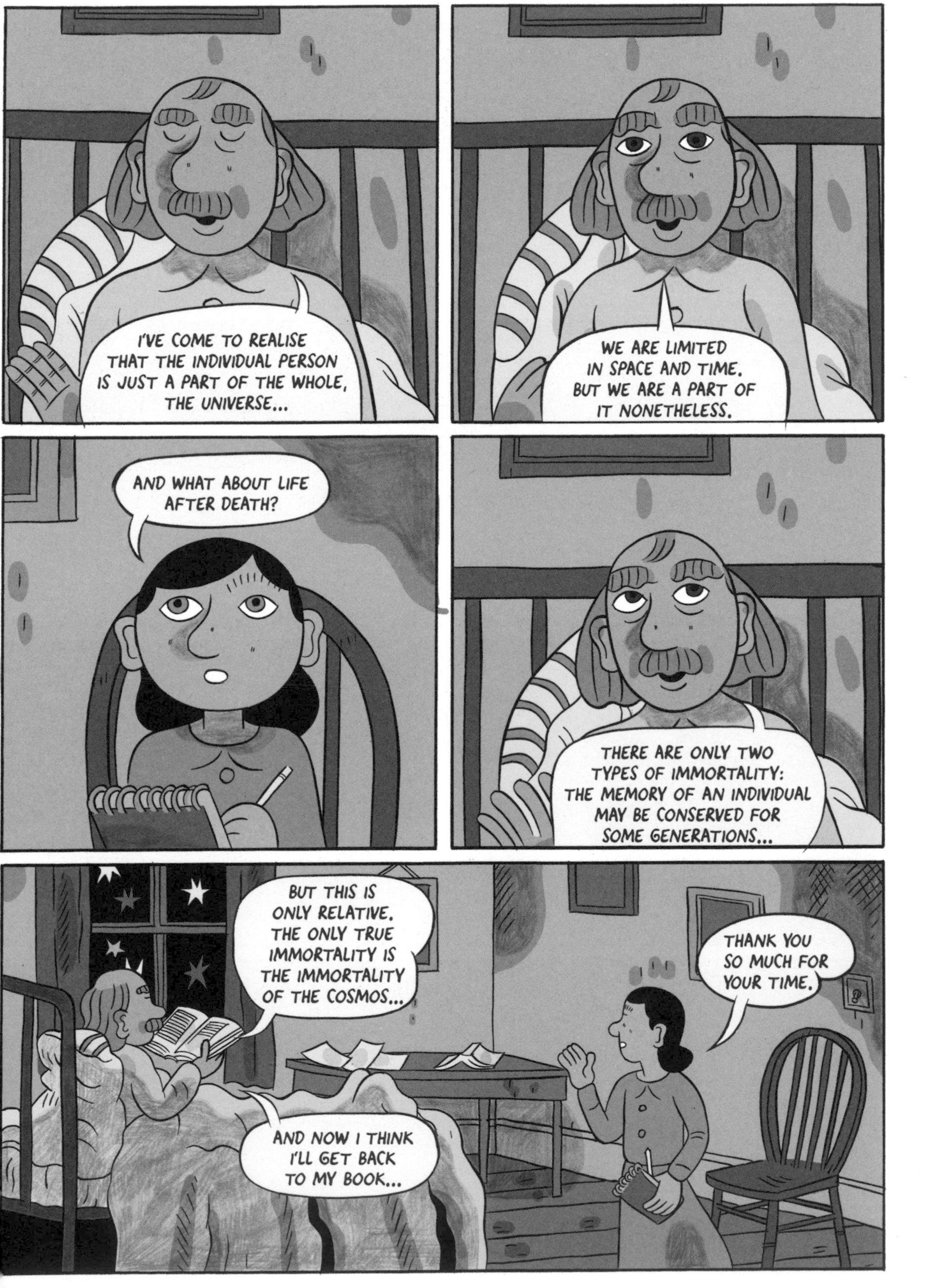
I'VE COME TO REALISE THAT THE INDIVIDUAL PERSON IS JUST A PART OF THE WHOLE, THE UNIVERSE...
WE ARE LIMITED IN SPACE AND TIME. BUT WE ARE A PART OF IT NONETHELESS.
AND WHAT ABOUT LIFE AFTER DEATH?
THERE ARE ONLY TWO TYPES OF IMMORTALITY: THE MEMORY OF AN INDIVIDUAL MAY BE CONSERVED FOR SOME GENERATIONS...
BUT THIS IS ONLY RELATIVE. THE ONLY TRUE IMMORTALITY IS THE IMMORTALITY OF THE COSMOS...
AND NOW I THINK I'LL GET BACK TO MY BOOK...
THANK YOU SO MUCH FOR YOUR TIME.

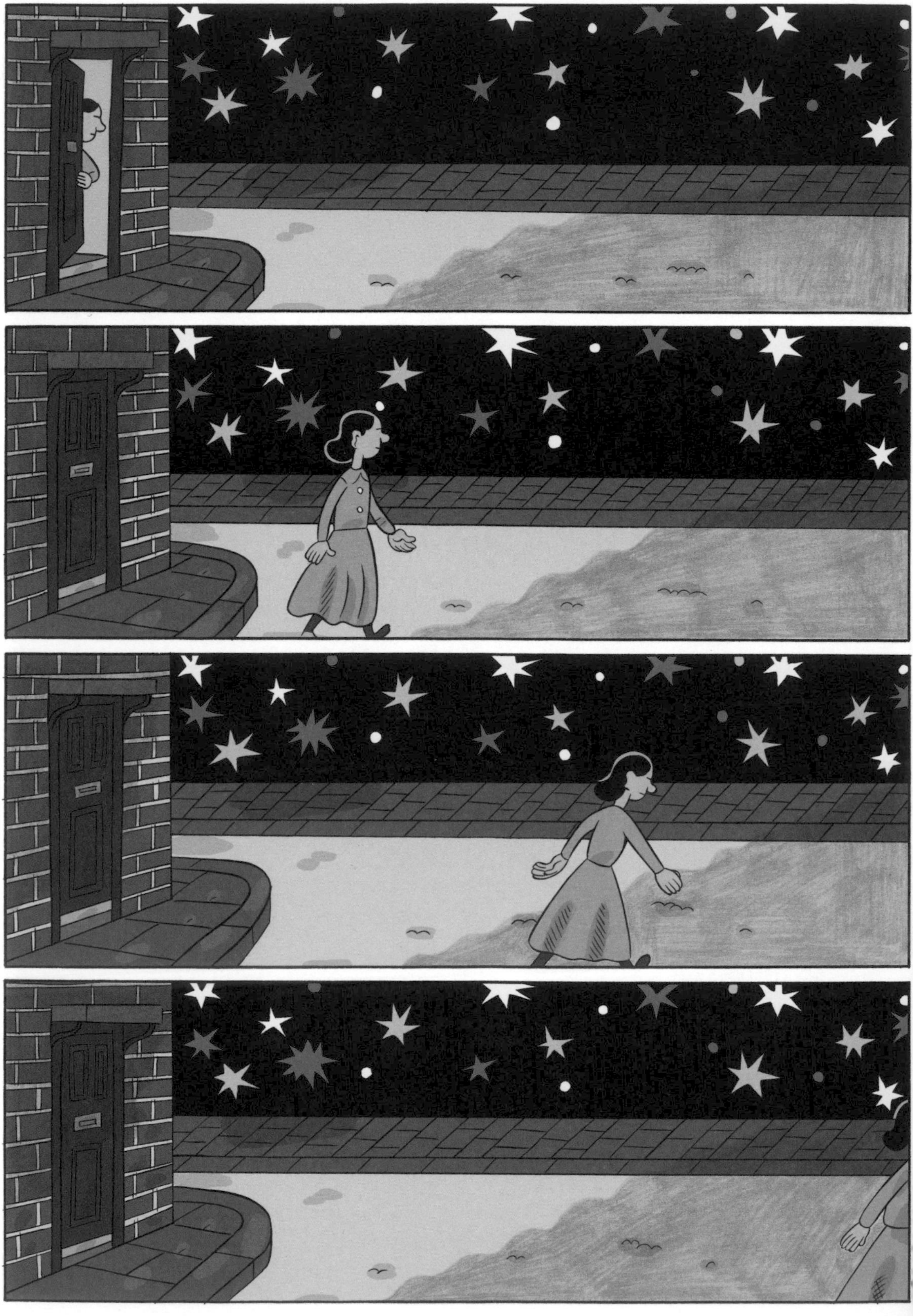

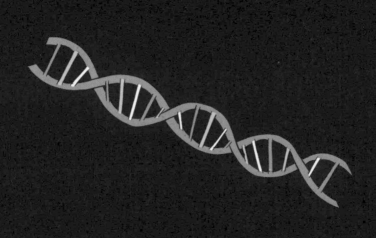

meanwhile...

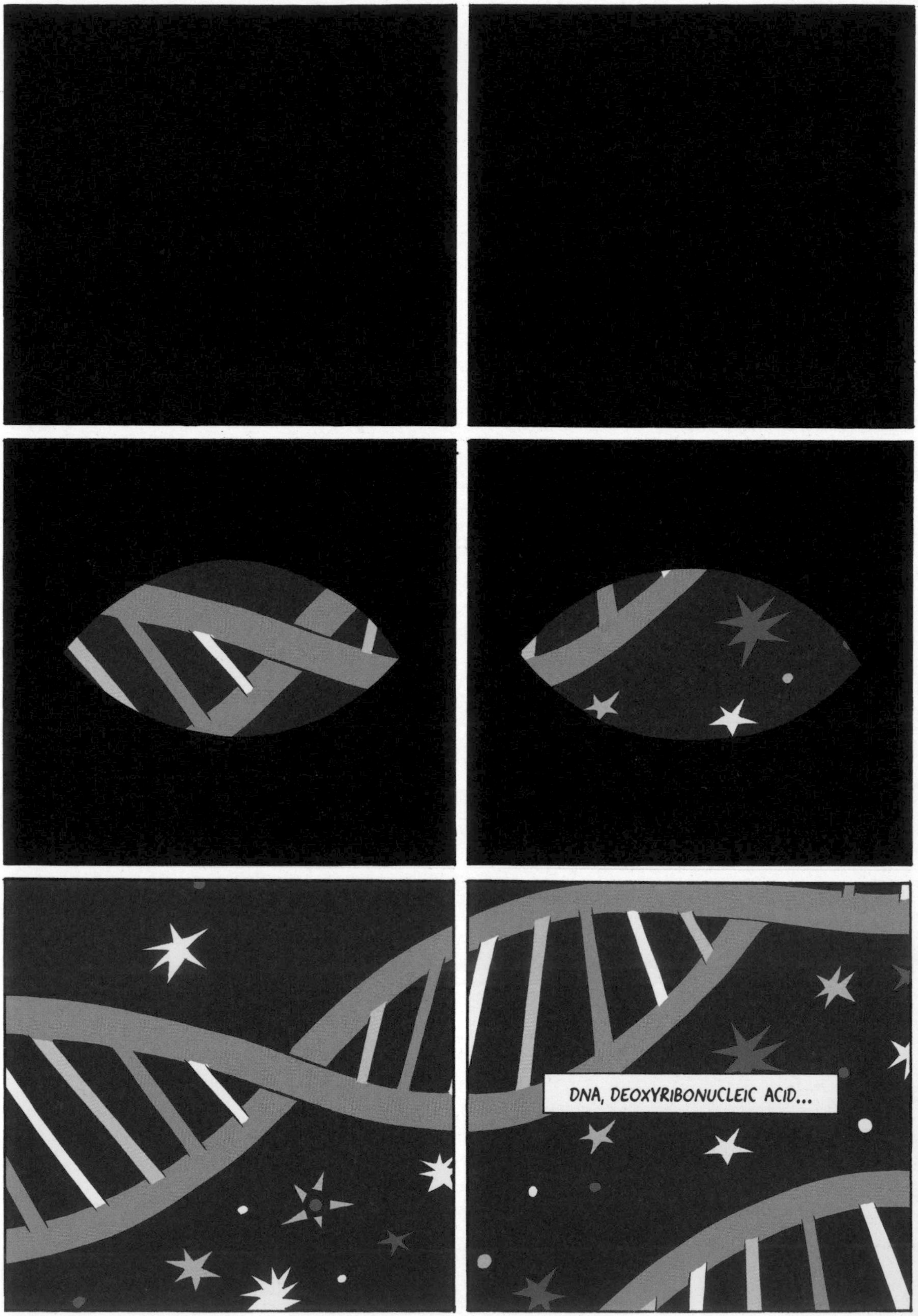
DNA, DEOXYRIBONUCLEIC ACID...

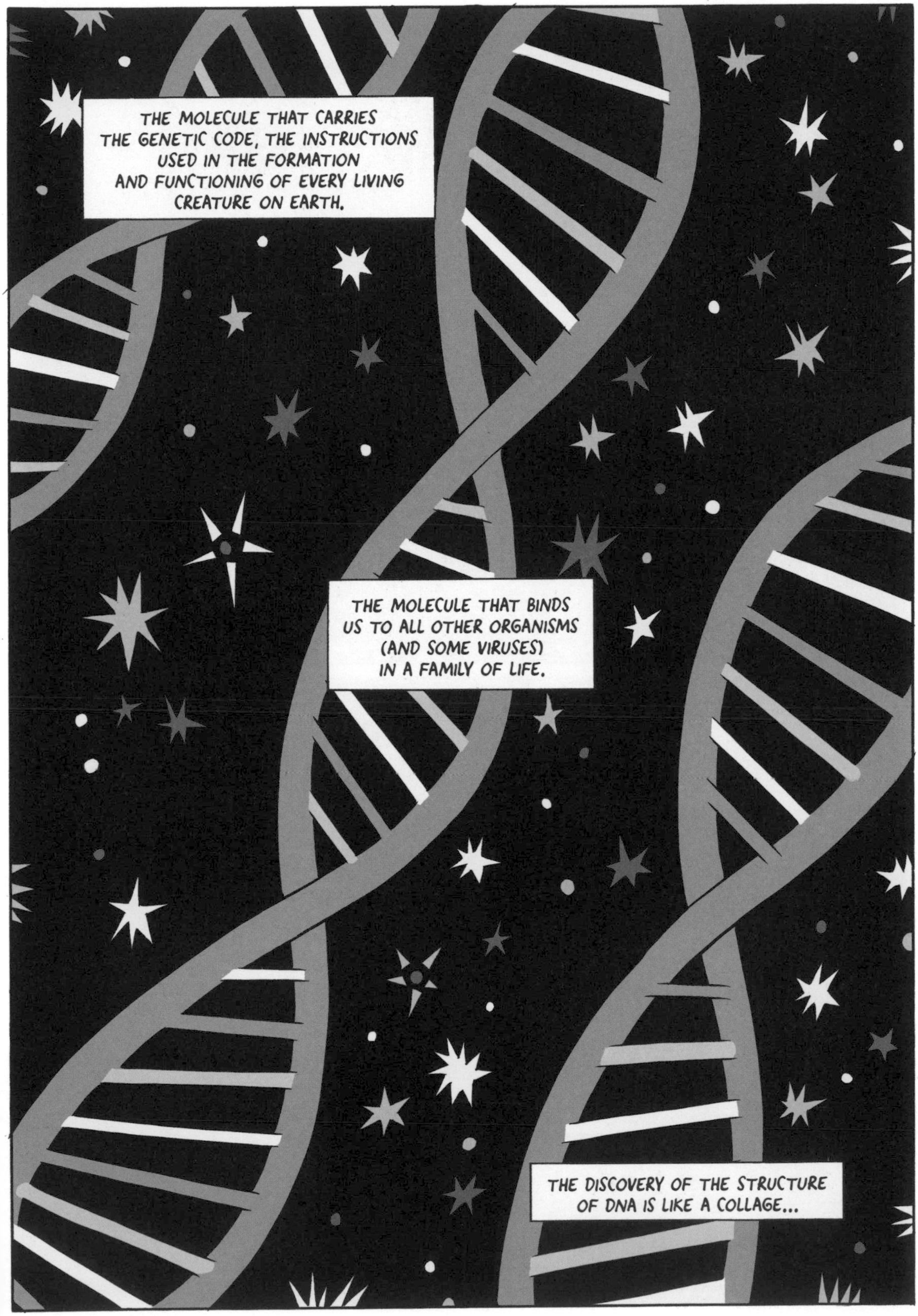
THE MOLECULE THAT CARRIES THE GENETIC CODE, THE INSTRUCTIONS USED IN THE FORMATION AND FUNCTIONING OF EVERY LIVING CREATURE ON EARTH.
THE MOLECULE THAT BINDS US TO ALL OTHER ORGANISMS (AND SOME VIRUSES) IN A FAMILY OF LIFE.
THE DISCOVERY OF THE STRUCTURE OF DNA IS LIKE A COLLAGE...

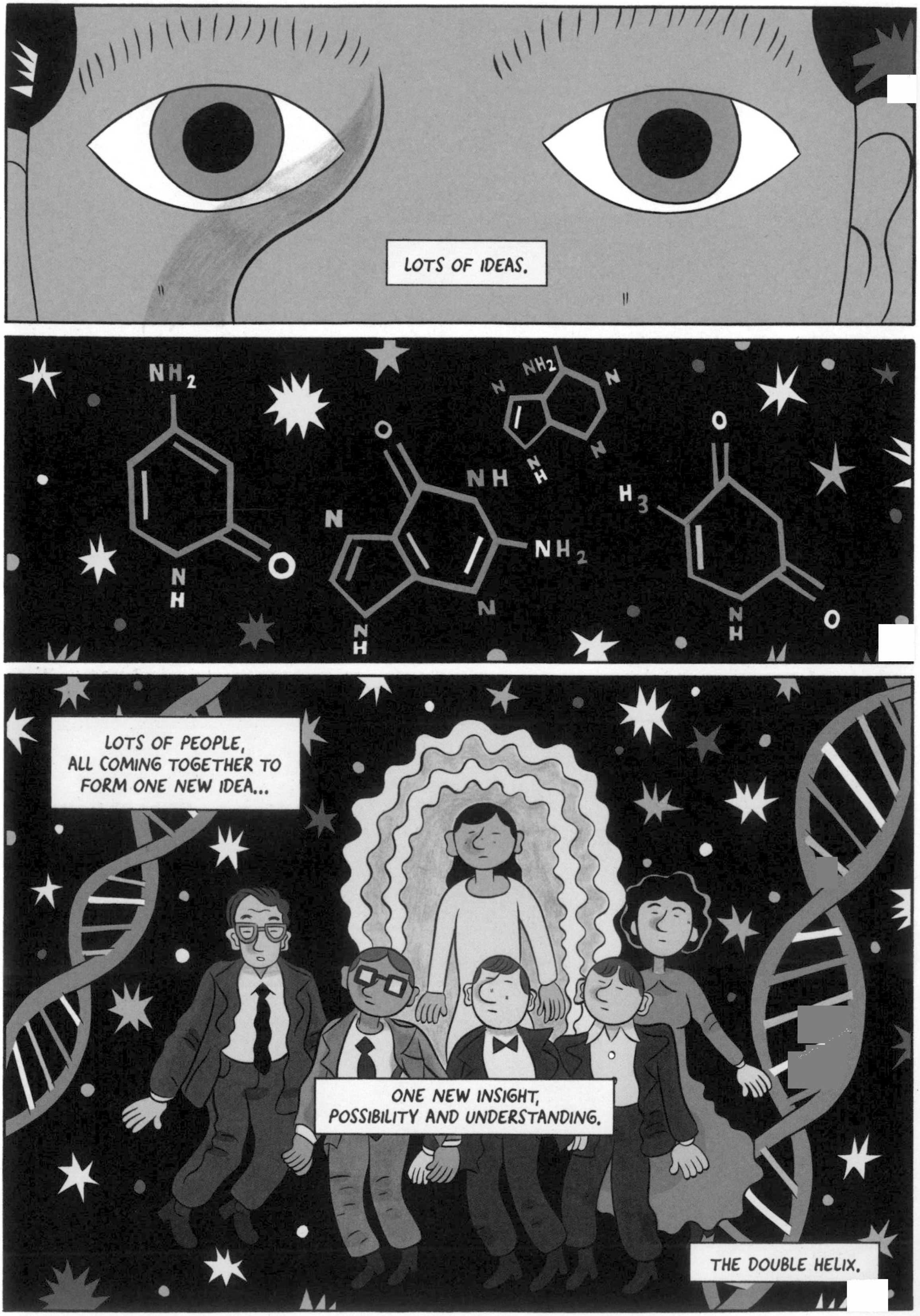
LOTS OF IDEAS.
LOTS OF PEOPLE, ALL COMING TOGETHER TO FORM ONE NEW IDEA...
ONE NEW INSIGHT, POSSIBILITY AND UNDERSTANDING.
THE DOUBLE HELIX.

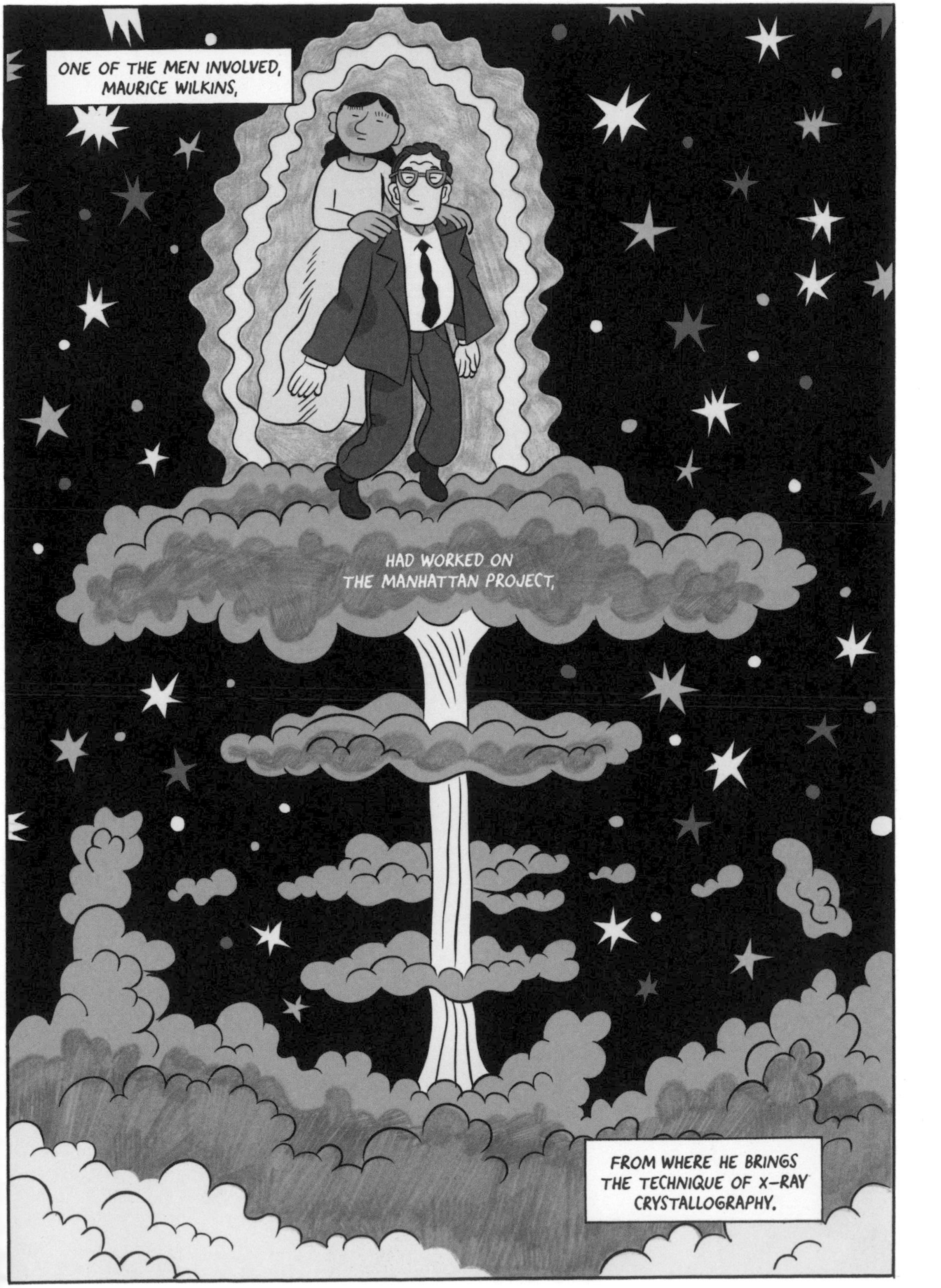
ONE OF THE MEN INVOLVED, MAURICE WILKINS,
HAD WORKED ON THE MANHATTAN PROJECT,
FROM WHERE HE BRINGS THE TECHNIQUE OF X-RAY CRYSTALLOGRAPHY.

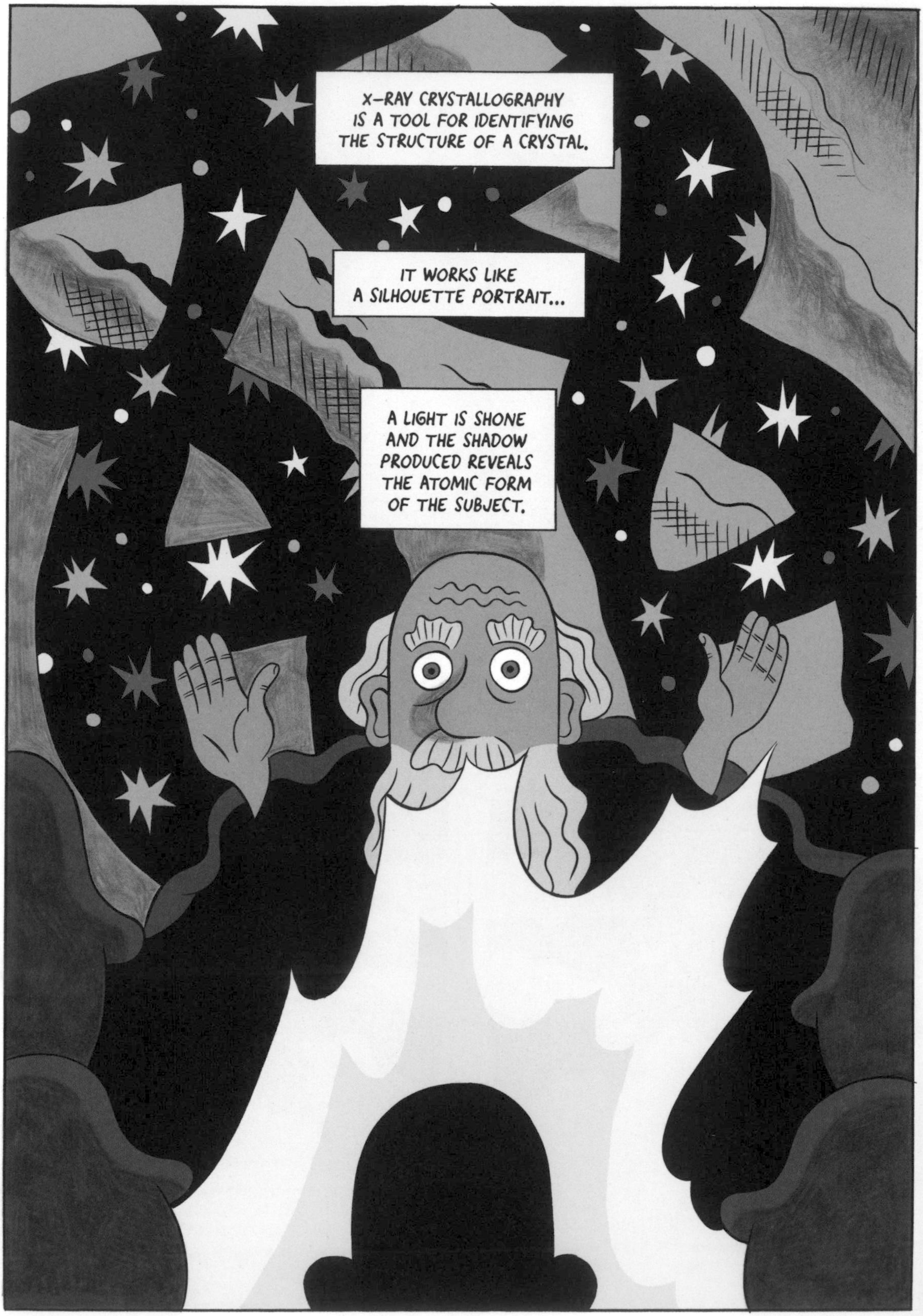
X-RAY CRYSTALLOGRAPHY IS A TOOL FOR IDENTIFYING THE STRUCTURE OF A CRYSTAL.
IT WORKS LIKE A SILHOUETTE PORTRAIT...
A LIGHT IS SHONE AND THE SHADOW PRODUCED REVEALS THE ATOMIC FORM OF THE SUBJECT.

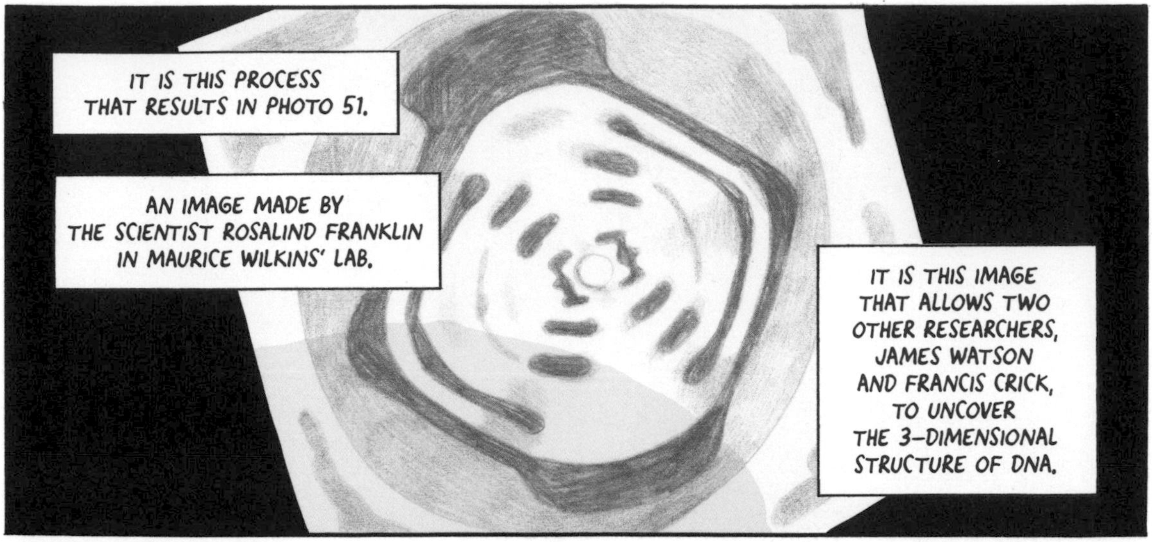
IT IS THIS PROCESS THAT RESULTS IN PHOTO 51.
AN IMAGE MADE BY THE SCIENTIST ROSALIND FRANKLIN IN MAURICE WILKINS' LAB.
IT IS THIS IMAGE THAT ALLOWS TWO OTHER RESEARCHERS, JAMES WATSON AND FRANCIS CRICK, TO UNCOVER THE 3-DIMENSIONAL STRUCTURE OF DNA.

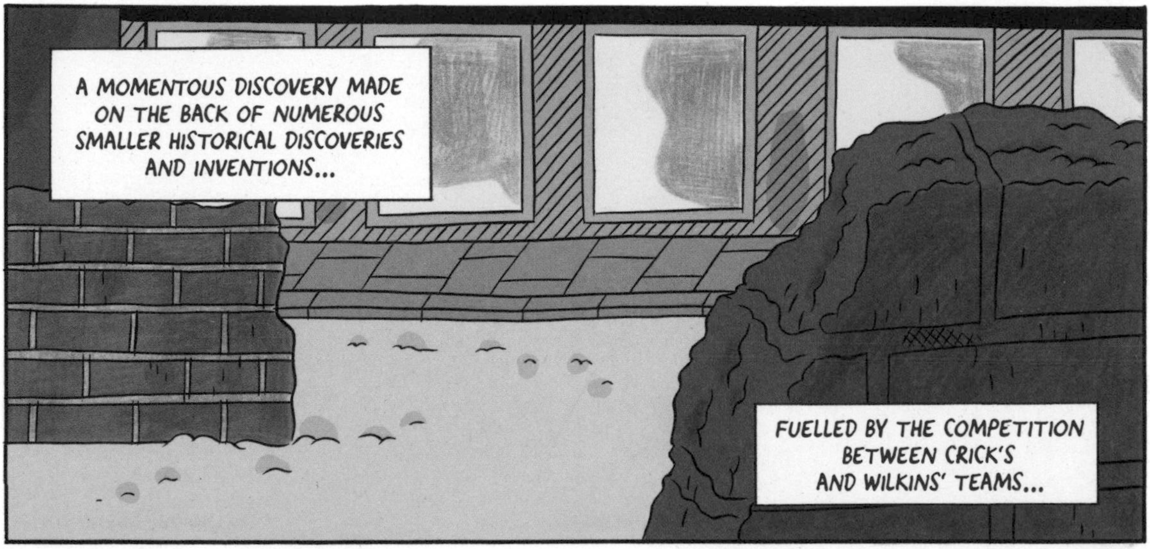
A MOMENTOUS DISCOVERY MADE ON THE BACK OF NUMEROUS SMALLER HISTORICAL DISCOVERIES AND INVENTIONS...
FUELLED BY THE COMPETITION BETWEEN CRICK'S AND WILKINS' TEAMS...

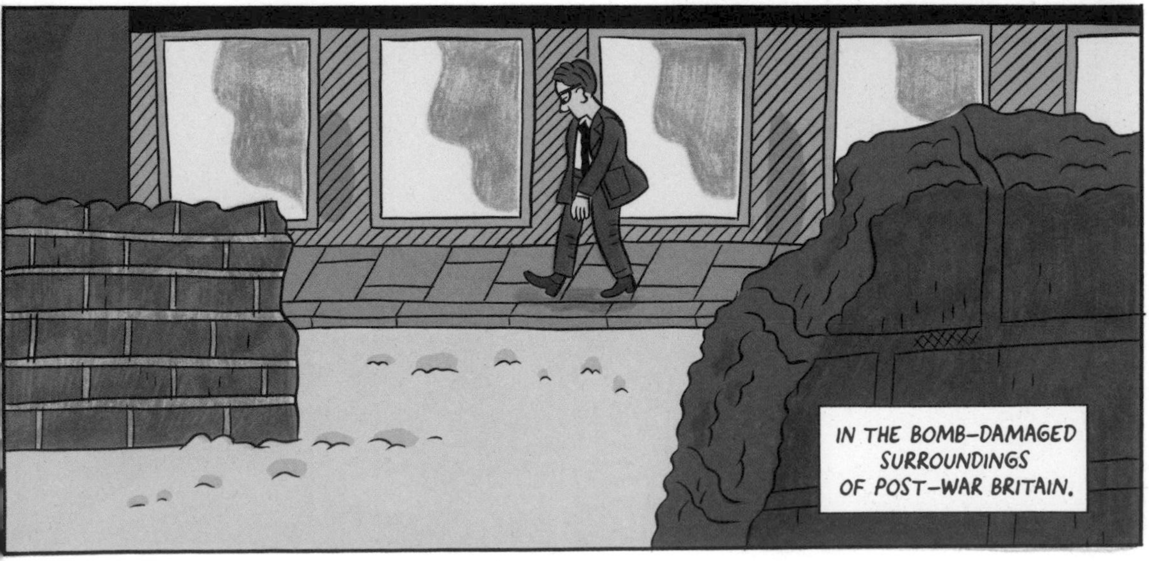
IN THE BOMB-DAMAGED SURROUNDINGS OF POST-WAR BRITAIN.

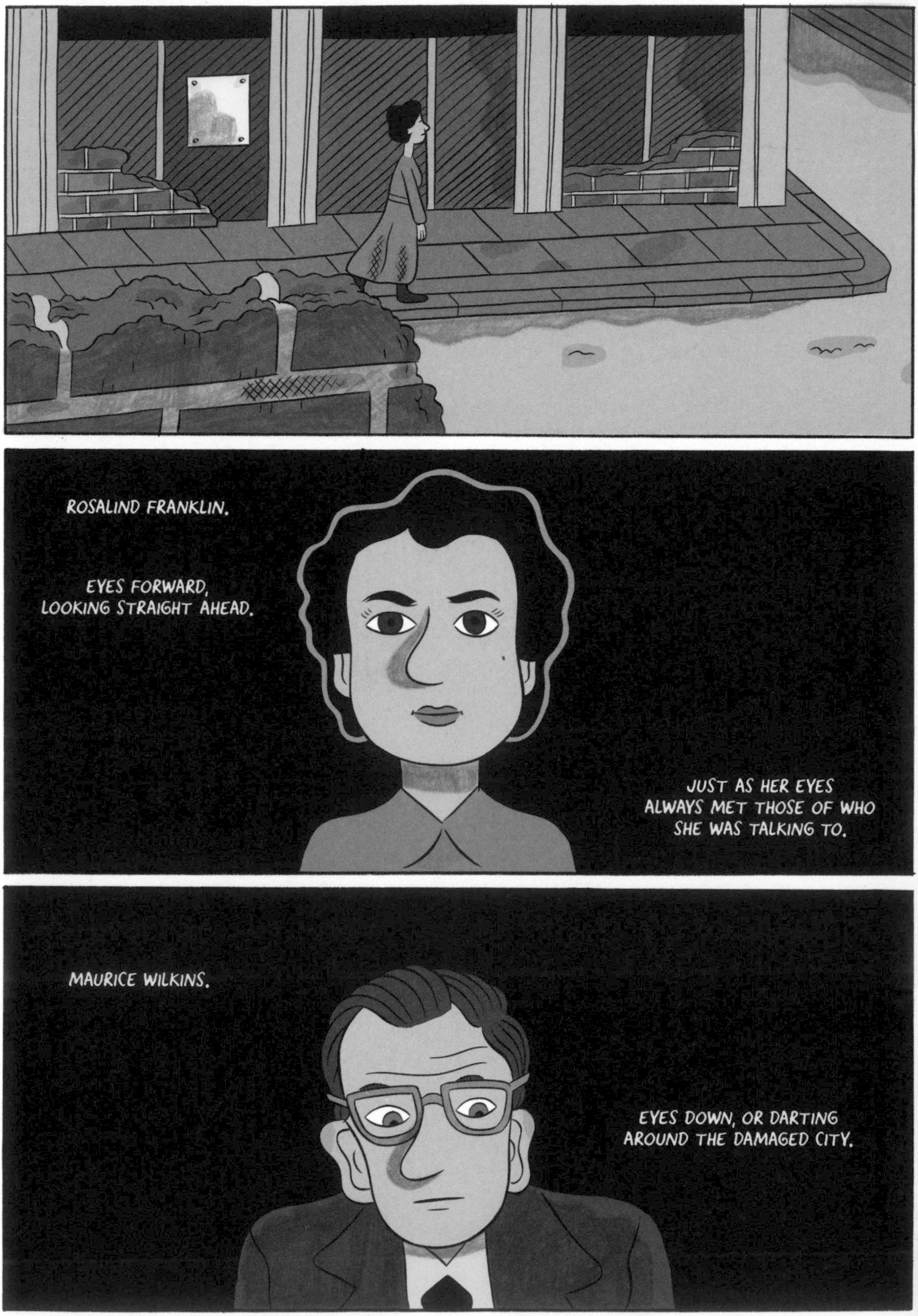
ROSALIND FRANKLIN.
EYES FORWARD, LOOKING STRAIGHT AHEAD.
JUST AS HER EYES ALWAYS MET THOSE OF WHO SHE WAS TALKING TO.
MAURICE WILKINS.
EYES DOWN, OR DARTING AROUND THE DAMAGED CITY.

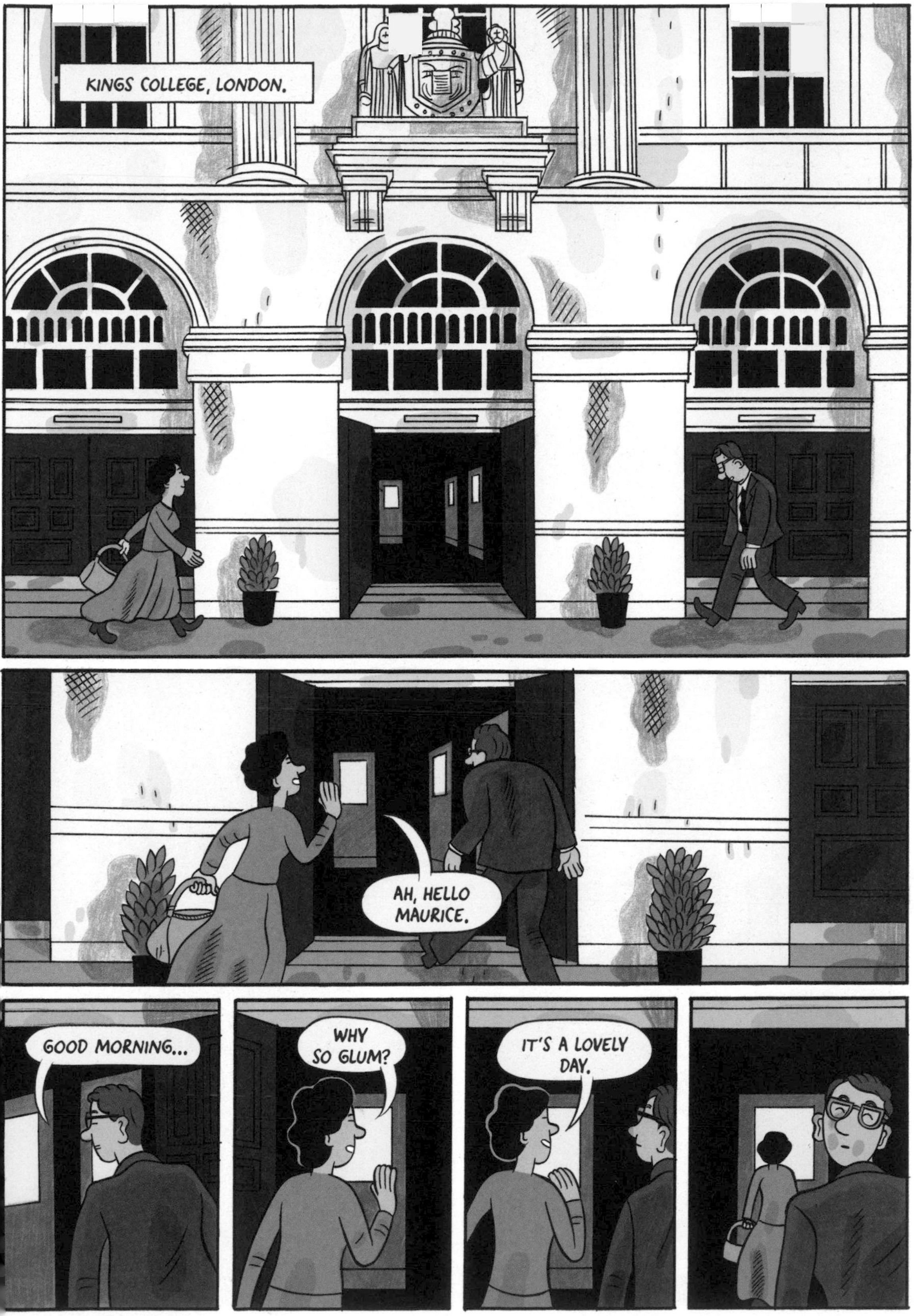
KINGS COLLEGE, LONDON.
AH, HELLO MAURICE.
GOOD MORNING...
WHY SO GLUM?
IT'S A LOVELY DAY.

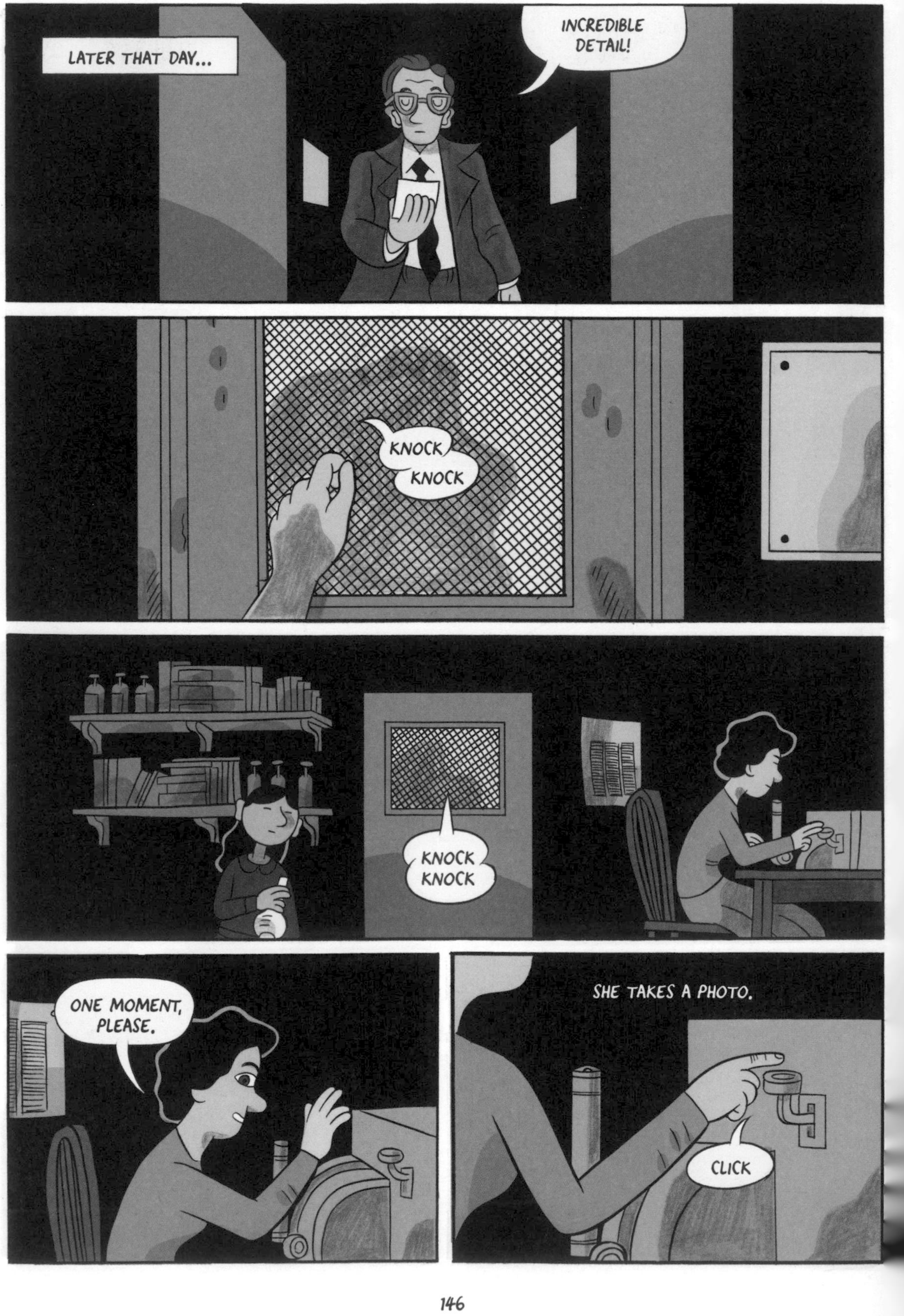
LATER THAT DAY...
INCREDIBLE DETAIL!
KNOCK KNOCK
KNOCK KNOCK
ONE MOMENT, PLEASE.
SHE TAKES A PHOTO.
CLICK

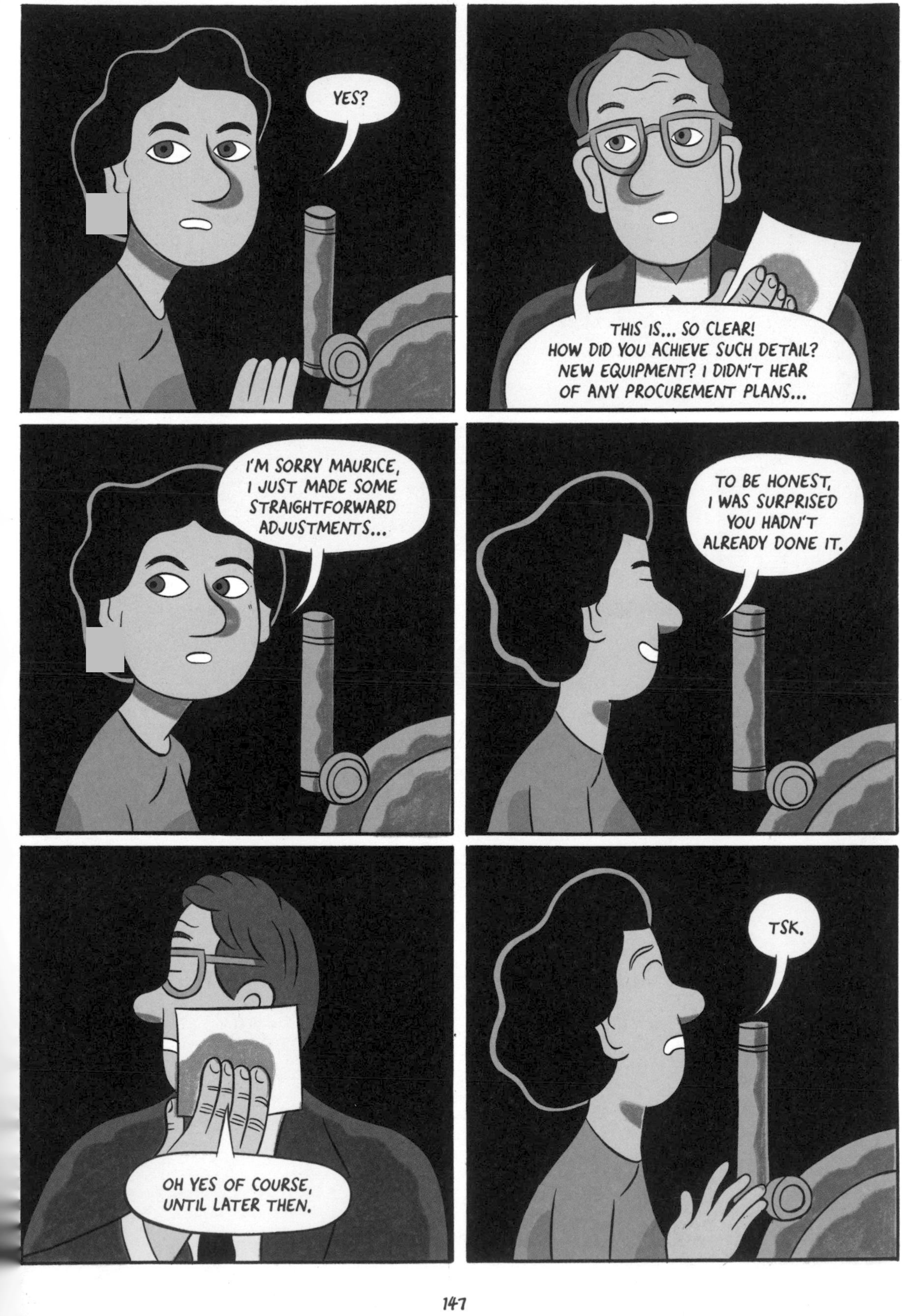
YES?
THIS IS... SO CLEAR! HOW DID YOU ACHIEVE SUCH DETAIL? NEW EQUIPMENT? I DIDN'T HEAR OF ANY PROCUREMENT PLANS...
I'M SORRY MAURICE, I JUST MADE SOME STRAIGHTFORWARD ADJUSTMENTS...
TO BE HONEST, I WAS SURPRISED YOU HADN'T ALREADY DONE IT.
OH YES OF COURSE, UNTIL LATER THEN.
TSK.

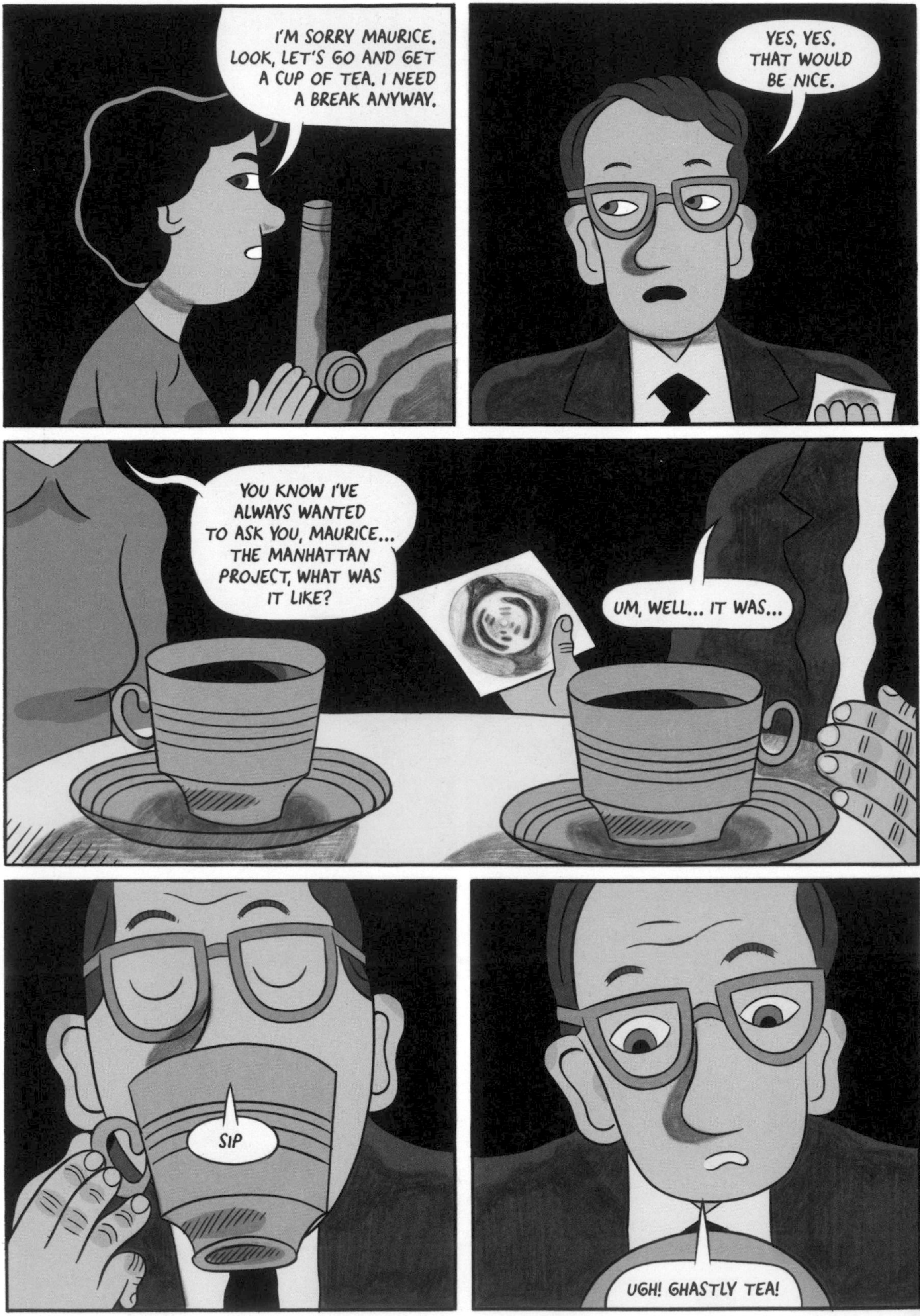
I'M SORRY MAURICE. LOOK, LET'S GO AND GET A CUP OF TEA. I NEED A BREAK ANYWAY.
YES, YES. THAT WOULD BE NICE.
YOU KNOW I'VE ALWAYS WANTED TO ASK YOU, MAURICE... THE MANHATTAN PROJECT, WHAT WAS IT LIKE?
UM, WELL... IT WAS...
SIP
UGH! GHASTLY TEA!

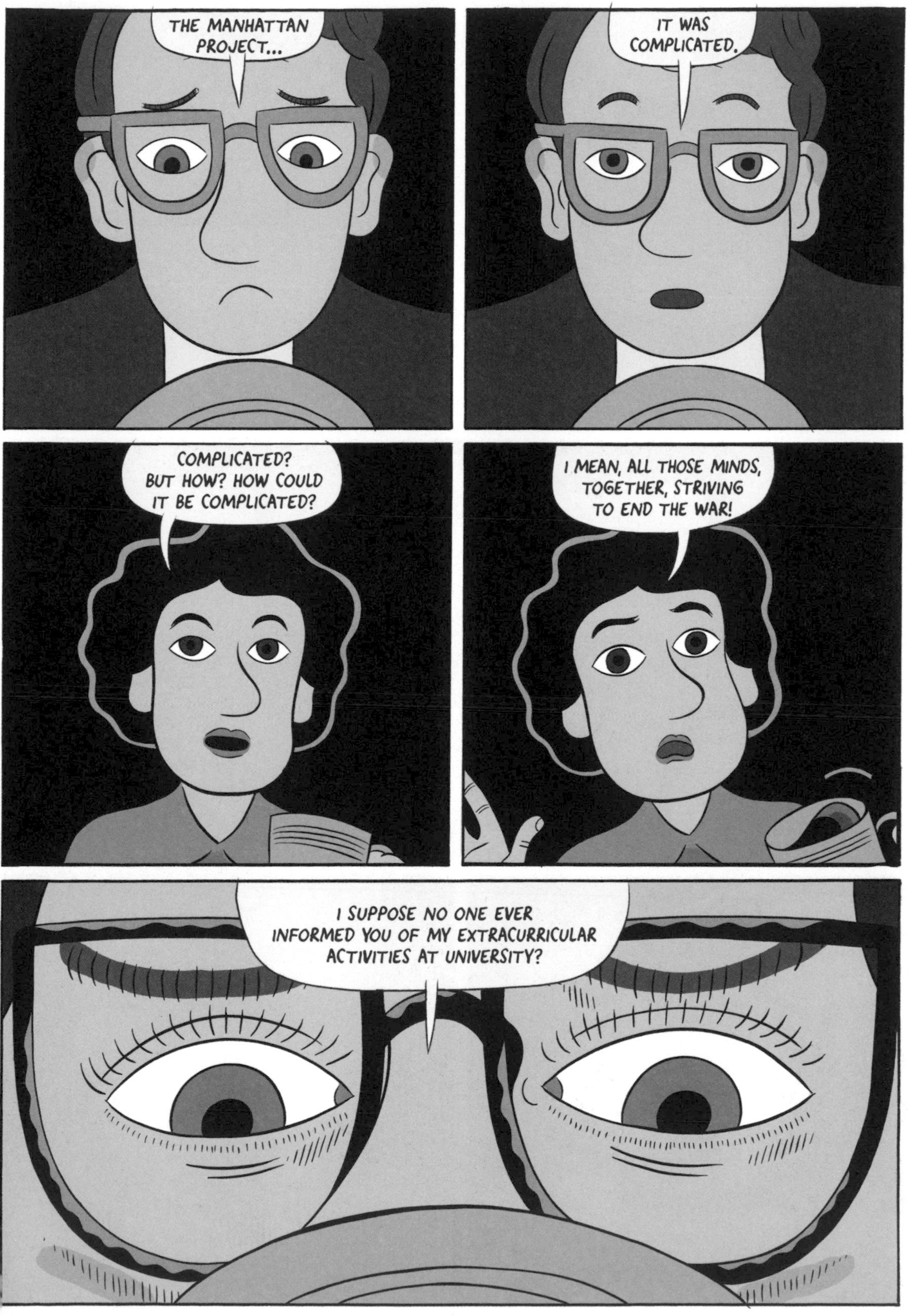
THE MANHATTAN PROJECT...
IT WAS COMPLICATED.
COMPLICATED? BUT HOW? HOW COULD IT BE COMPLICATED?
I MEAN, ALL THOSE MINDS, TOGETHER, STRIVING TO END THE WAR!
I SUPPOSE NO ONE EVER INFORMED YOU OF MY EXTRACURRICULAR ACTIVITIES AT UNIVERSITY?

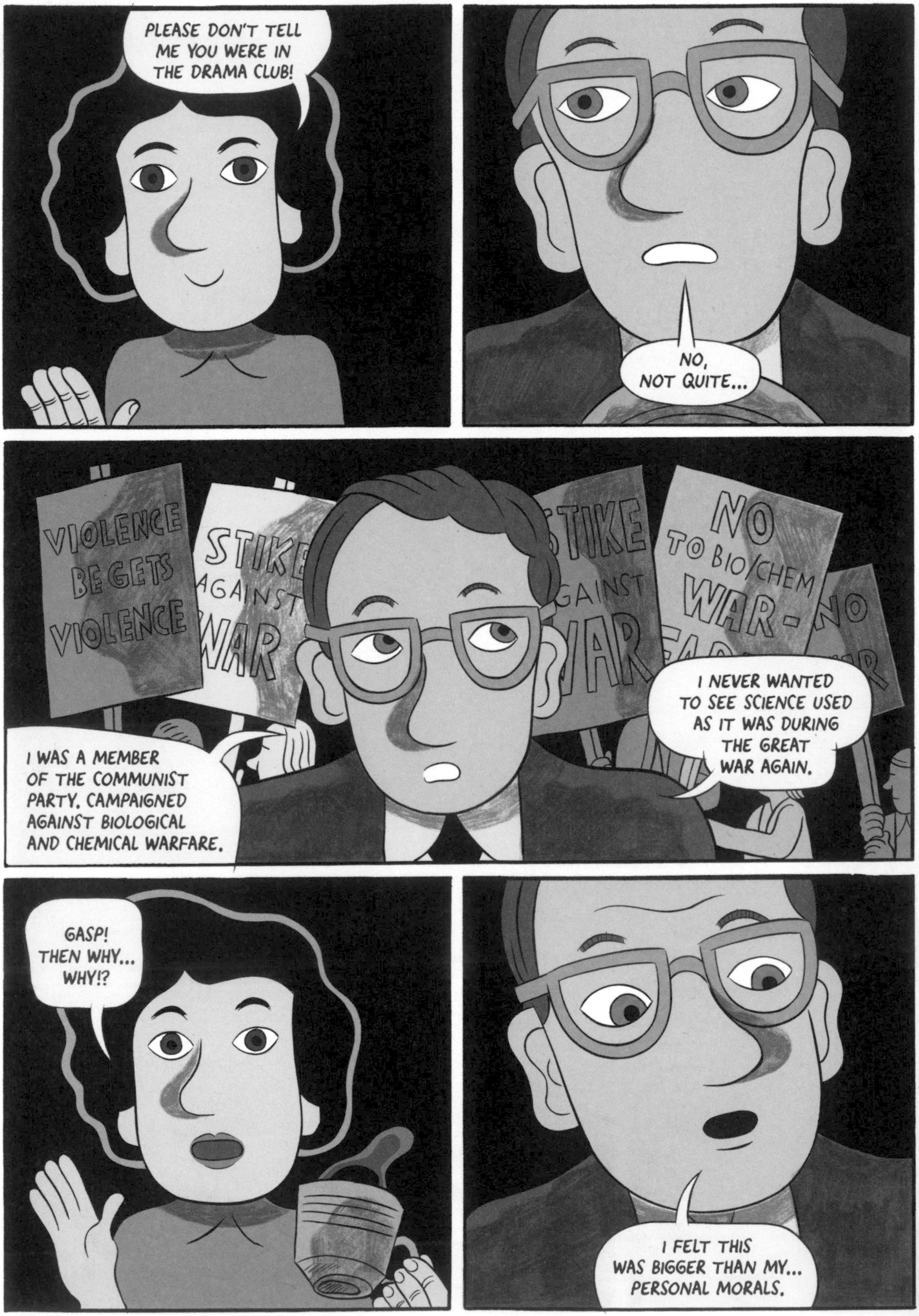
PLEASE DON'T TELL ME YOU WERE IN THE DRAMA CLUB!
NO, NOT QUITE...
VIOLENCE BEGETS VIOLENCE
STIKE AGAINST WAR
STIKE AGAINST WAR
NO TO BIO/CHEM WAR-
NO
I WAS A MEMBER OF THE COMMUNIST PARTY. CAMPAIGNED AGAINST BIOLOGICAL AND CHEMICAL WARFARE.
I NEVER WANTED TO SEE SCIENCE USED AS IT WAS DURING THE GREAT WAR AGAIN.
GASP! THEN WHY... WHY!?
I FELT THIS WAS BIGGER THAN MY... PERSONAL MORALS.

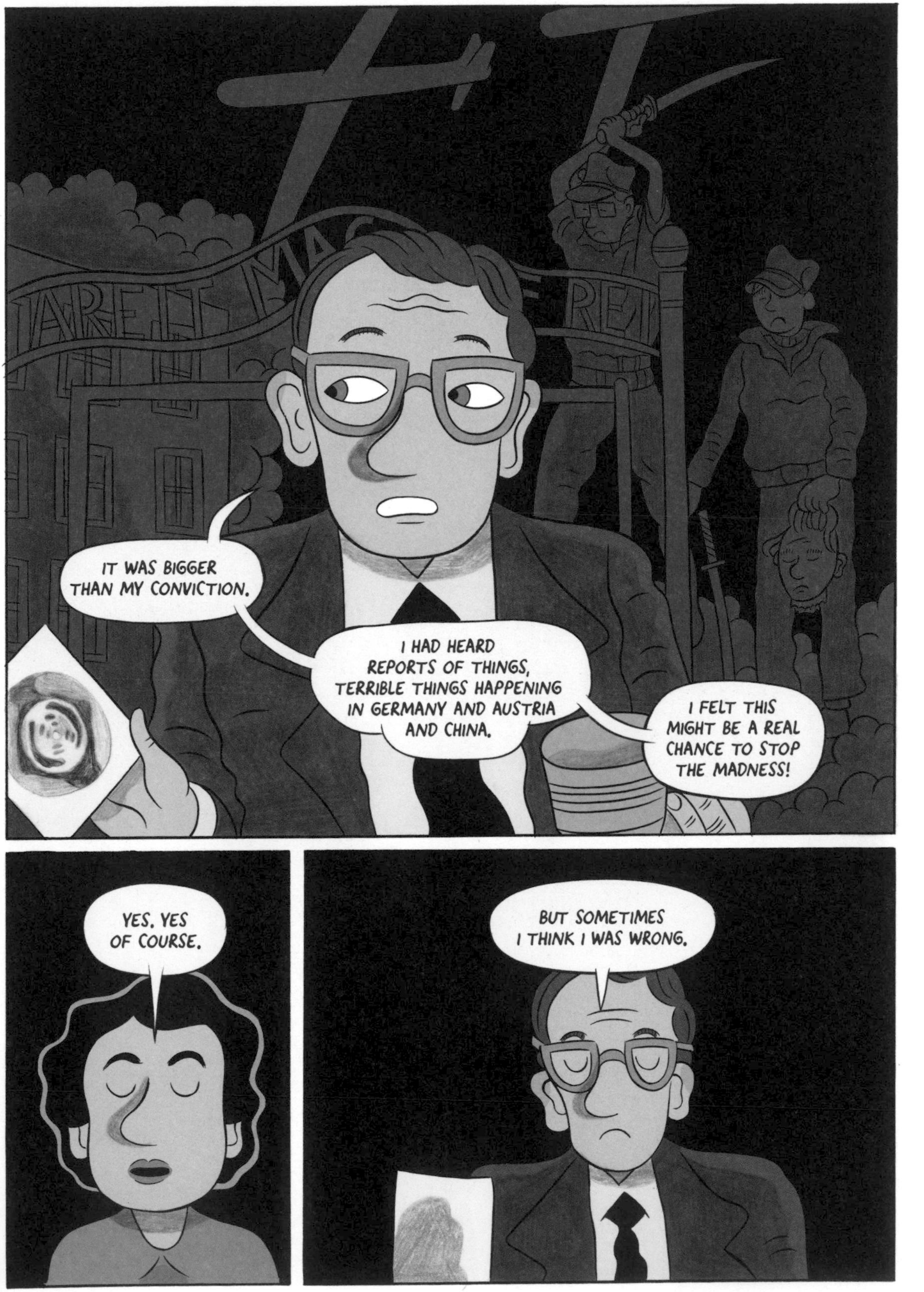
IT WAS BIGGER THAN MY CONVICTION.
I HAD HEARD REPORTS OF THINGS, TERRIBLE THINGS HAPPENING IN GERMANY AND AUSTRIA AND CHINA.
I FELT THIS MIGHT BE A REAL CHANCE TO STOP THE MADNESS!
YES, YES OF COURSE.
BUT SOMETIMES I THINK I WAS WRONG.

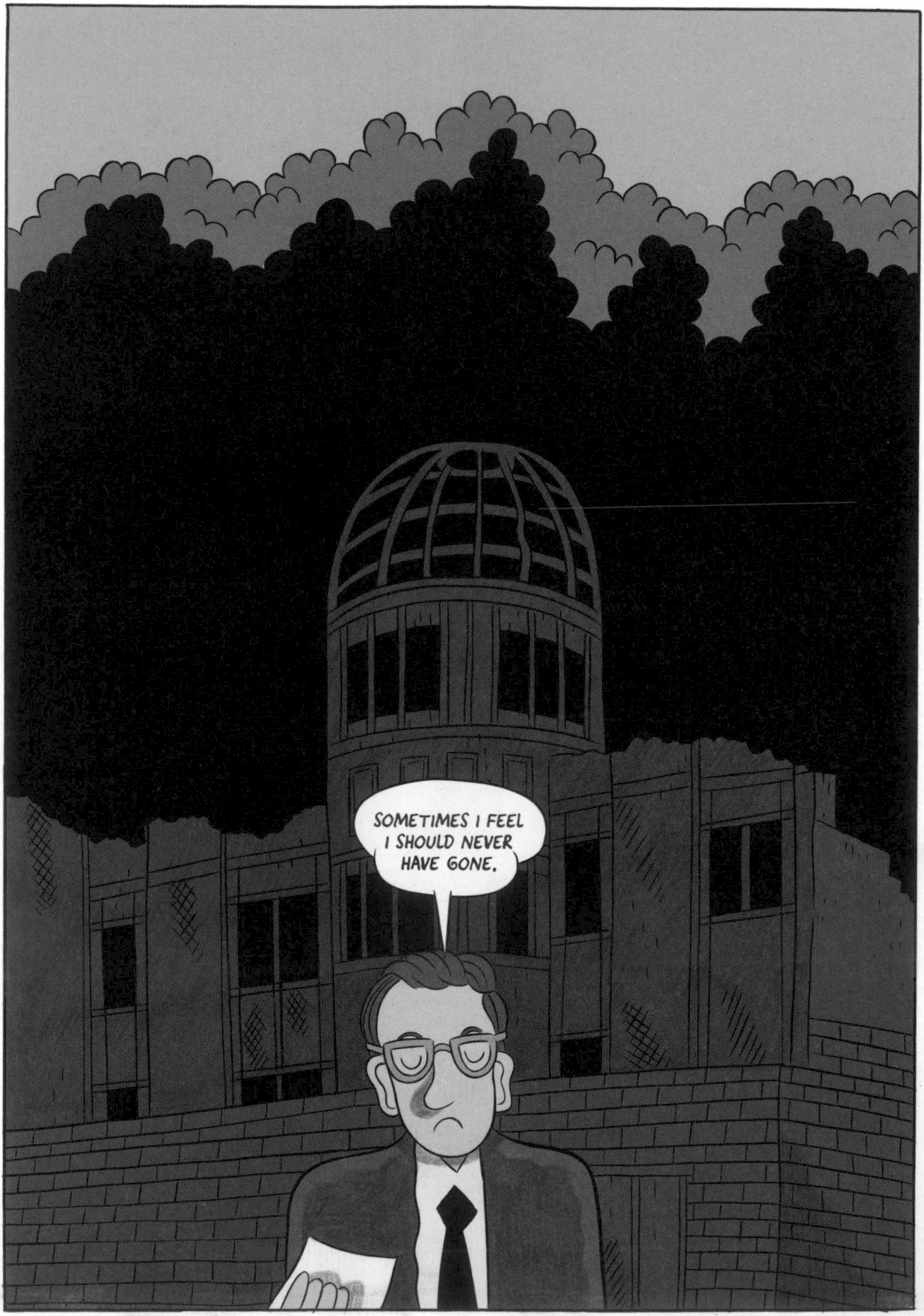
SOMETIMES I FEEL I SHOULD NEVER HAVE GONE.

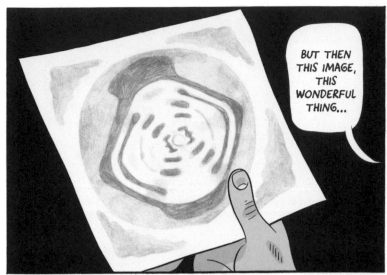

BUT THEN THIS IMAGE, THIS WONDERFUL THING...

...WOULD NOT EXIST.

HMM, NOW, I MUST BE GETTING BACK TO THE LAB...

IT WAS NICE TO CHAT TO YOU MAURICE...

WELL, THEN ...

WHERE WERE WE?

I THINK YOU NEEDED 15 ML?

AH YES.

CLICK

153

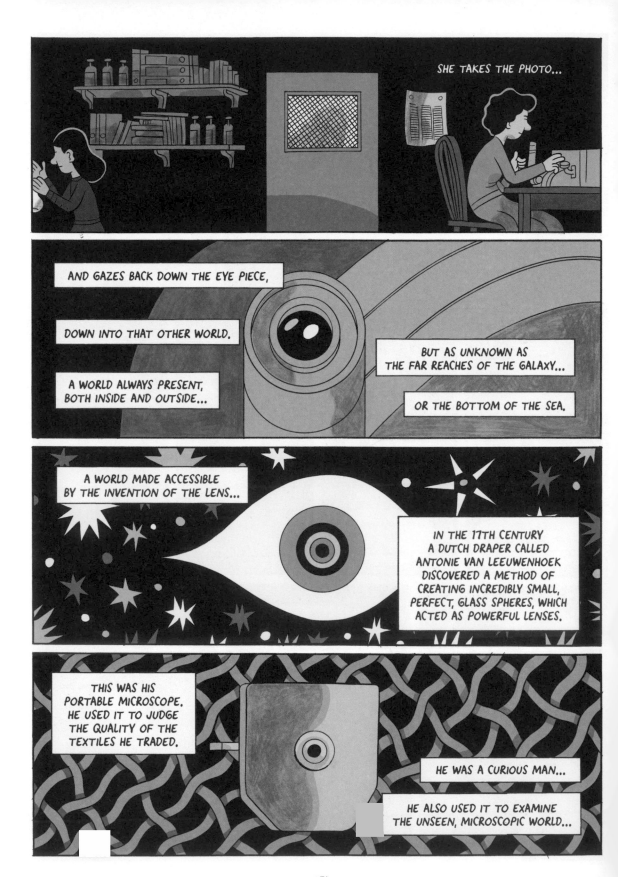

SHE TAKES THE PHOTO...

AND GAZES BACK DOWN THE EYE PIECE,

DOWN INTO THAT OTHER WORLD.

A WORLD ALWAYS PRESENT, BOTH INSIDE AND OUTSIDE...

BUT AS UNKNOWN AS THE FAR REACHES OF THE GALAXY...

OR THE BOTTOM OF THE SEA.

A WORLD MADE ACCESSIBLE BY THE INVENTION OF THE LENS...

IN THE 17TH CENTURY A DUTCH DRAPER CALLED ANTONIE VAN LEEUWENHOEK DISCOVERED A METHOD OF CREATING INCREDIBLY SMALL, PERFECT, GLASS SPHERES, WHICH ACTED AS POWERFUL LENSES.

THIS WAS HIS PORTABLE MICROSCOPE. HE USED IT TO JUDGE THE QUALITY OF THE TEXTILES HE TRADED.

HE WAS A CURIOUS MAN...

HE ALSO USED IT TO EXAMINE THE UNSEEN, MICROSCOPIC WORLD...

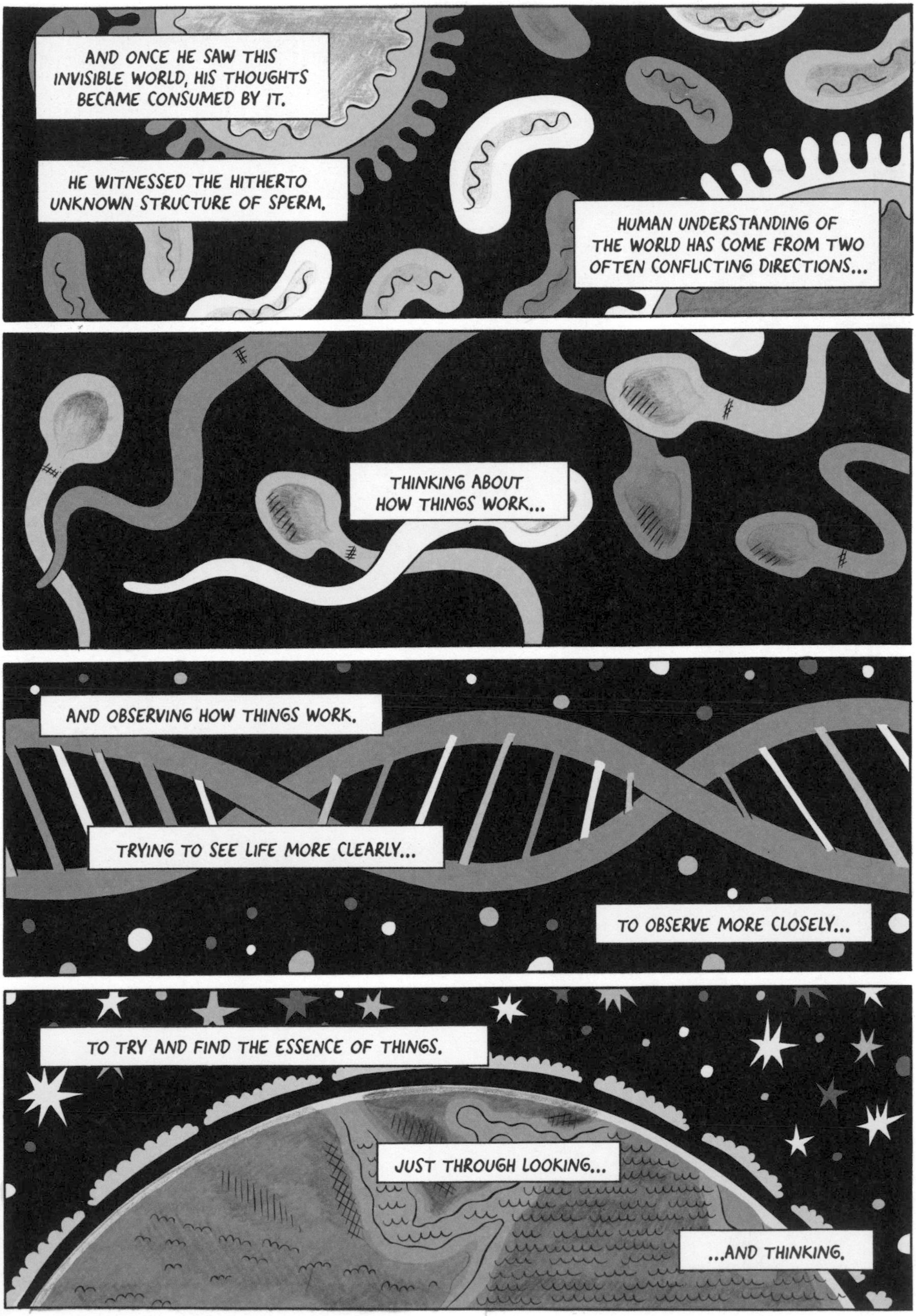
AND ONCE HE SAW THIS INVISIBLE WORLD, HIS THOUGHTS BECAME CONSUMED BY IT.
HE WITNESSED THE HITHERTO UNKNOWN STRUCTURE OF SPERM.
HUMAN UNDERSTANDING OF THE WORLD HAS COME FROM TWO OFTEN CONFLICTING DIRECTIONS...
THINKING ABOUT HOW THINGS WORK...
AND OBSERVING HOW THINGS WORK.
TRYING TO SEE LIFE MORE CLEARLY...
TO OBSERVE MORE CLOSELY...
TO TRY AND FIND THE ESSENCE OF THINGS.
JUST THROUGH LOOKING...
...AND THINKING.

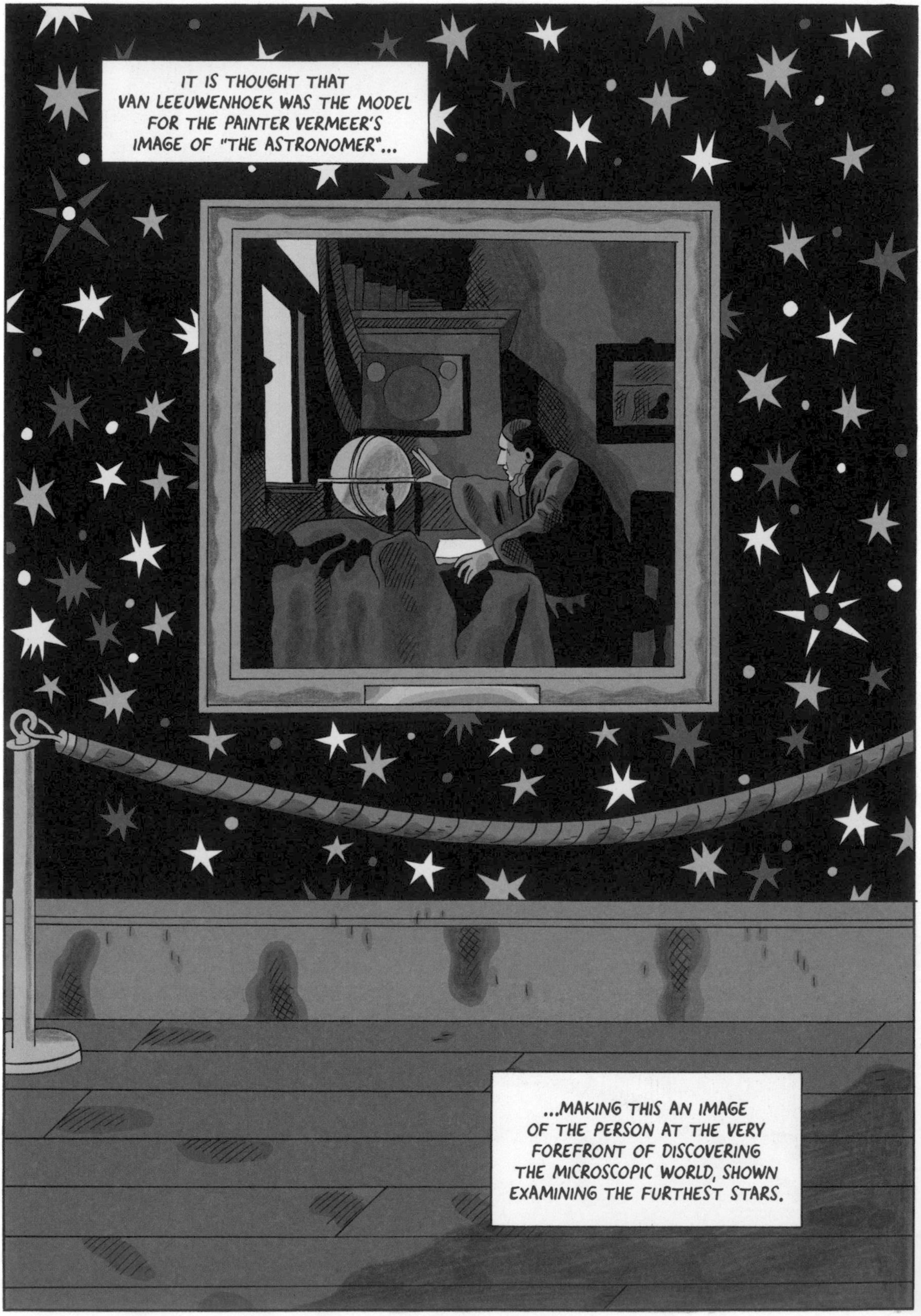
IT IS THOUGHT THAT VAN LEEUWENHOEK WAS THE MODEL FOR THE PAINTER VERMEER'S IMAGE OF "THE ASTRONOMER"...
...MAKING THIS AN IMAGE OF THE PERSON AT THE VERY FOREFRONT OF DISCOVERING THE MICROSCOPIC WORLD, SHOWN EXAMINING THE FURTHEST STARS.

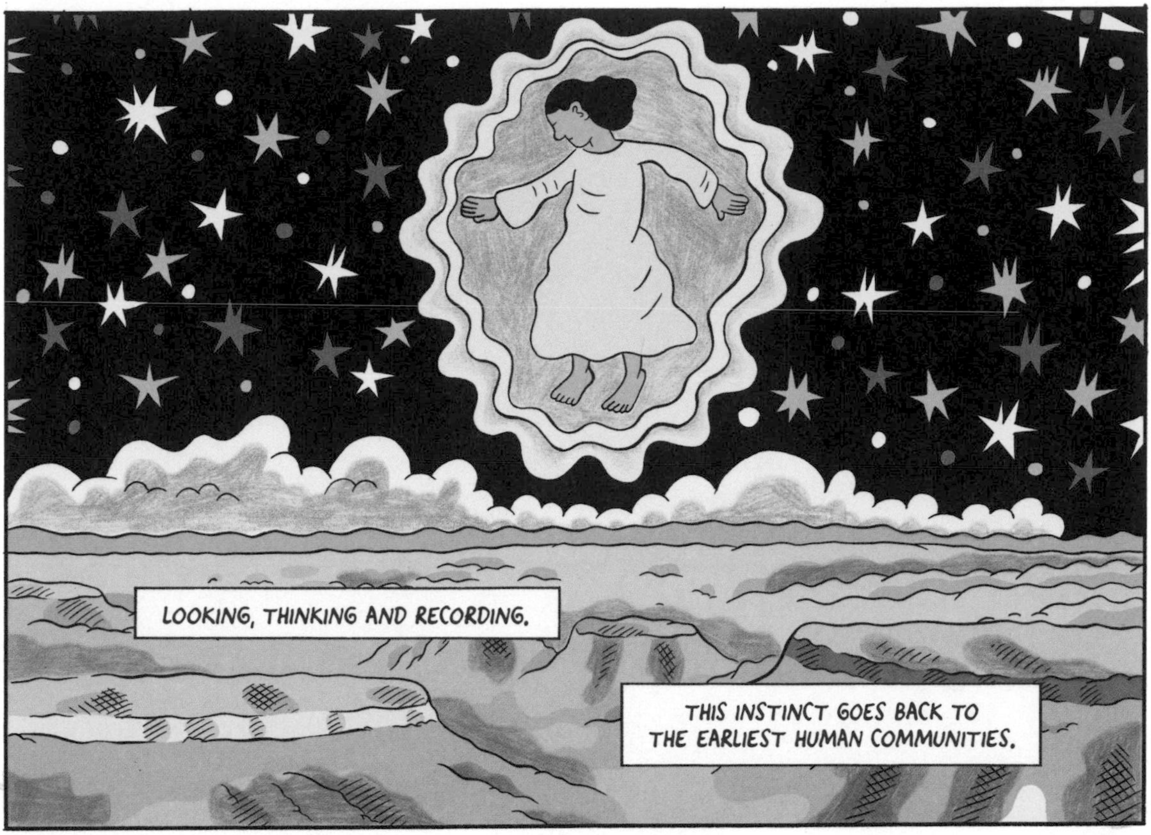
LOOKING, THINKING AND RECORDING.
THIS INSTINCT GOES BACK TO THE EARLIEST HUMAN COMMUNITIES.

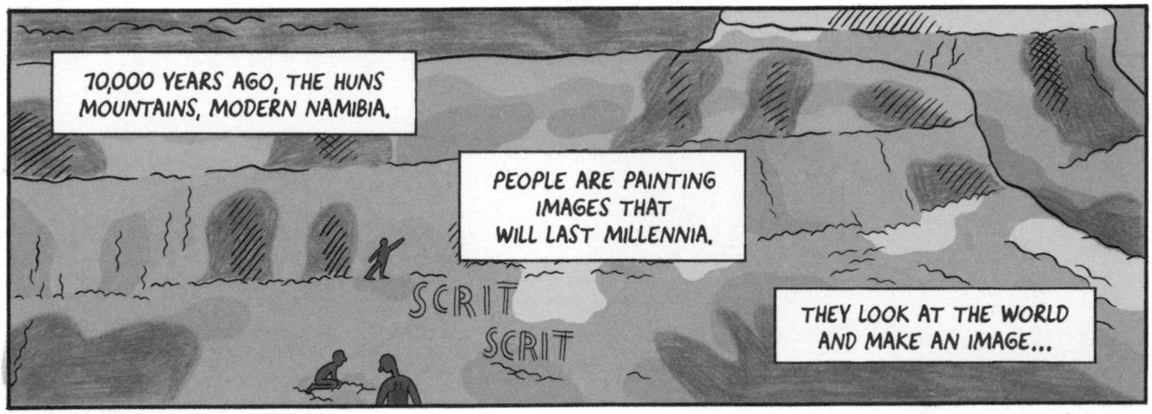
70,000 YEARS AGO, THE HUNS MOUNTAINS, MODERN NAMIBIA.
PEOPLE ARE PAINTING IMAGES THAT WILL LAST MILLENNIA.
SCRIT
SCRIT
THEY LOOK AT THE WORLD AND MAKE AN IMAGE...

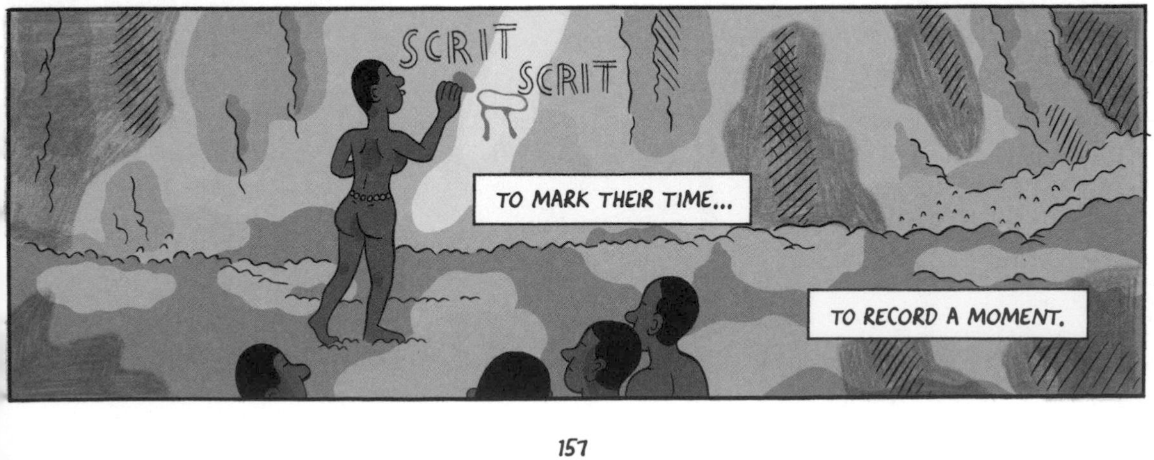
SCRIT
SCRIT
TO MARK THEIR TIME...
TO RECORD A MOMENT.

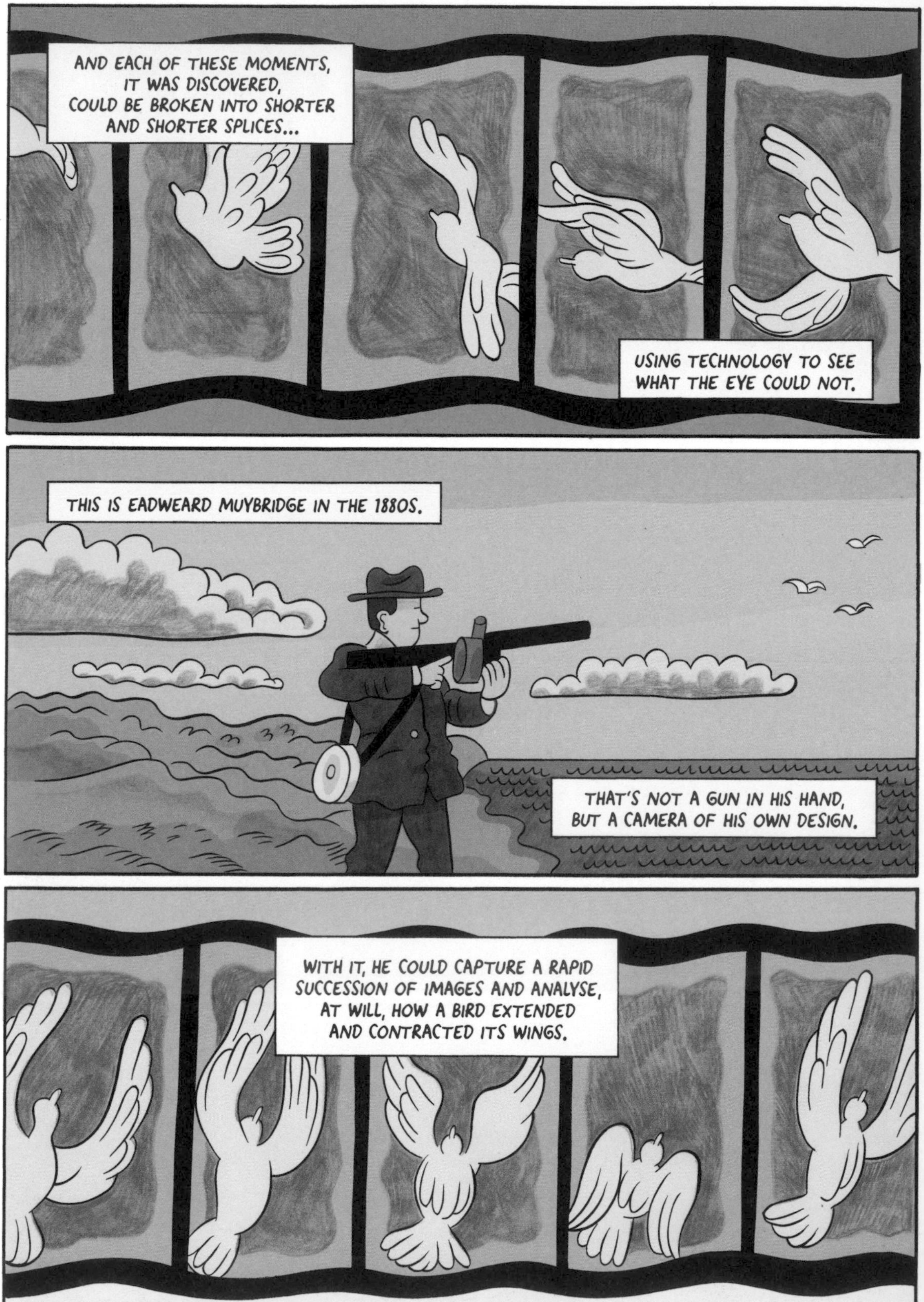
AND EACH OF THESE MOMENTS, IT WAS DISCOVERED, COULD BE BROKEN INTO SHORTER AND SHORTER SPLICES...
USING TECHNOLOGY TO SEE WHAT THE EYE COULD NOT.
THIS IS EADWEARD MUYBRIDGE IN THE 1880S.
THAT'S NOT A GUN IN HIS HAND, BUT A CAMERA OF HIS OWN DESIGN.
WITH IT, HE COULD CAPTURE A RAPID SUCCESSION OF IMAGES AND ANALYSE, AT WILL, HOW A BIRD EXTENDED AND CONTRACTED ITS WINGS.

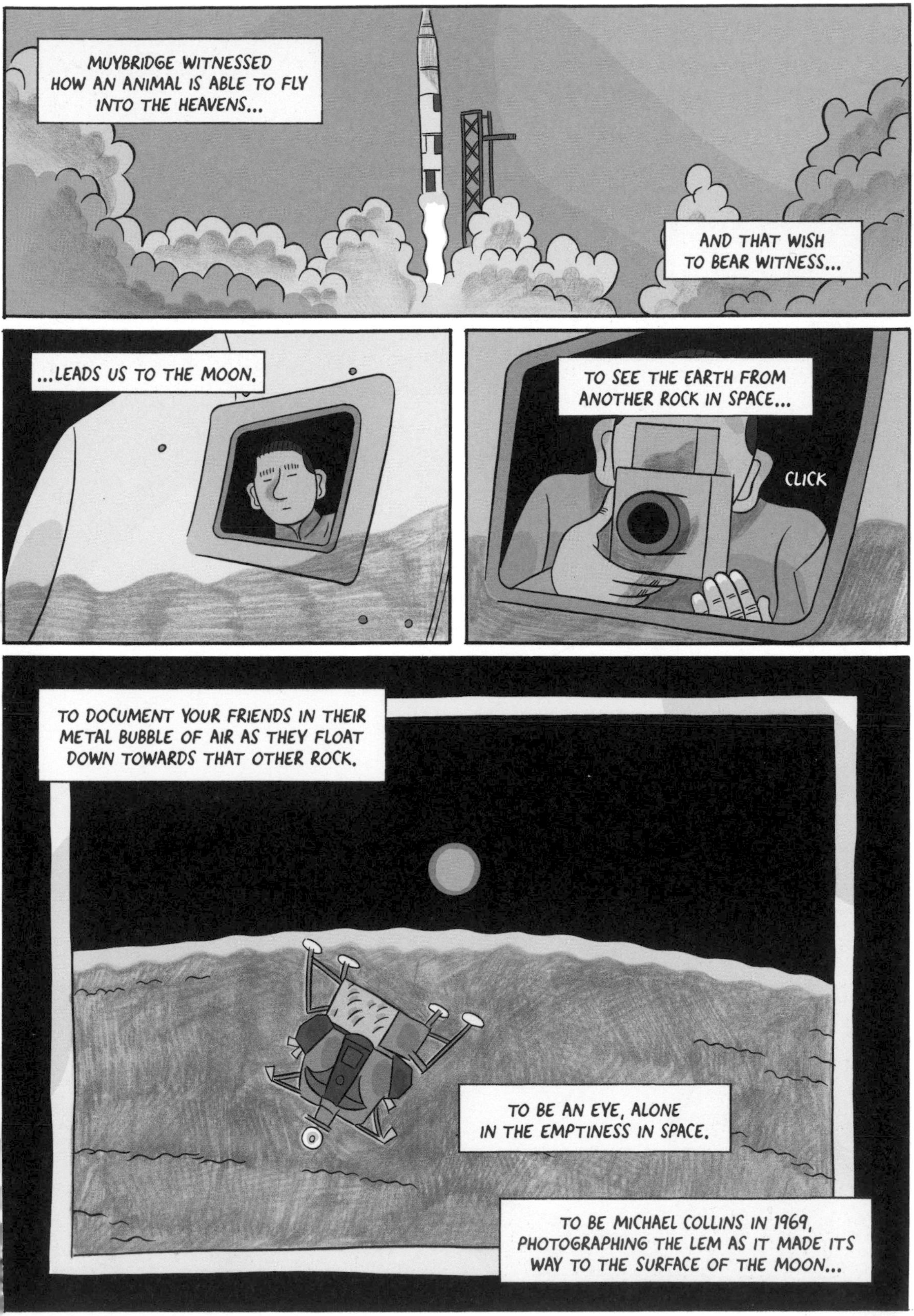
MUYBRIDGE WITNESSED HOW AN ANIMAL IS ABLE TO FLY INTO THE HEAVENS...
AND THAT WISH TO BEAR WITNESS...
...LEADS US TO THE MOON.
TO SEE THE EARTH FROM ANOTHER ROCK IN SPACE...
CLICK
TO DOCUMENT YOUR FRIENDS IN THEIR METAL BUBBLE OF AIR AS THEY FLOAT DOWN TOWARDS THAT OTHER ROCK.
TO BE AN EYE, ALONE IN THE EMPTINESS IN SPACE.
TO BE MICHAEL COLLINS IN 1969, PHOTOGRAPHING THE LEM AS IT MADE ITS WAY TO THE SURFACE OF THE MOON...

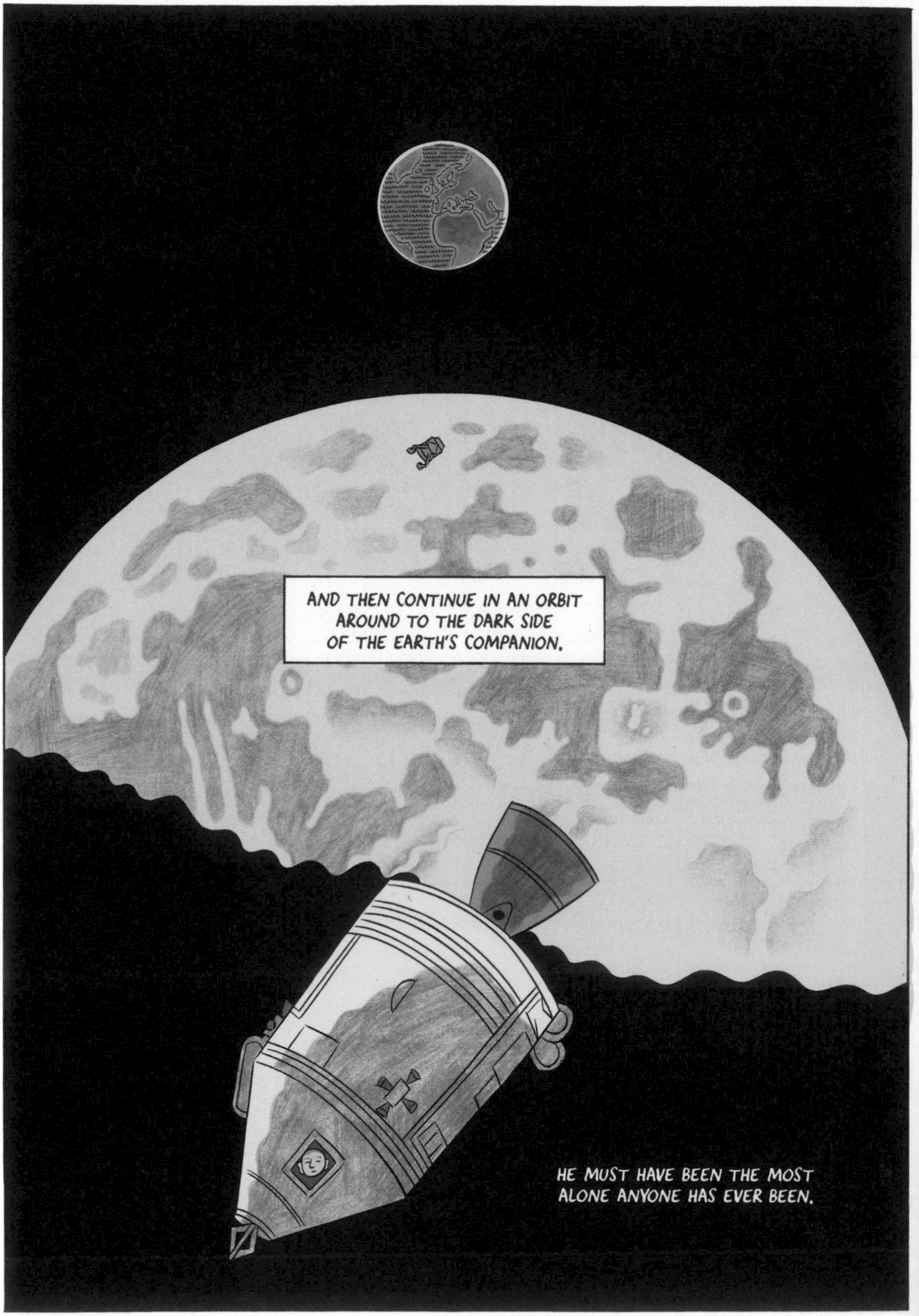
AND THEN CONTINUE IN AN ORBIT
AROUND TO THE DARK SIDE
OF THE EARTH'S COMPANION.
HE MUST HAVE BEEN THE MOST
ALONE ANYONE HAS EVER BEEN.

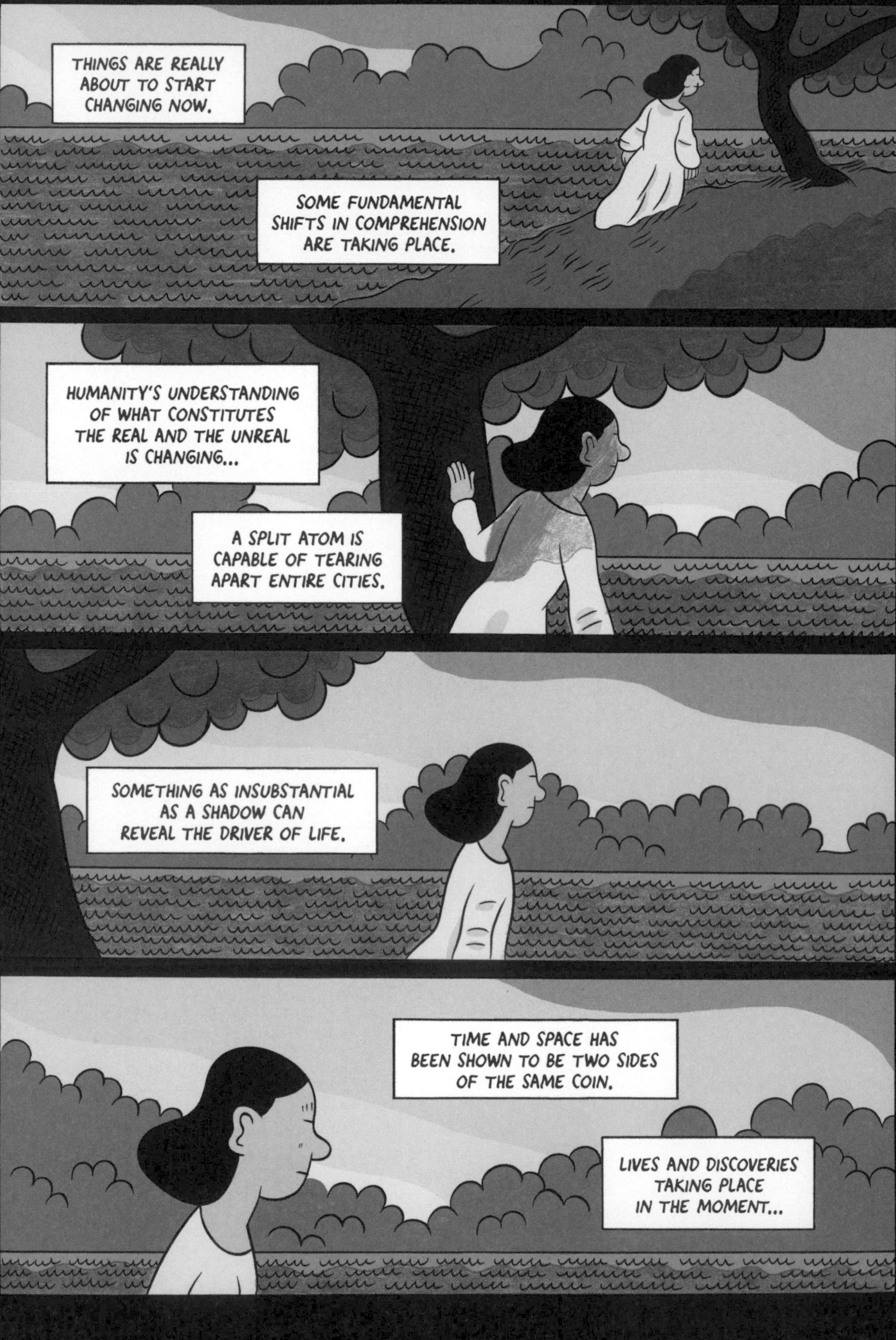
THINGS ARE REALLY ABOUT TO START CHANGING NOW.
SOME FUNDAMENTAL SHIFTS IN COMPREHENSION ARE TAKING PLACE.
HUMANITY'S UNDERSTANDING OF WHAT CONSTITUTES THE REAL AND THE UNREAL IS CHANGING...
A SPLIT ATOM IS CAPABLE OF TEARING APART ENTIRE CITIES.
SOMETHING AS INSUBSTANTIAL AS A SHADOW CAN REVEAL THE DRIVER OF LIFE.
TIME AND SPACE HAS BEEN SHOWN TO BE TWO SIDES OF THE SAME COIN.
LIVES AND DISCOVERIES TAKING PLACE IN THE MOMENT...

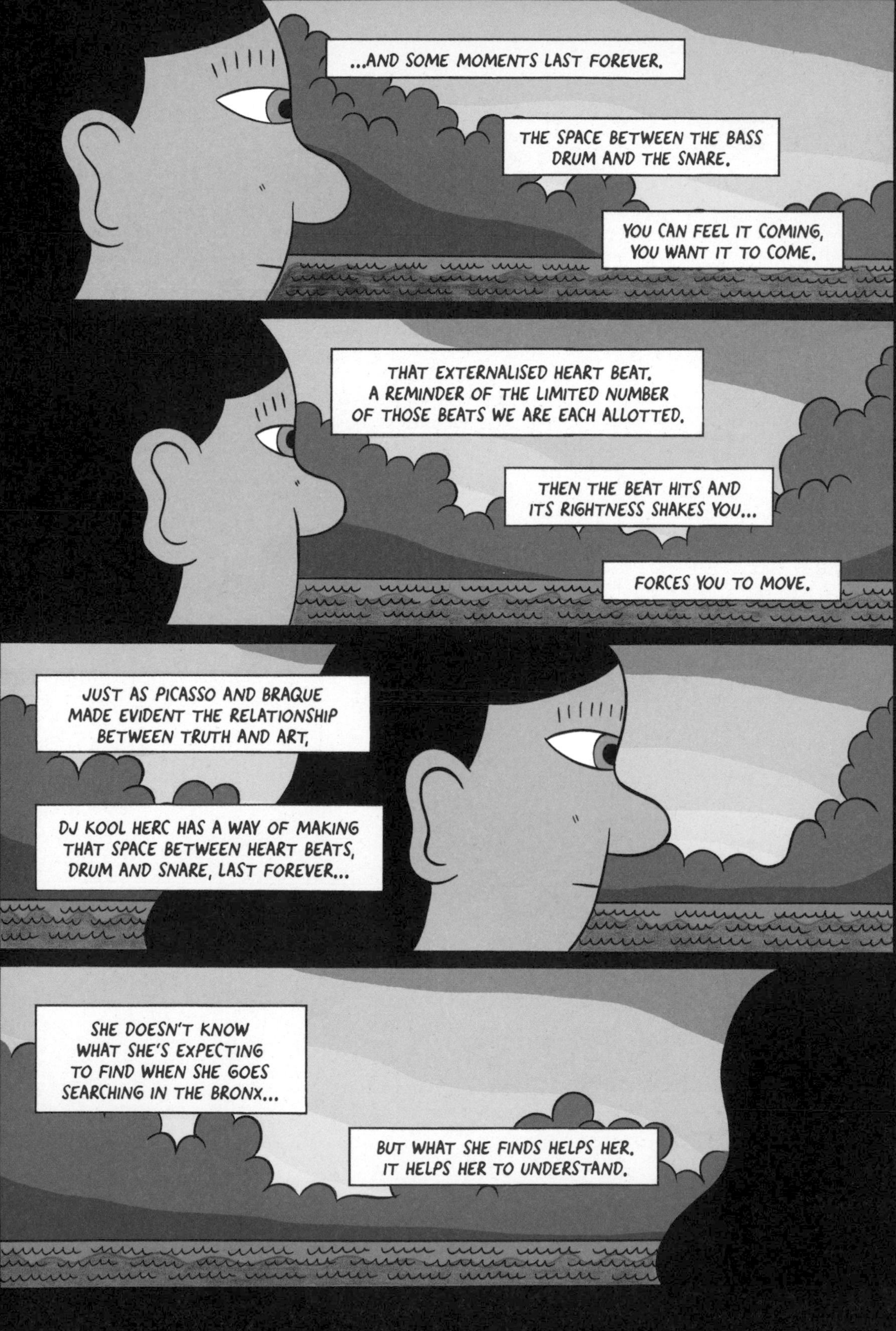
...AND SOME MOMENTS LAST FOREVER.
THE SPACE BETWEEN THE BASS DRUM AND THE SNARE.
YOU CAN FEEL IT COMING, YOU WANT IT TO COME.
THAT EXTERNALISED HEART BEAT. A REMINDER OF THE LIMITED NUMBER OF THOSE BEATS WE ARE EACH ALLOTTED.
THEN THE BEAT HITS AND ITS RIGHTNESS SHAKES YOU...
FORCES YOU TO MOVE.
JUST AS PICASSO AND BRAQUE MADE EVIDENT THE RELATIONSHIP BETWEEN TRUTH AND ART,
DJ KOOL HERC HAS A WAY OF MAKING THAT SPACE BETWEEN HEART BEATS, DRUM AND SNARE, LAST FOREVER...
SHE DOESN'T KNOW WHAT SHE'S EXPECTING TO FIND WHEN SHE GOES SEARCHING IN THE BRONX...
BUT WHAT SHE FINDS HELPS HER. IT HELPS HER TO UNDERSTAND.

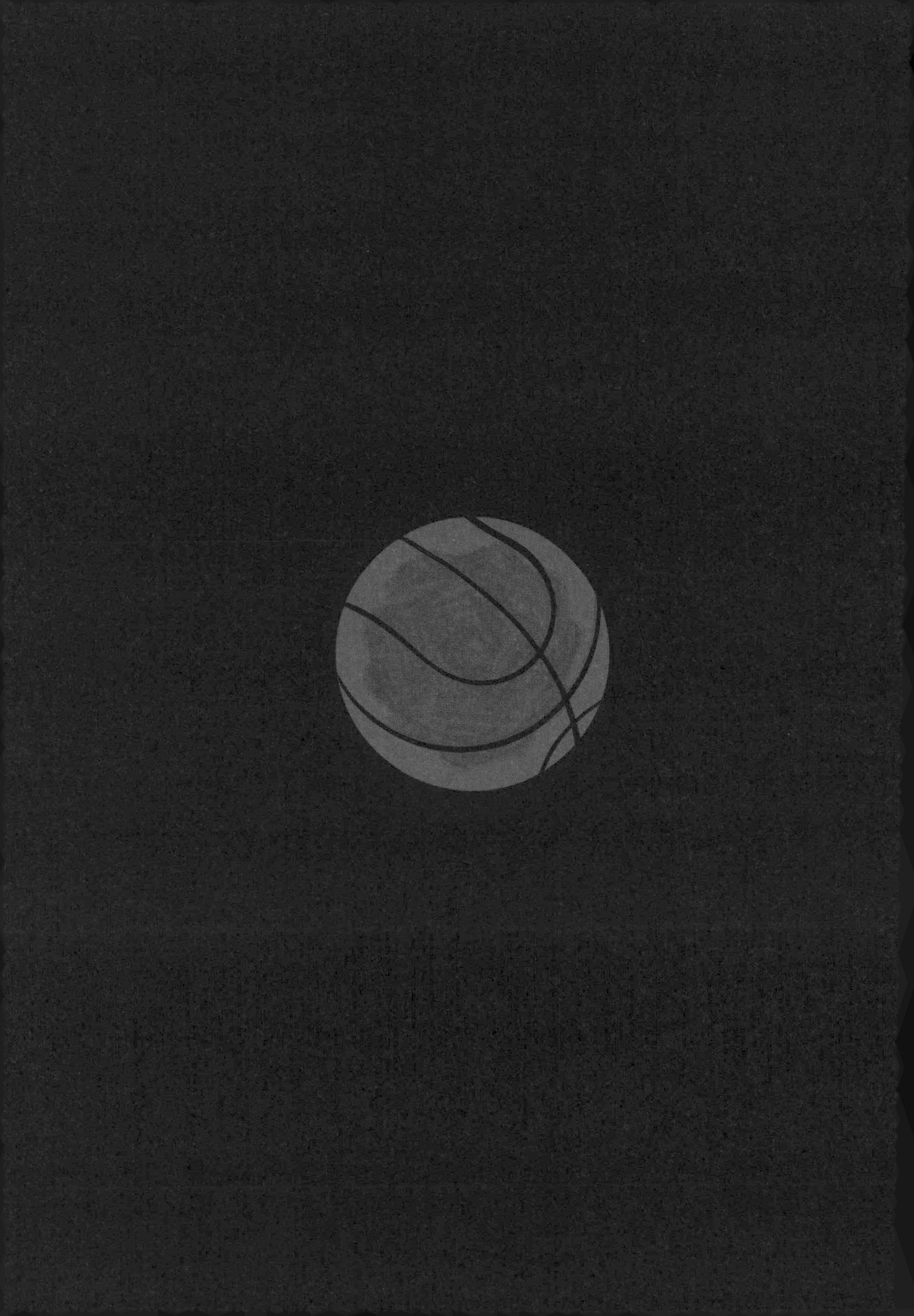

a little while
later...

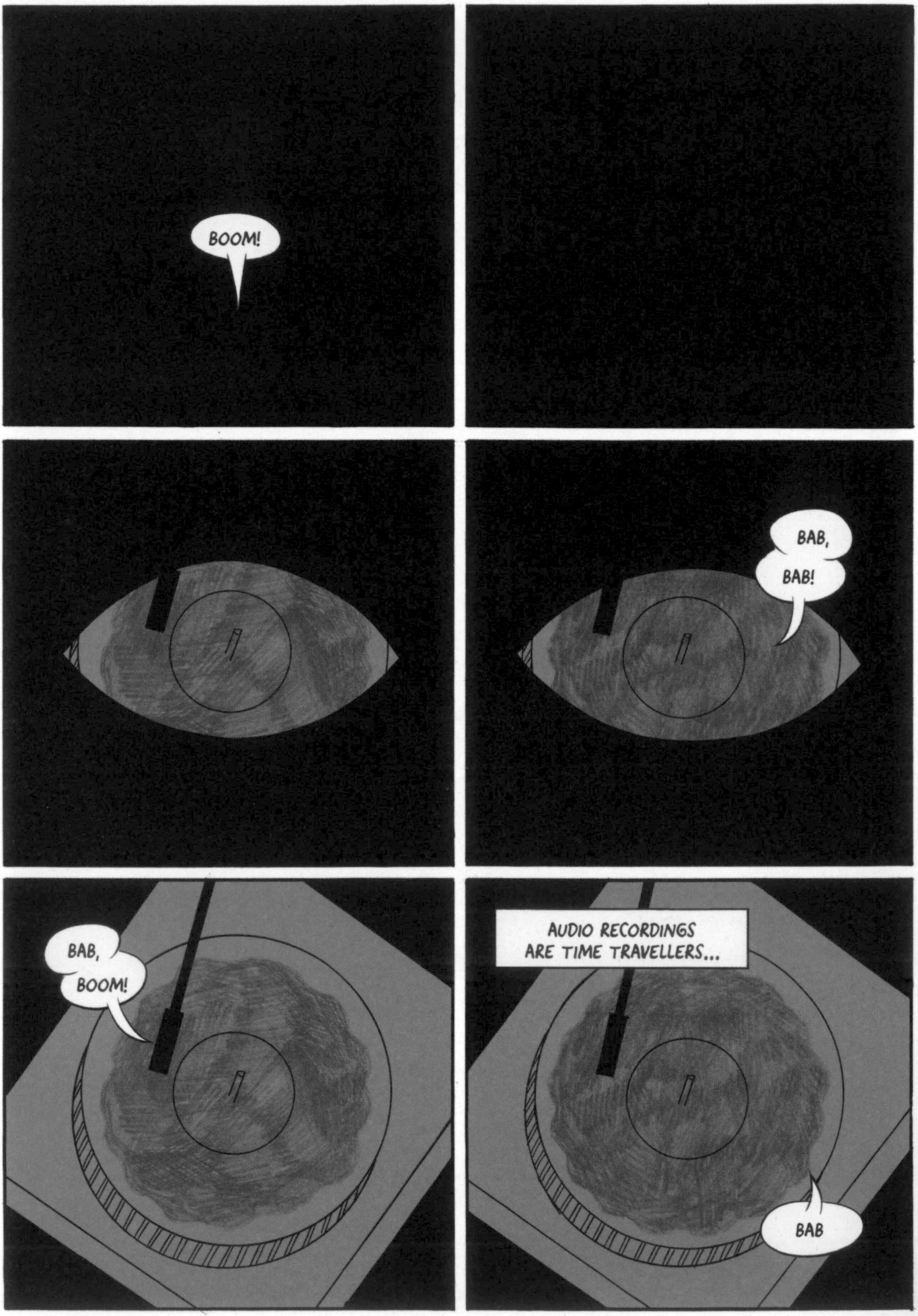
BOOM!
BAB,
BAB!
BAB,
BOOM!
AUDIO RECORDINGS ARE TIME TRAVELLERS...
BAB

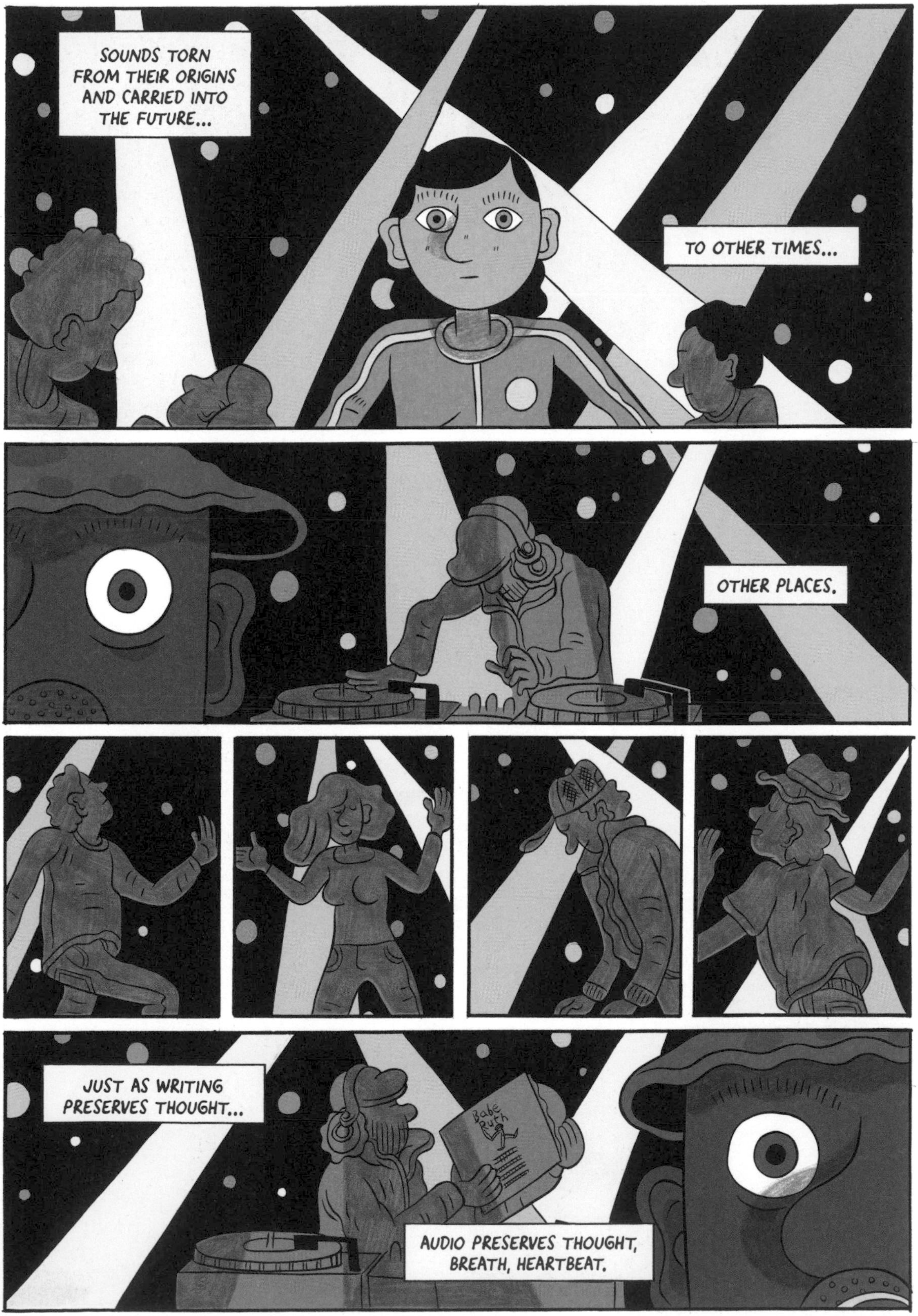
SOUNDS TORN FROM THEIR ORIGINS AND CARRIED INTO THE FUTURE...
TO OTHER TIMES...
OTHER PLACES.
JUST AS WRITING PRESERVES THOUGHT...
Babe Ruth
AUDIO PRESERVES THOUGHT, BREATH, HEARTBEAT.

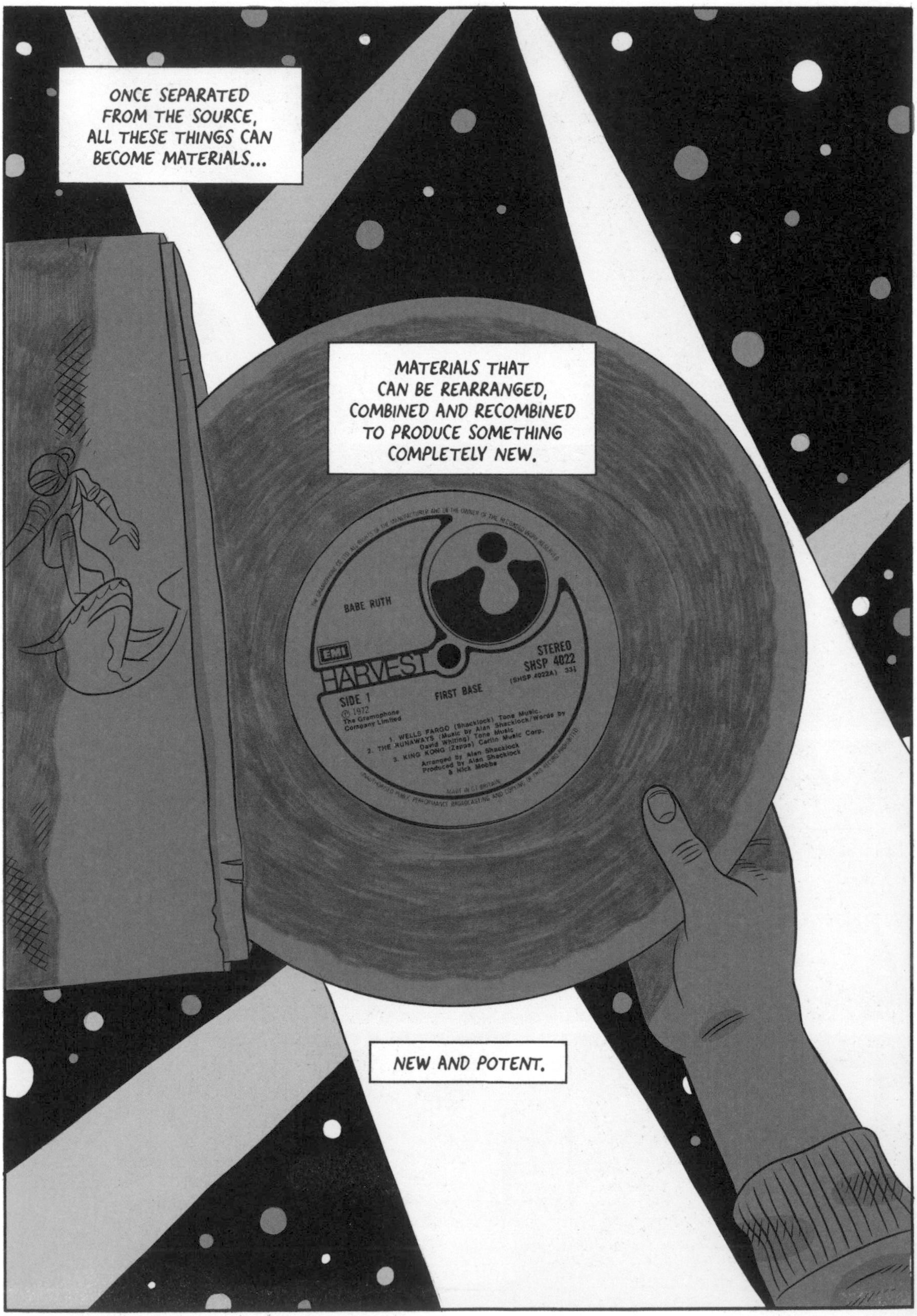
ONCE SEPARATED FROM THE SOURCE, ALL THESE THINGS CAN BECOME MATERIALS...
MATERIALS THAT CAN BE REARRANGED, COMBINED AND RECOMBINED TO PRODUCE SOMETHING COMPLETELY NEW.
BABE RUTH
EMI
HARVEST
STEREO
SHSP 4022
(SHSP 4022A) 33⅓
SIDE 1
℗ 1972
The Gramophone Company Limited
FIRST BASE
1. WELLS FARGO (Shacklock) Tone Music.
2. THE RUNAWAYS (Music by Alan Shacklock/Words by David Whiting) Tone Music
3. KING KONG (Zappa) Carlin Music Corp.
Arranged by Alan Shacklock
Produced by Alan Shacklock & Nick Mobbs
NEW AND POTENT.

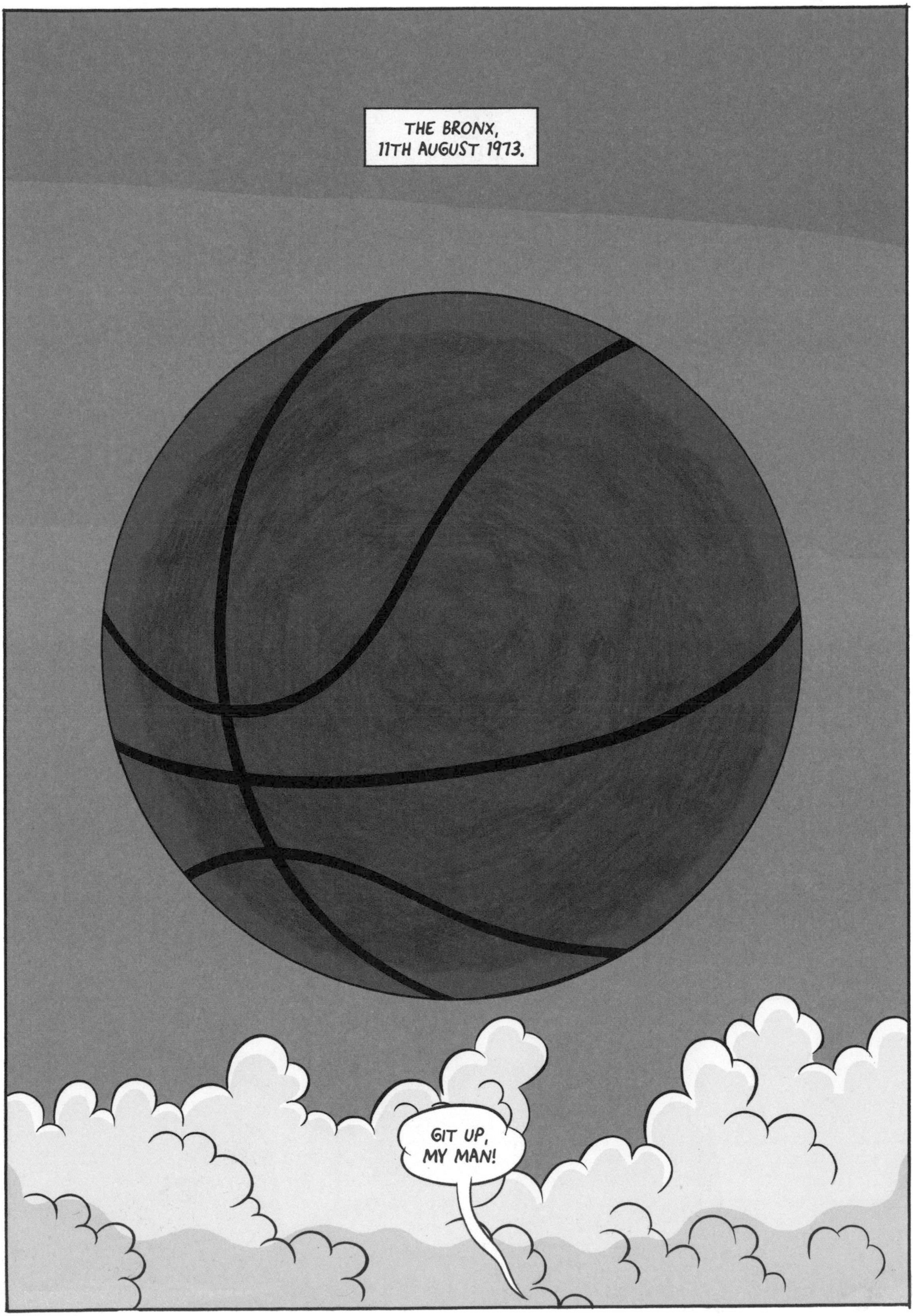
THE BRONX,
11TH AUGUST 1973.
GIT UP,
MY MAN!

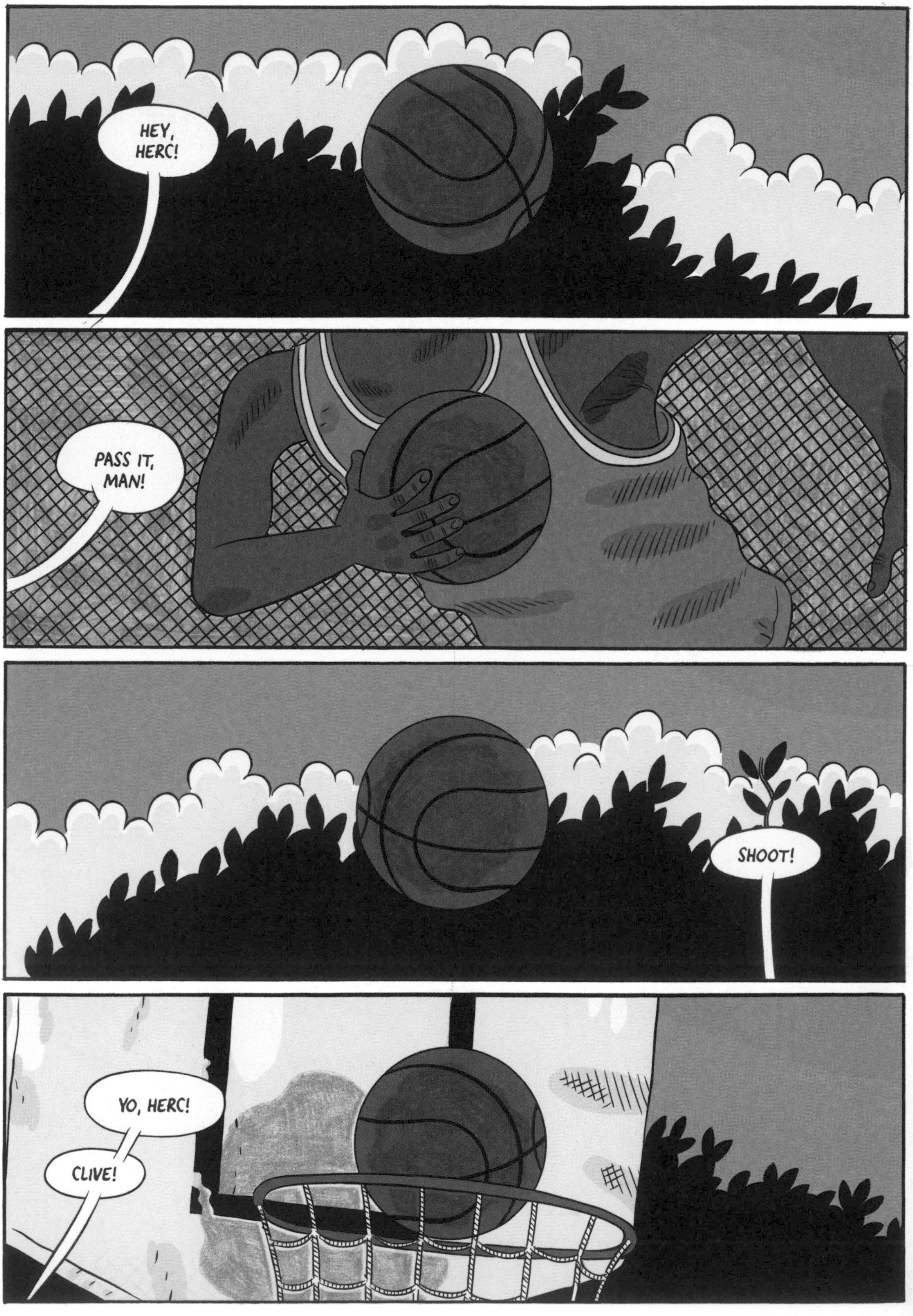
HEY, HERC!
PASS IT, MAN!
SHOOT!
YO, HERC!
CLIVE!

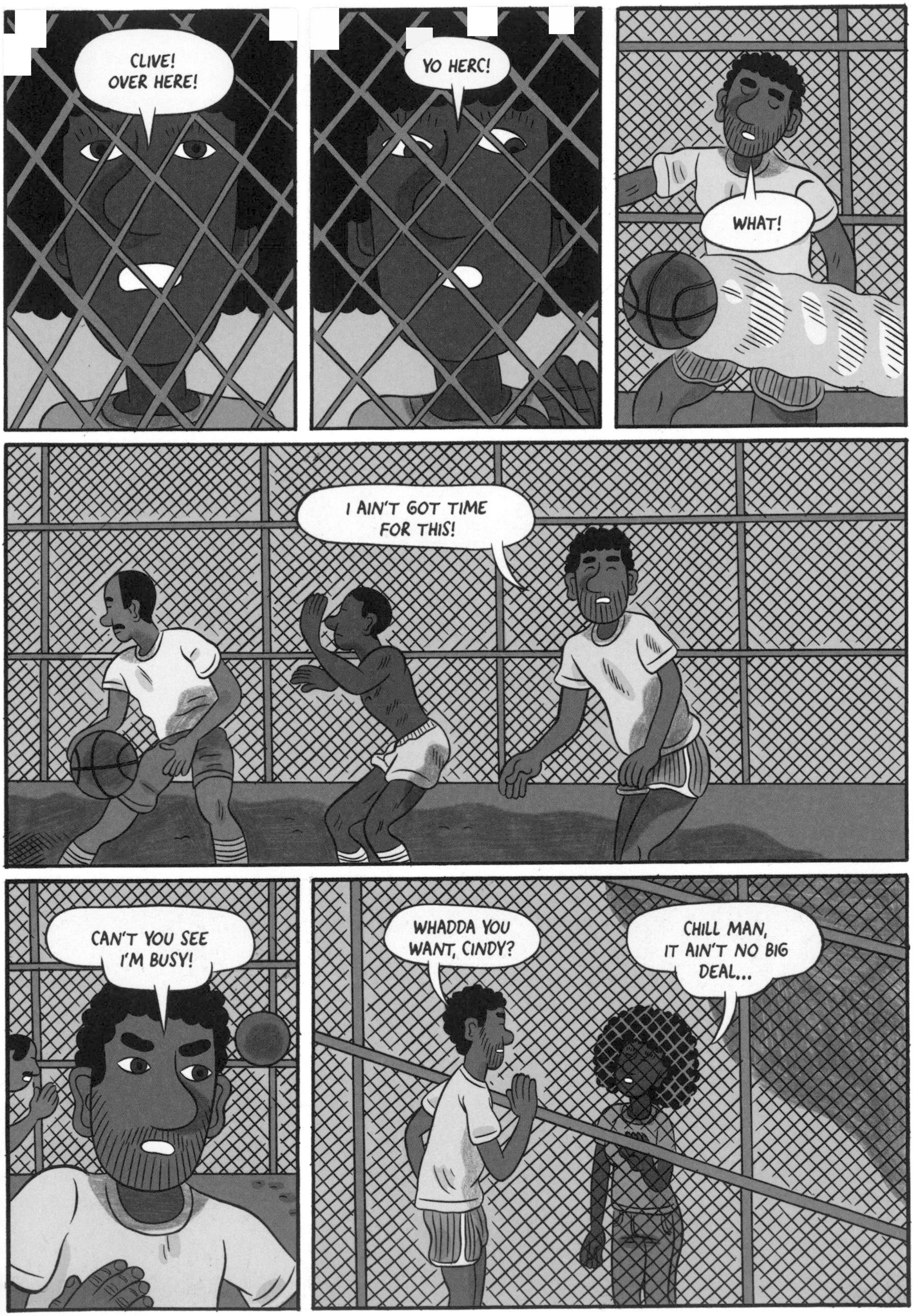
CLIVE! OVER HERE!
YO HERC!
WHAT!
I AIN'T GOT TIME FOR THIS!
CAN'T YOU SEE I'M BUSY!
WHADDA YOU WANT, CINDY?
CHILL MAN, IT AIN'T NO BIG DEAL...

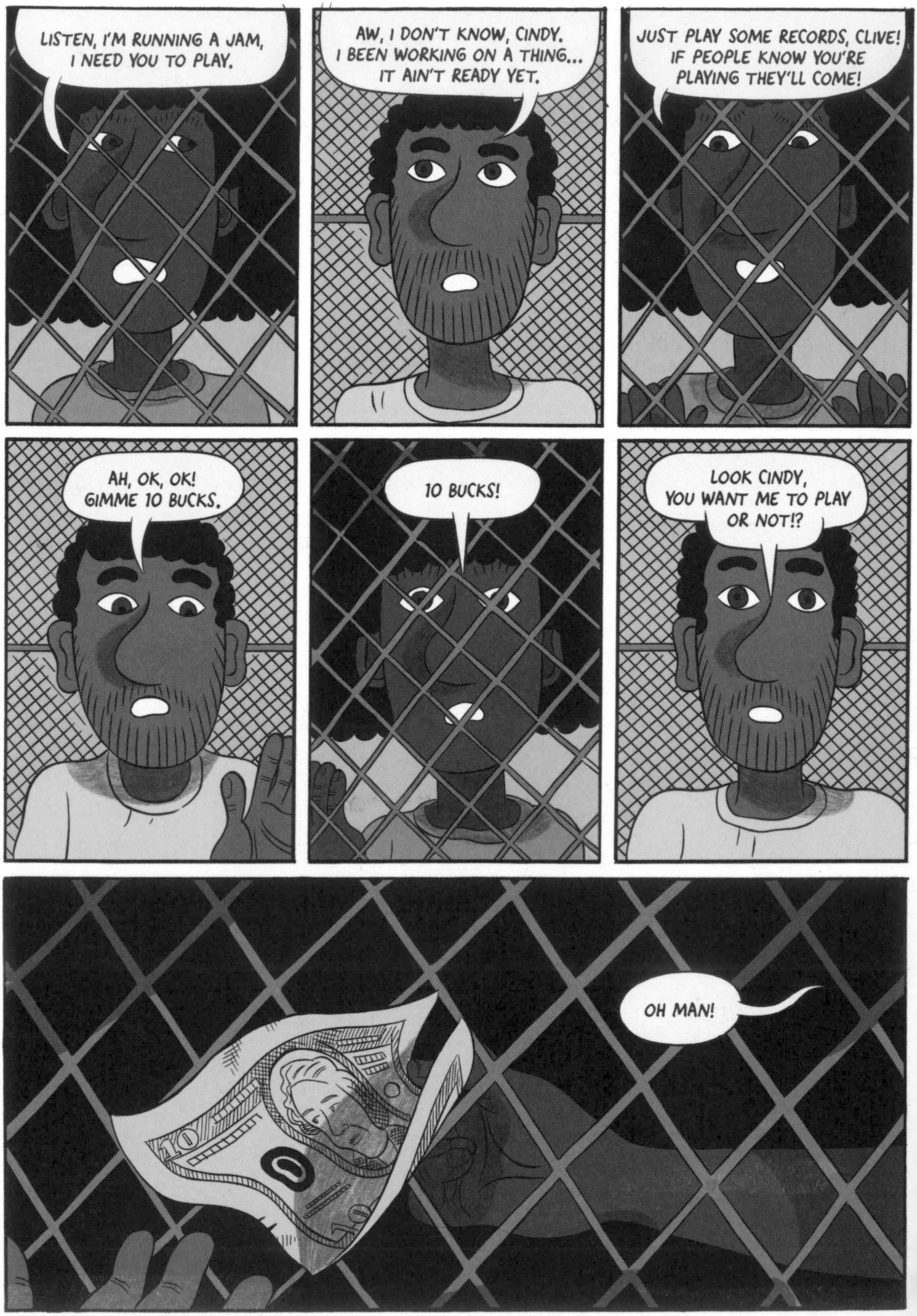
LISTEN, I'M RUNNING A JAM, I NEED YOU TO PLAY.
AW, I DON'T KNOW, CINDY. I BEEN WORKING ON A THING... IT AIN'T READY YET.
JUST PLAY SOME RECORDS, CLIVE! IF PEOPLE KNOW YOU'RE PLAYING THEY'LL COME!
AH, OK, OK! GIMME 10 BUCKS.
10 BUCKS!
LOOK CINDY, YOU WANT ME TO PLAY OR NOT!?
OH MAN!

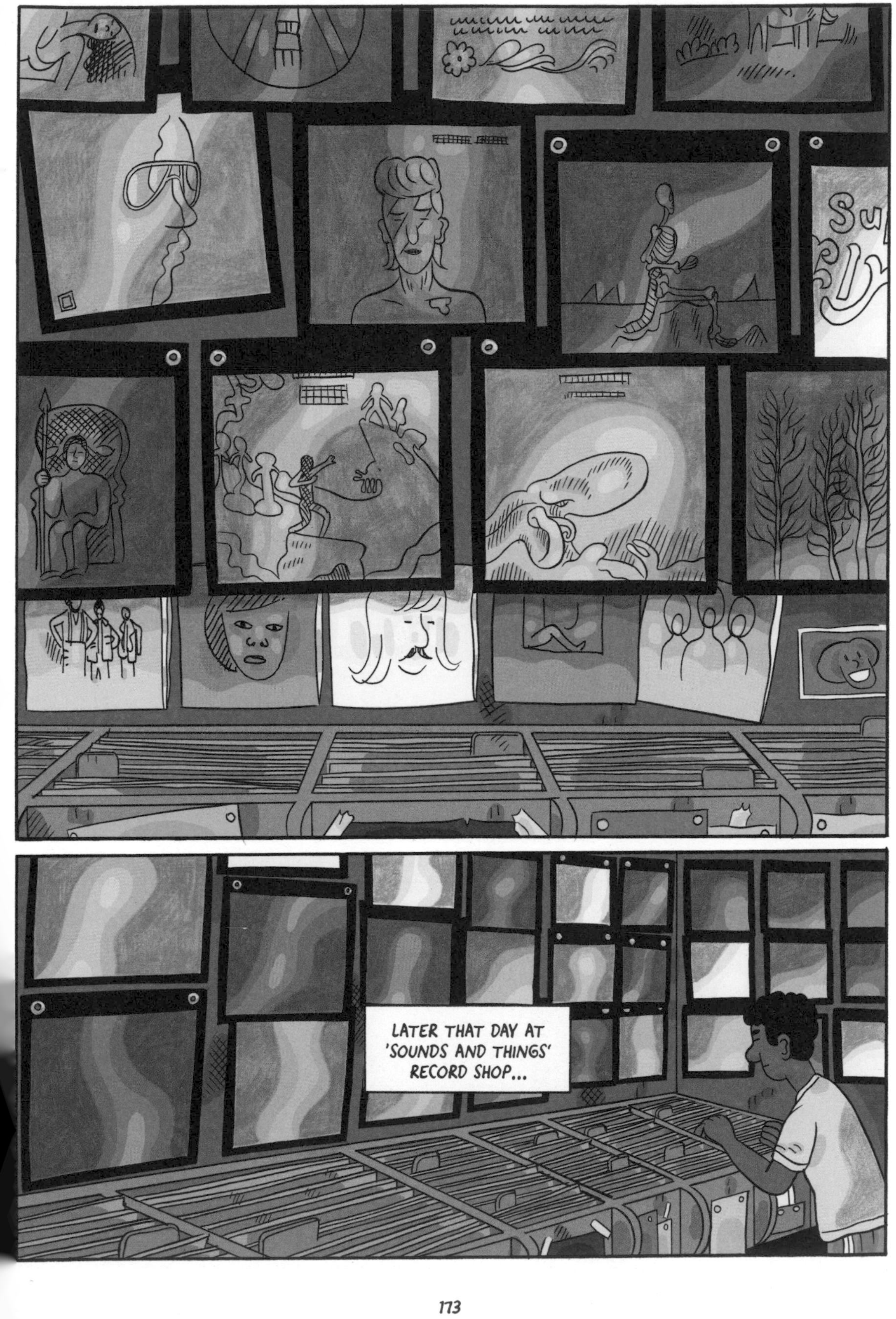
LATER THAT DAY AT 'SOUNDS AND THINGS' RECORD SHOP...

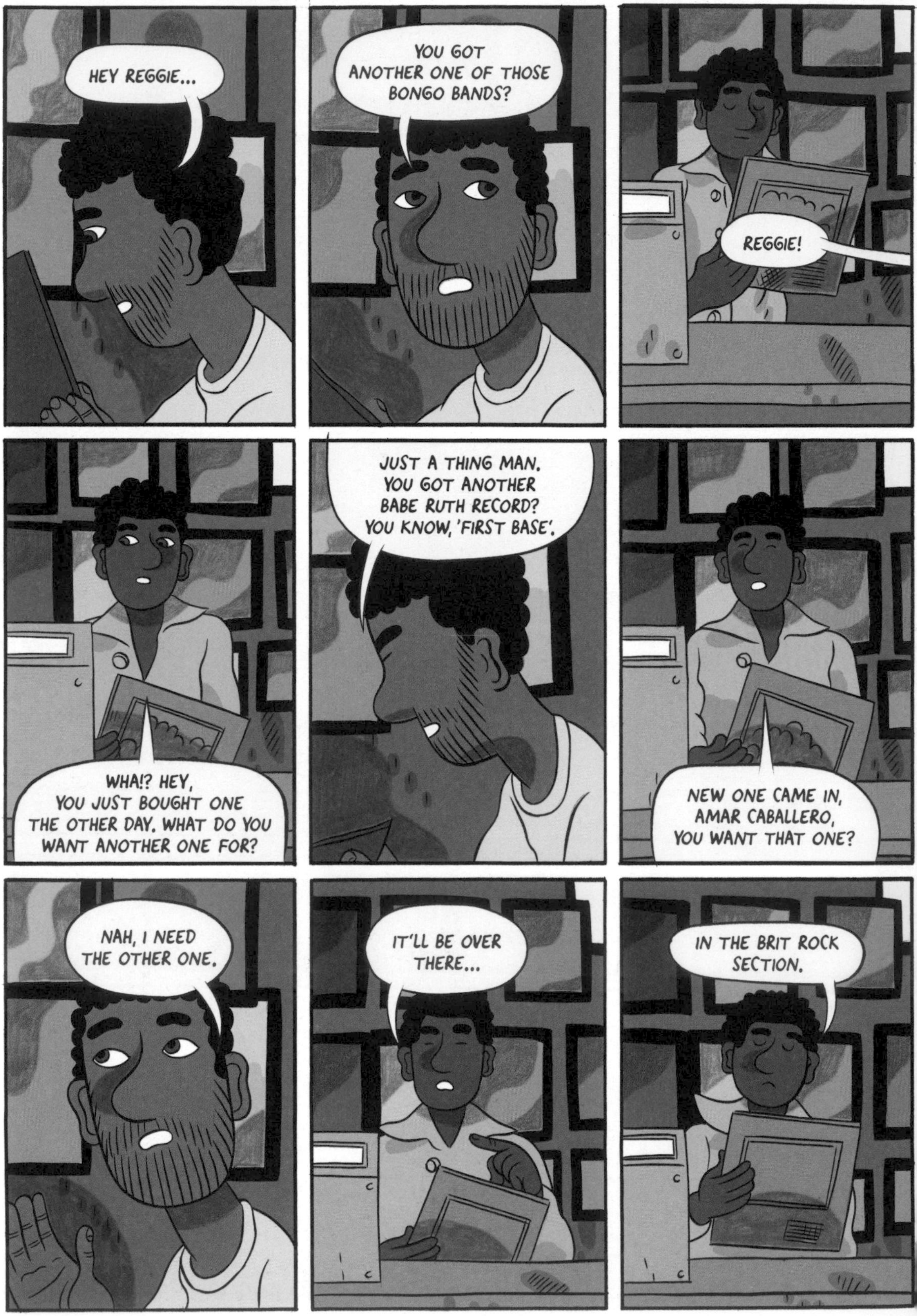
HEY REGGIE...
YOU GOT ANOTHER ONE OF THOSE BONGO BANDS?
REGGIE!
WHA!? HEY, YOU JUST BOUGHT ONE THE OTHER DAY. WHAT DO YOU WANT ANOTHER ONE FOR?
JUST A THING MAN. YOU GOT ANOTHER BABE RUTH RECORD? YOU KNOW, 'FIRST BASE'.
NEW ONE CAME IN, AMAR CABALLERO, YOU WANT THAT ONE?
NAH, I NEED THE OTHER ONE.
IT'LL BE OVER THERE...
IN THE BRIT ROCK SECTION.

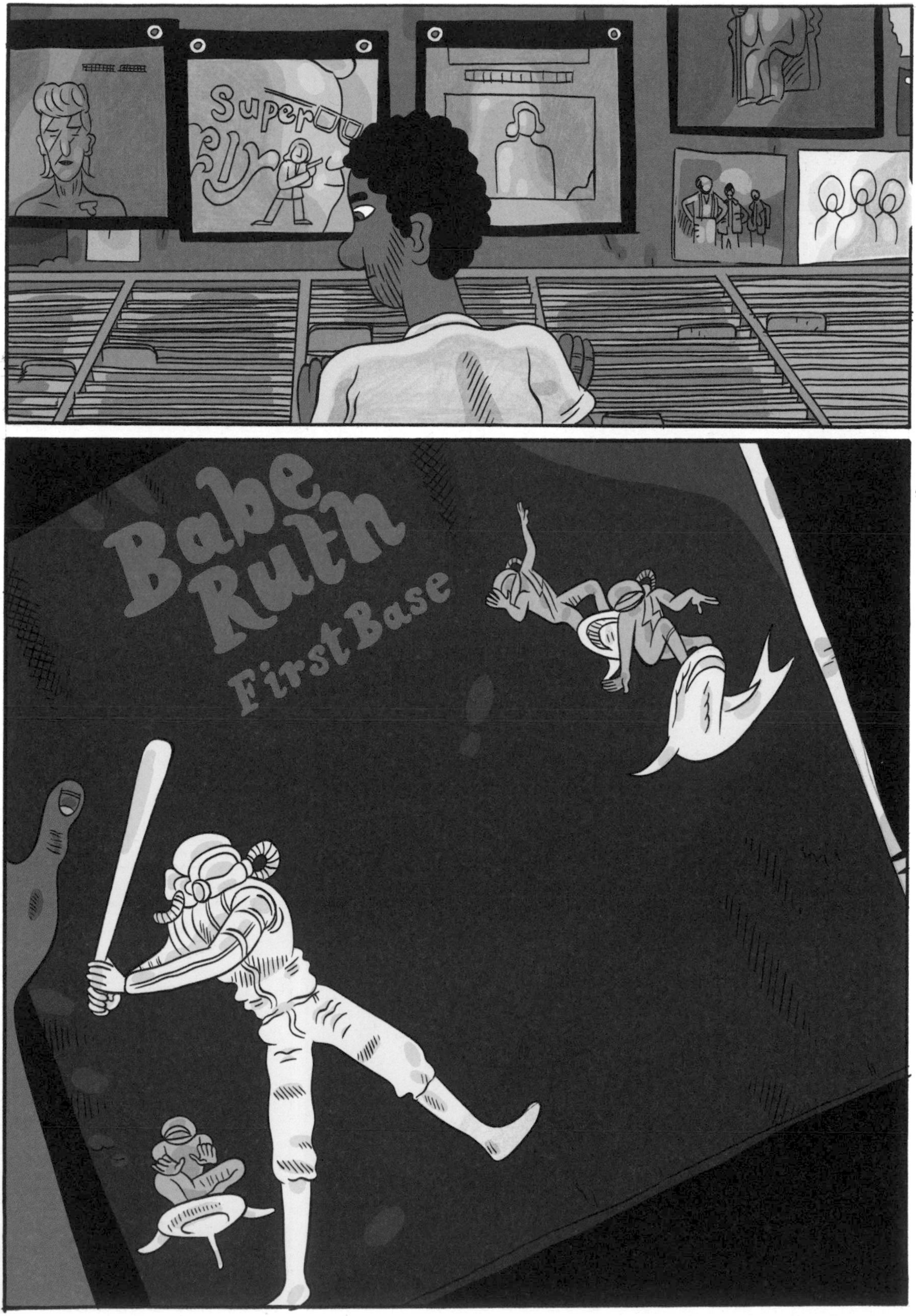
Super
Babe Ruth
First Base

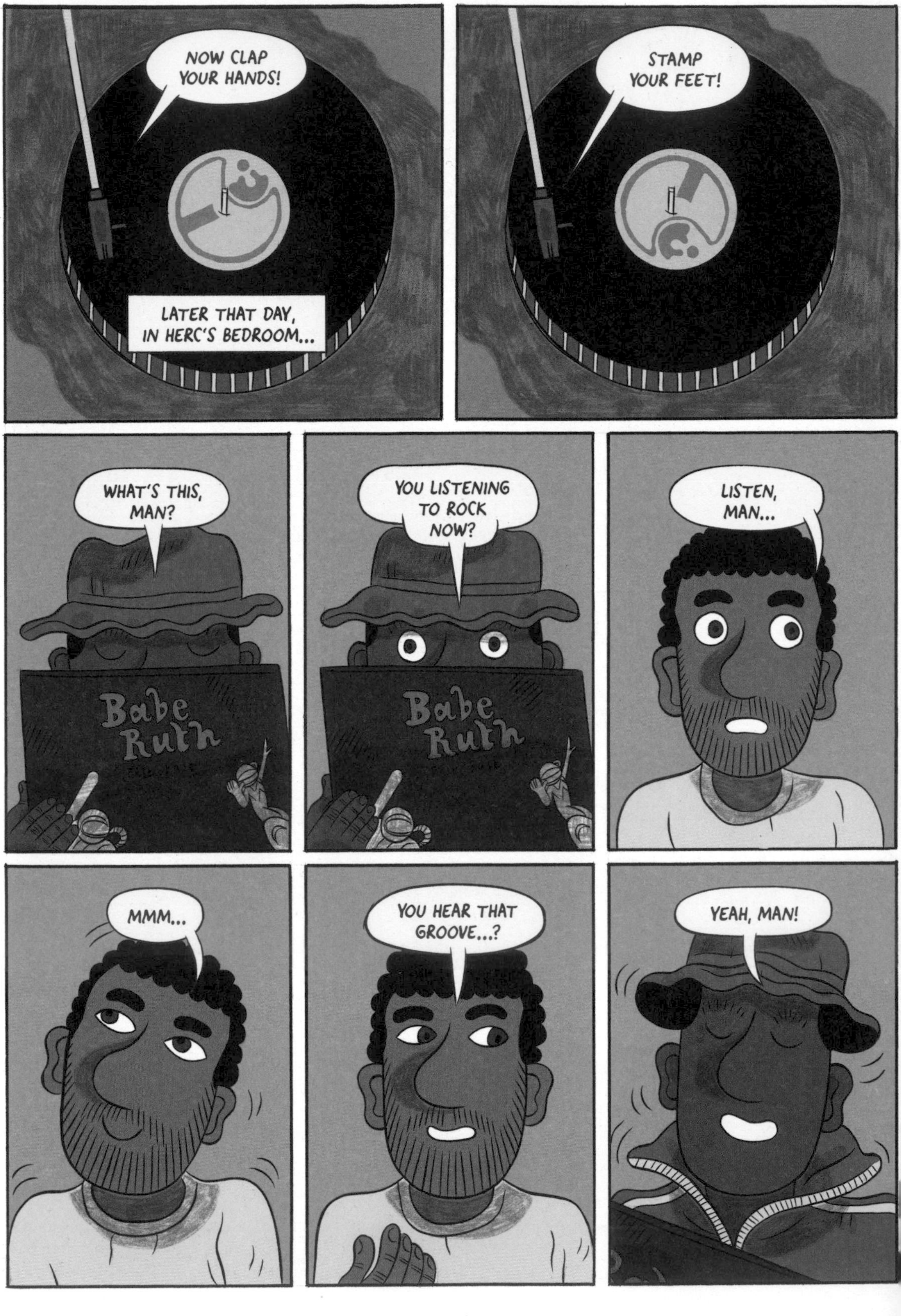
NOW CLAP YOUR HANDS!
LATER THAT DAY, IN HERC'S BEDROOM...
STAMP YOUR FEET!
WHAT'S THIS, MAN?
Babe Ruth
YOU LISTENING TO ROCK NOW?
Babe Ruth
LISTEN, MAN...
MMM...
YOU HEAR THAT GROOVE...?
YEAH, MAN!

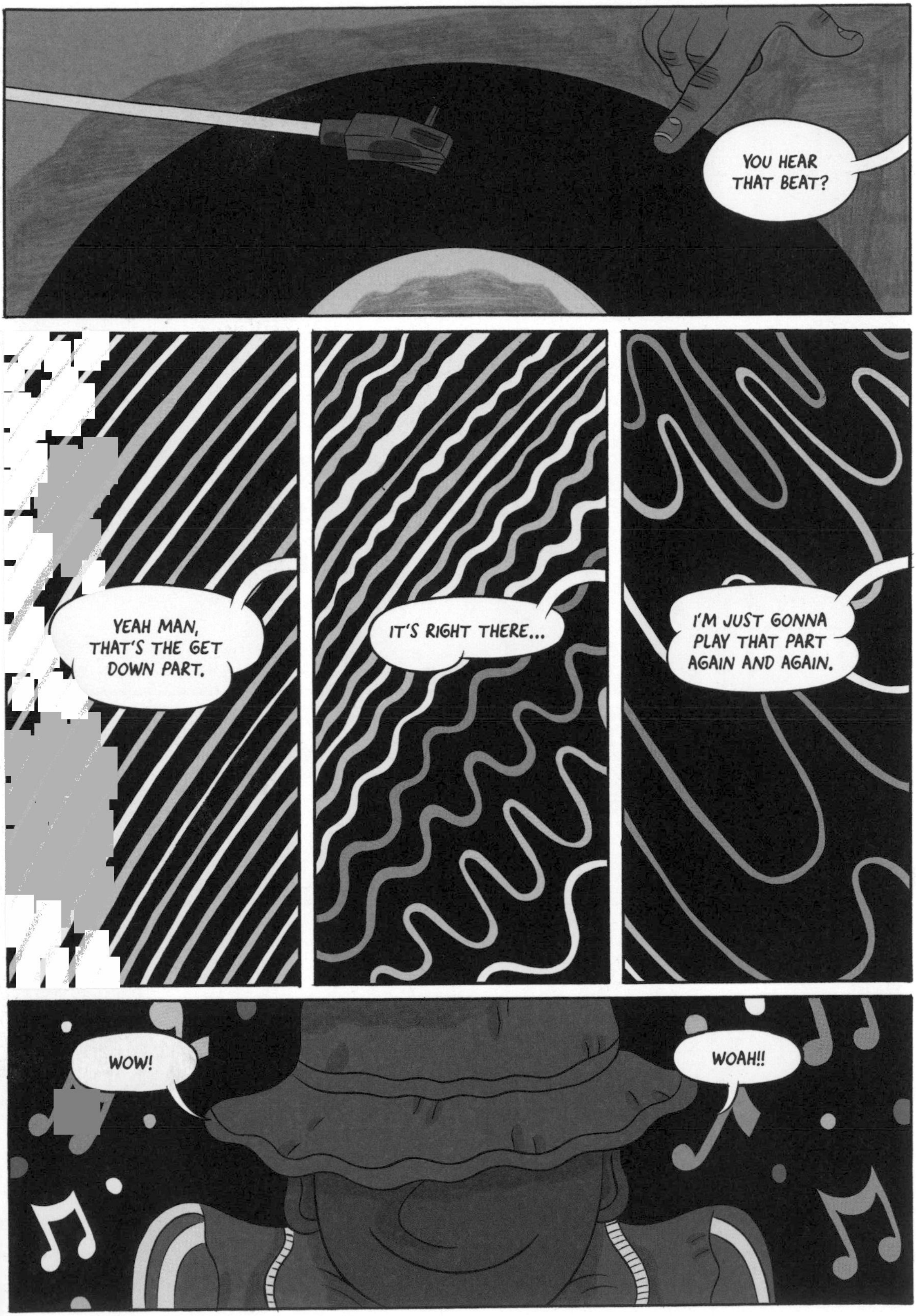
YOU HEAR THAT BEAT?
YEAH MAN, THAT'S THE GET DOWN PART.
IT'S RIGHT THERE...
I'M JUST GONNA PLAY THAT PART AGAIN AND AGAIN.
WOW!
WOAH!!

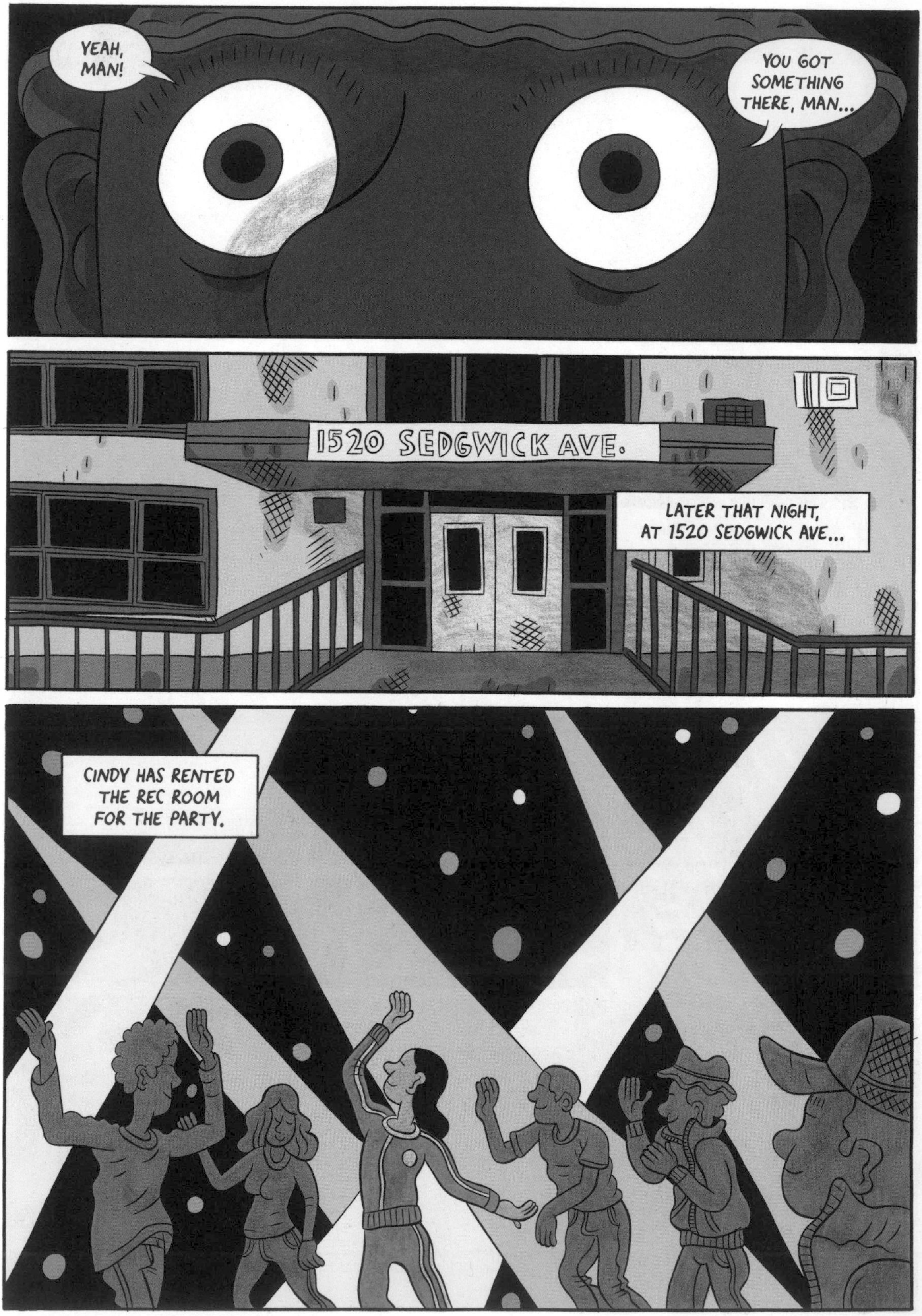
YEAH, MAN!
YOU GOT SOMETHING THERE, MAN...
1520 SEDGWICK AVE.
LATER THAT NIGHT, AT 1520 SEDGWICK AVE...
CINDY HAS RENTED THE REC ROOM FOR THE PARTY.

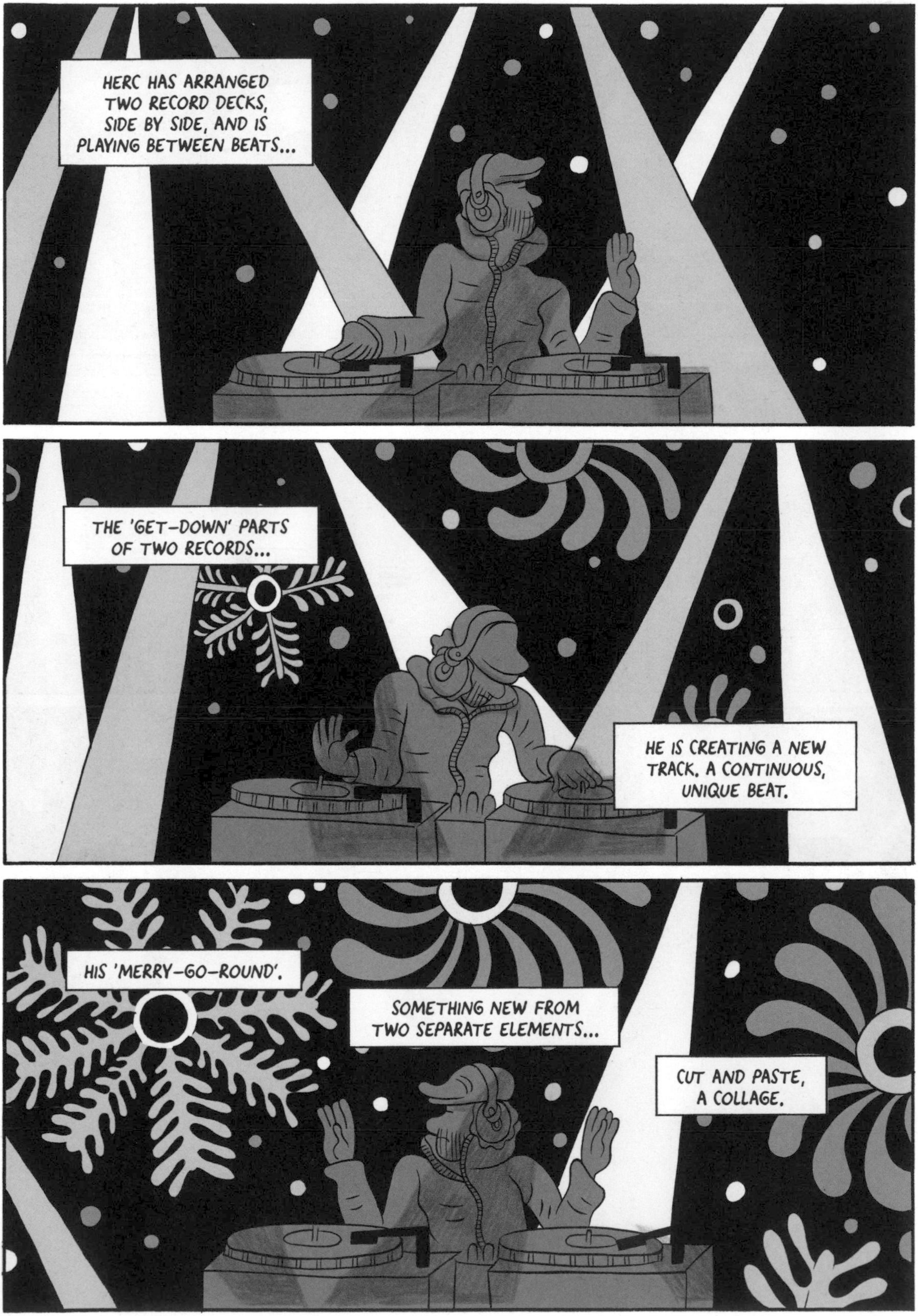
HERC HAS ARRANGED TWO RECORD DECKS, SIDE BY SIDE, AND IS PLAYING BETWEEN BEATS...
THE 'GET-DOWN' PARTS OF TWO RECORDS...
HE IS CREATING A NEW TRACK. A CONTINUOUS, UNIQUE BEAT.
HIS 'MERRY-GO-ROUND'.
SOMETHING NEW FROM TWO SEPARATE ELEMENTS...
CUT AND PASTE, A COLLAGE.

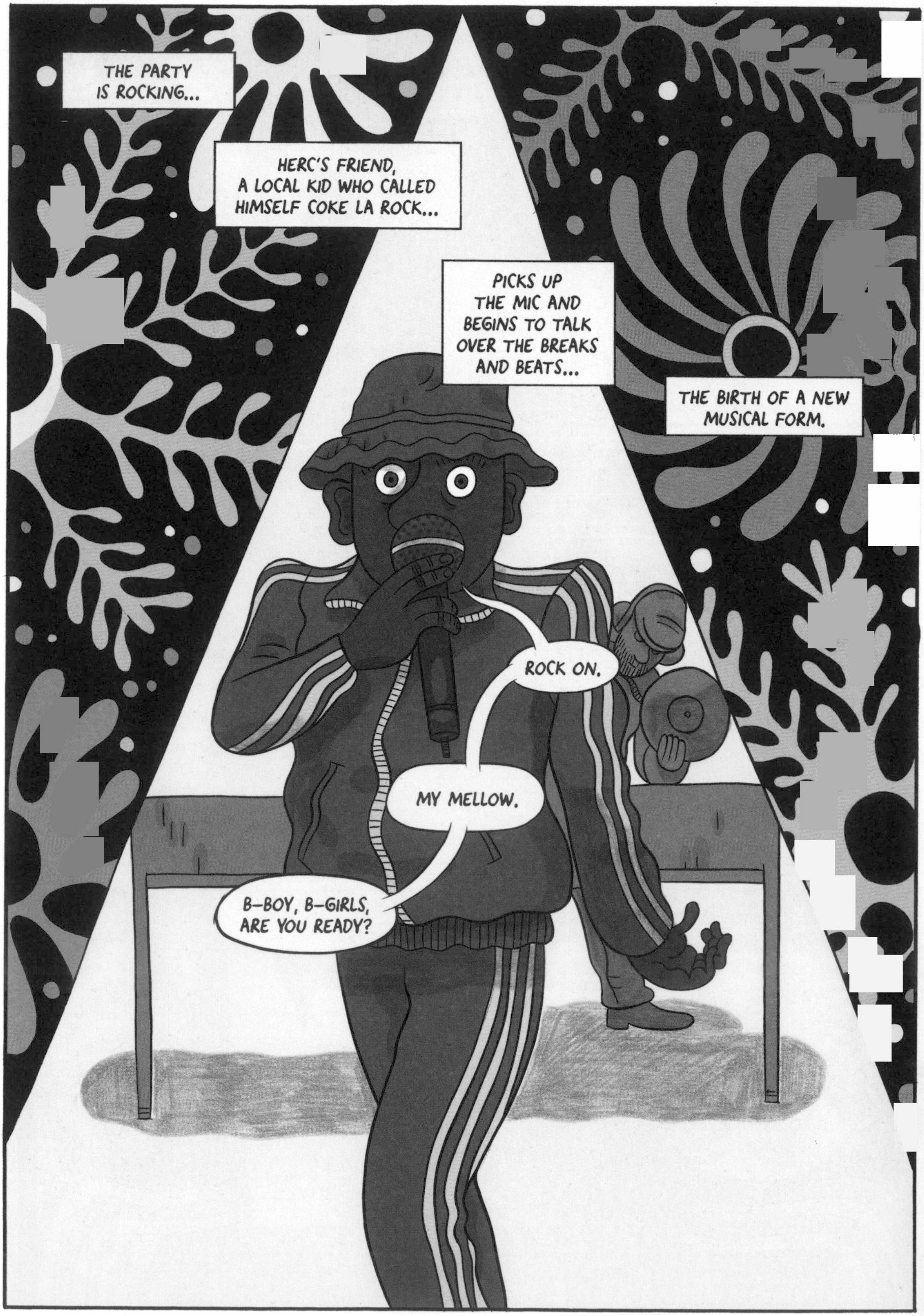
THE PARTY IS ROCKING...
HERC'S FRIEND, A LOCAL KID WHO CALLED HIMSELF COKE LA ROCK...
PICKS UP THE MIC AND BEGINS TO TALK OVER THE BREAKS AND BEATS...
THE BIRTH OF A NEW MUSICAL FORM.
ROCK ON.
MY MELLOW.
B-BOY, B-GIRLS, ARE YOU READY?

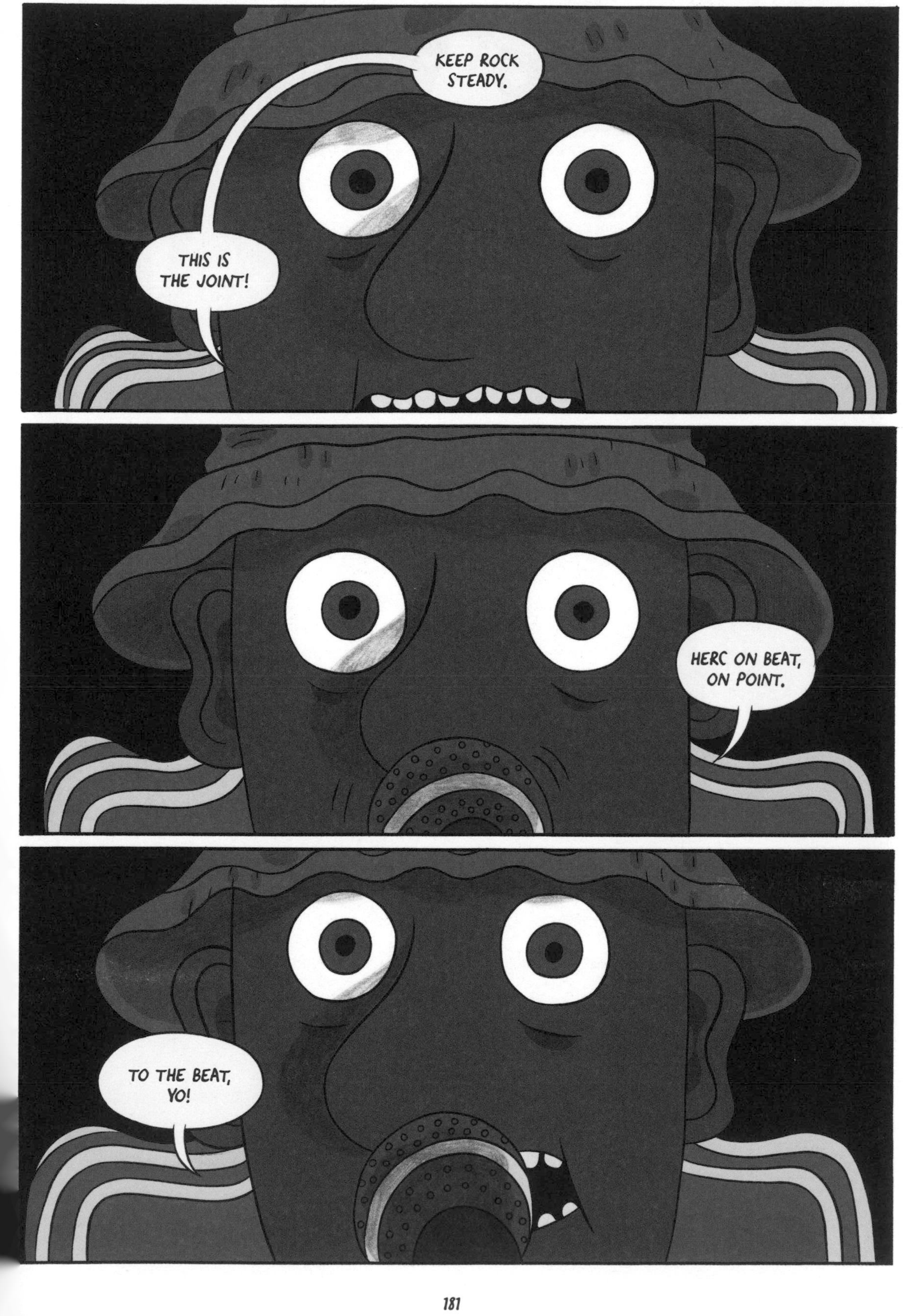
THIS IS THE JOINT!
KEEP ROCK STEADY.
HERC ON BEAT, ON POINT.
TO THE BEAT, YO!

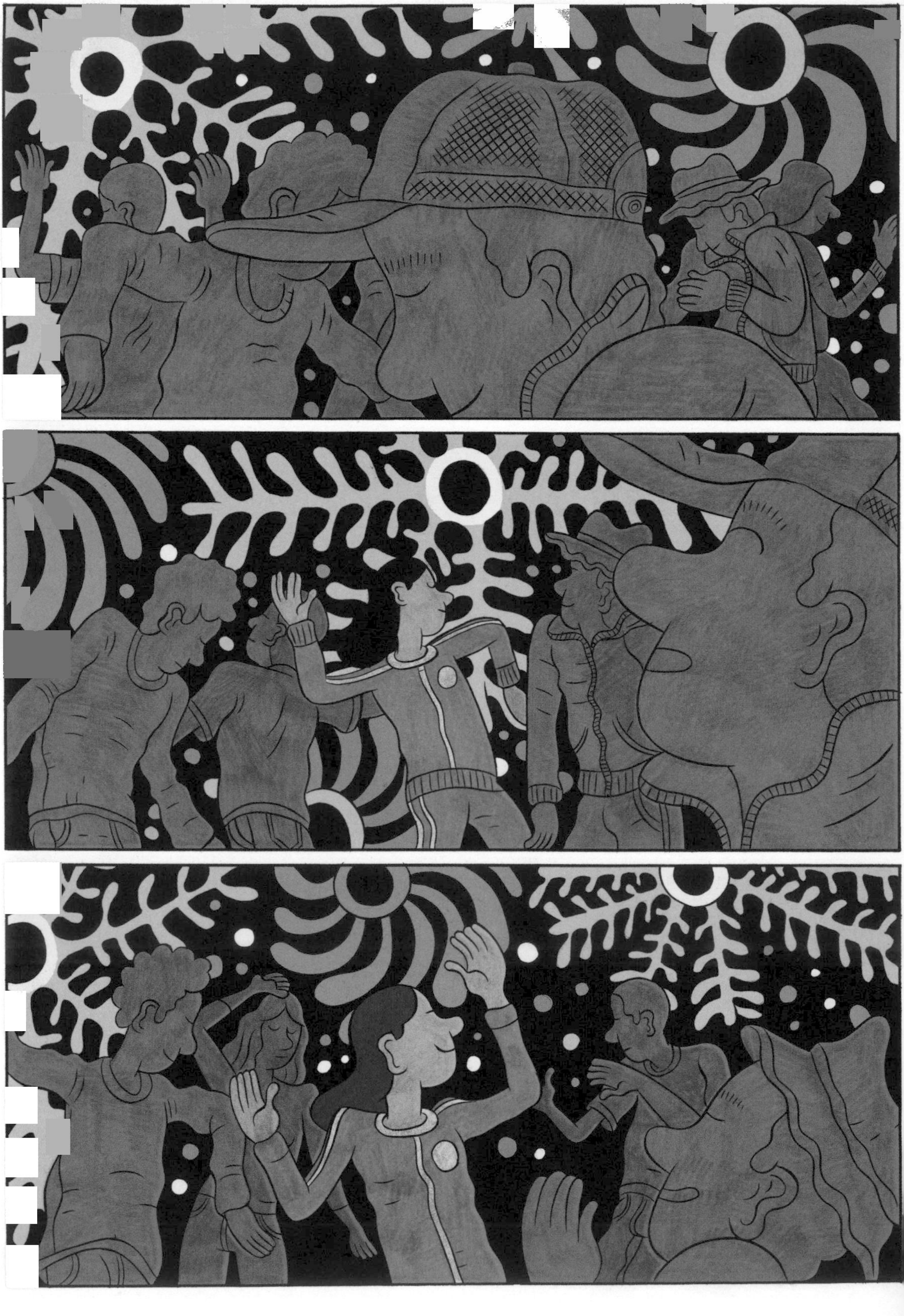

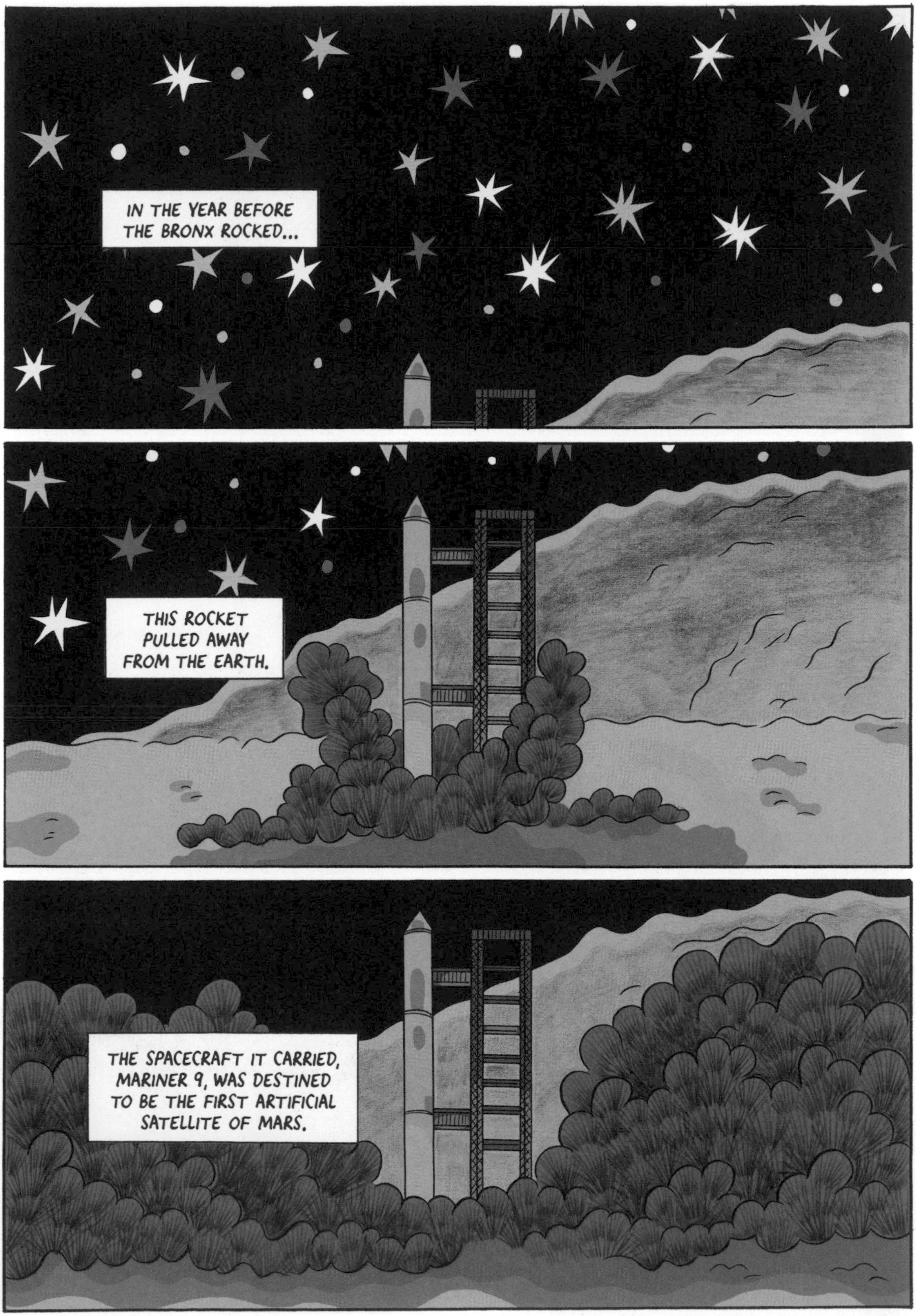
IN THE YEAR BEFORE THE BRONX ROCKED...
THIS ROCKET PULLED AWAY FROM THE EARTH.
THE SPACECRAFT IT CARRIED, MARINER 9, WAS DESTINED TO BE THE FIRST ARTIFICIAL SATELLITE OF MARS.

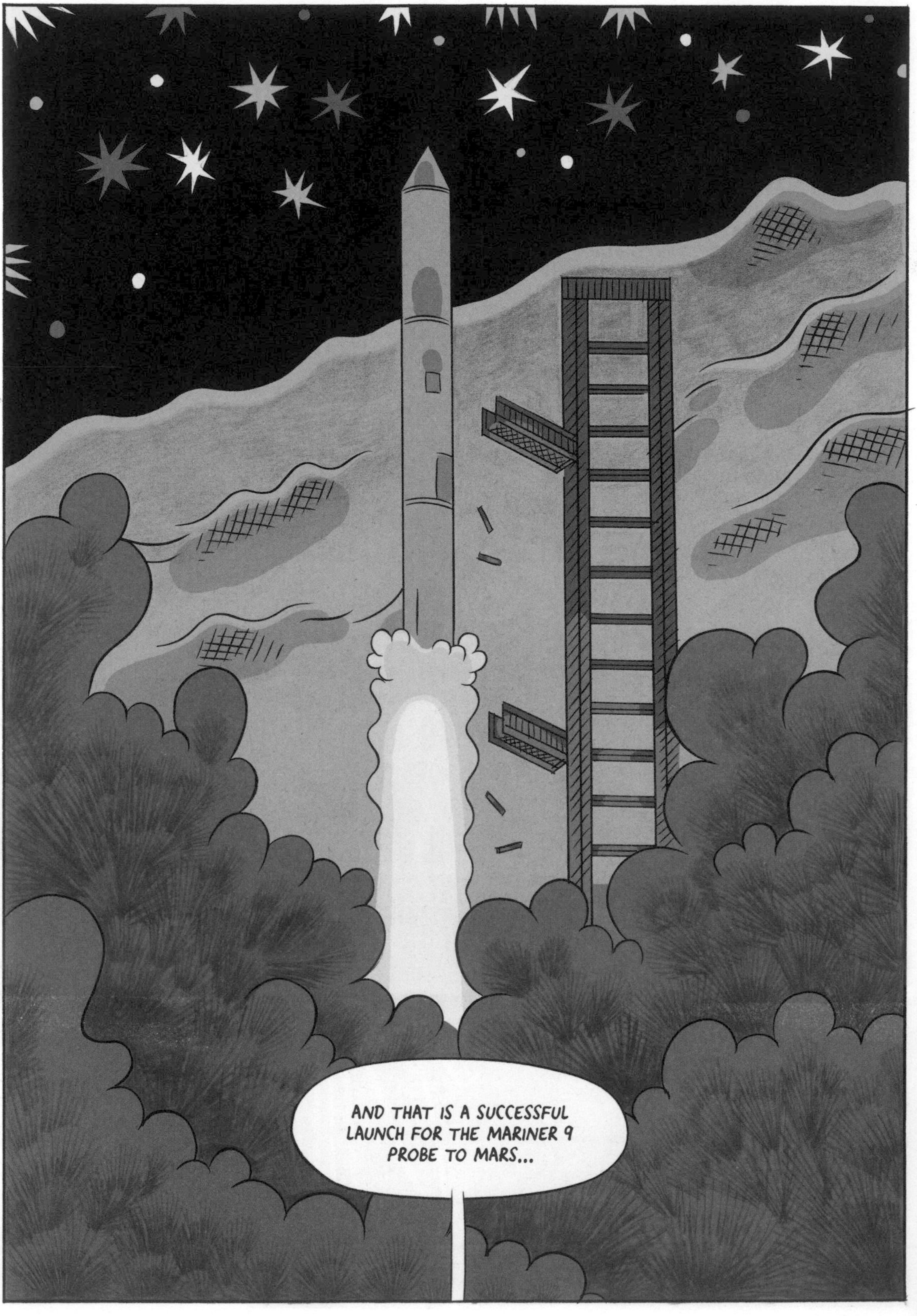
AND THAT IS A SUCCESSFUL LAUNCH FOR THE MARINER 9 PROBE TO MARS...

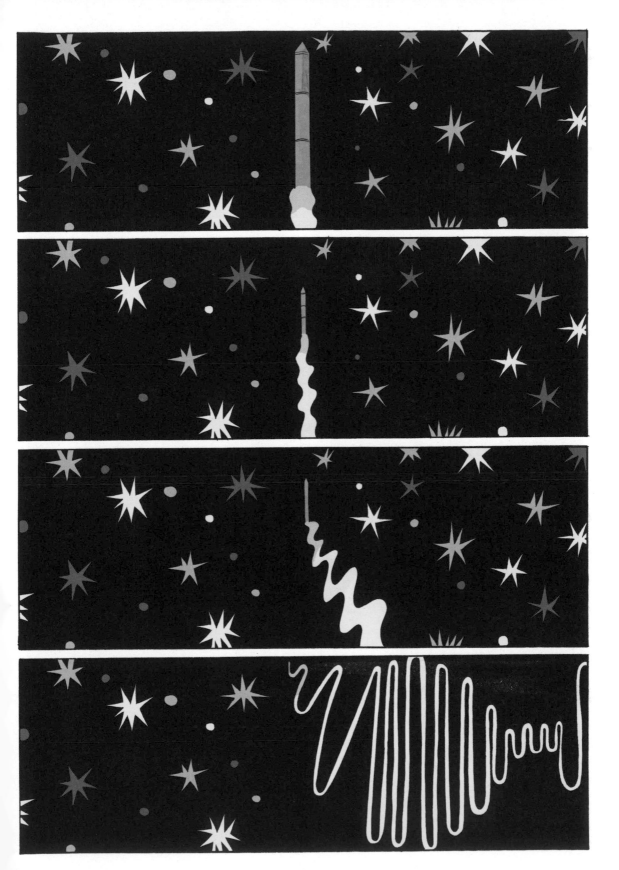

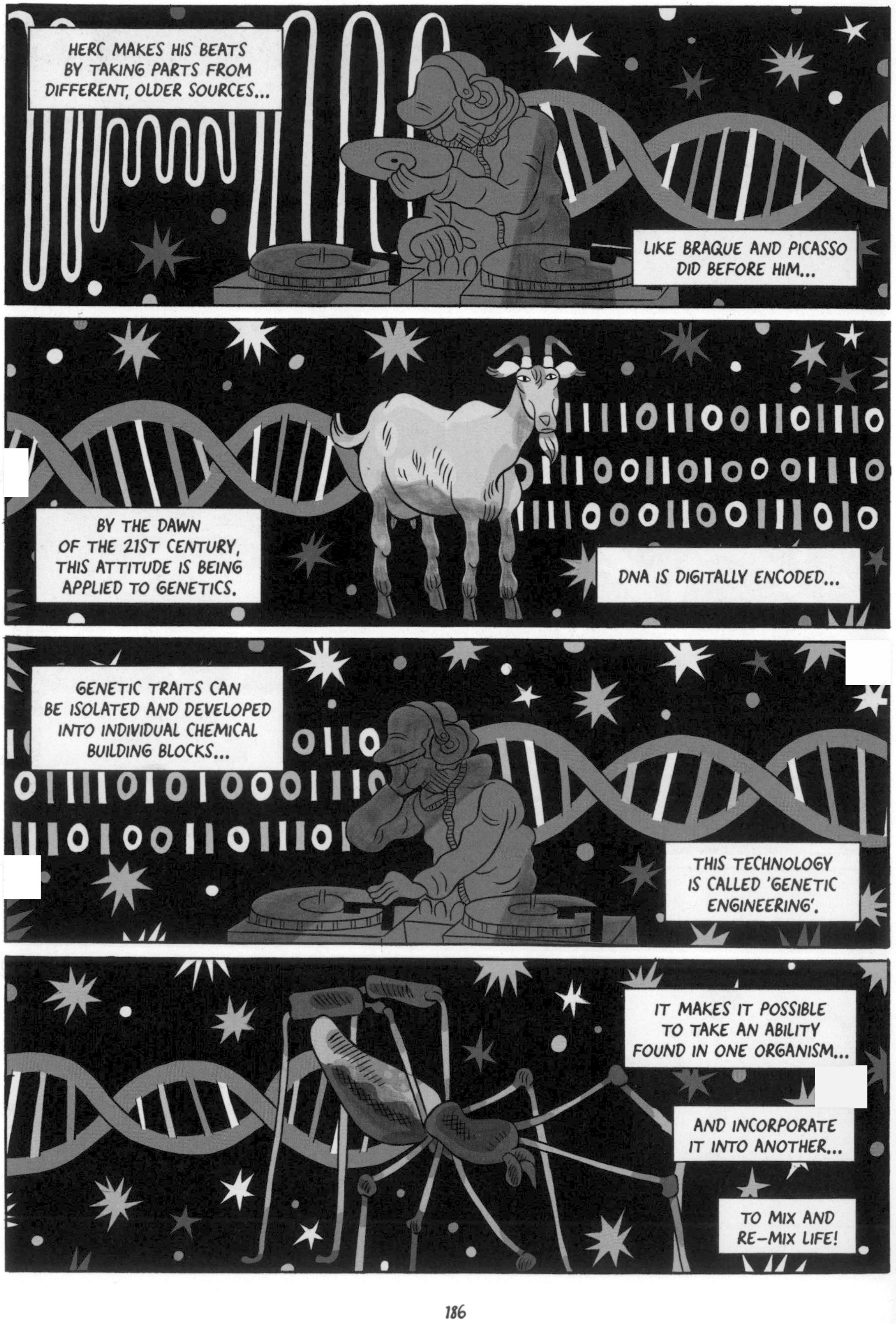
HERC MAKES HIS BEATS BY TAKING PARTS FROM DIFFERENT, OLDER SOURCES...
LIKE BRAQUE AND PICASSO DID BEFORE HIM...
BY THE DAWN OF THE 21ST CENTURY, THIS ATTITUDE IS BEING APPLIED TO GENETICS.
DNA IS DIGITALLY ENCODED...
GENETIC TRAITS CAN BE ISOLATED AND DEVELOPED INTO INDIVIDUAL CHEMICAL BUILDING BLOCKS...
THIS TECHNOLOGY IS CALLED 'GENETIC ENGINEERING'.
IT MAKES IT POSSIBLE TO TAKE AN ABILITY FOUND IN ONE ORGANISM...
AND INCORPORATE IT INTO ANOTHER...
TO MIX AND RE-MIX LIFE!

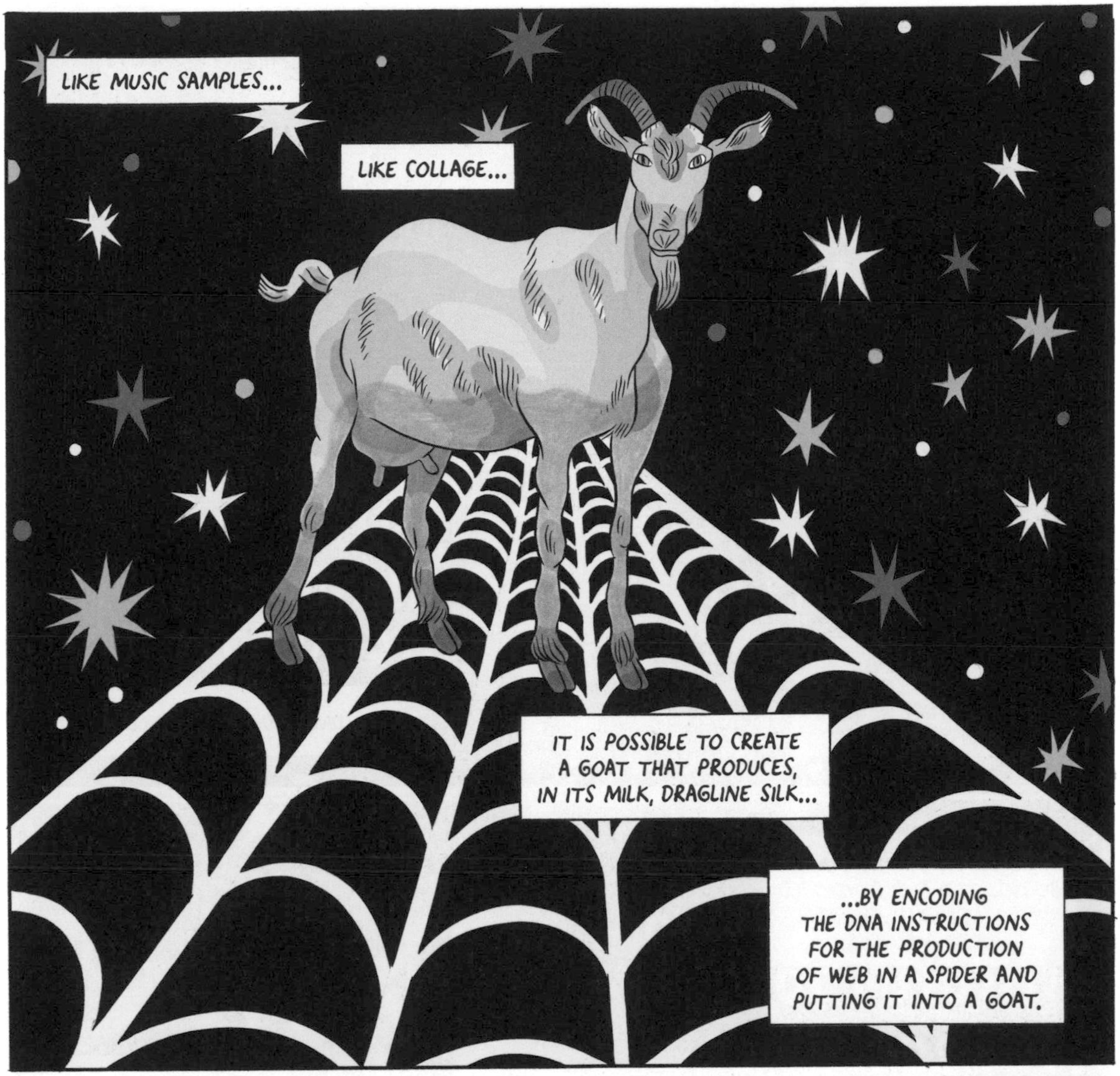
LIKE MUSIC SAMPLES...
LIKE COLLAGE...
IT IS POSSIBLE TO CREATE A GOAT THAT PRODUCES, IN ITS MILK, DRAGLINE SILK...
...BY ENCODING THE DNA INSTRUCTIONS FOR THE PRODUCTION OF WEB IN A SPIDER AND PUTTING IT INTO A GOAT.

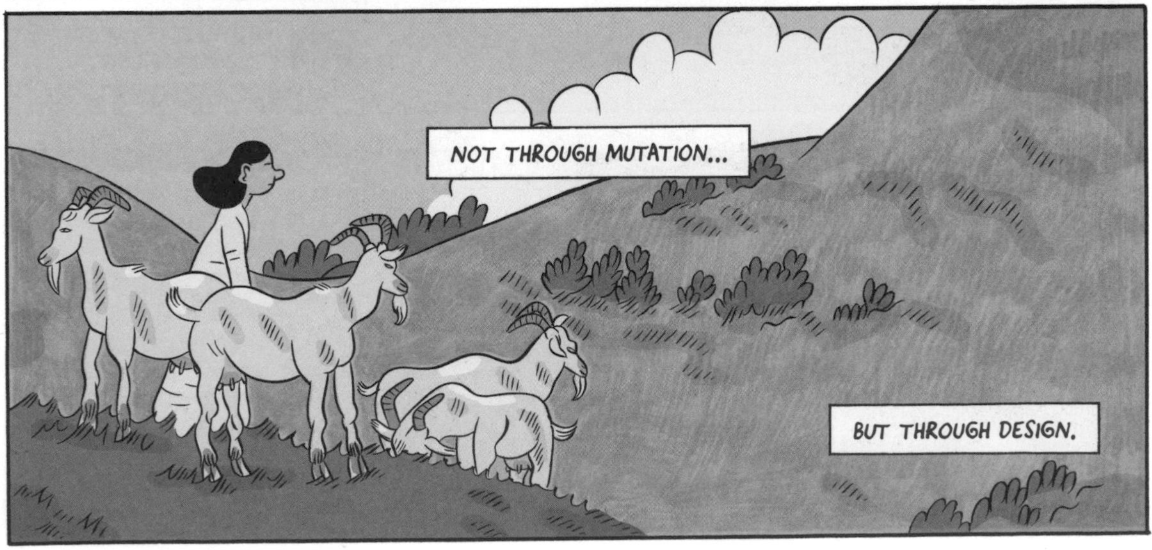
NOT THROUGH MUTATION...
BUT THROUGH DESIGN.

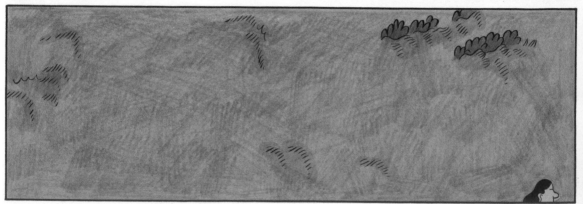

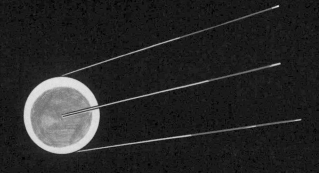

16
years later...

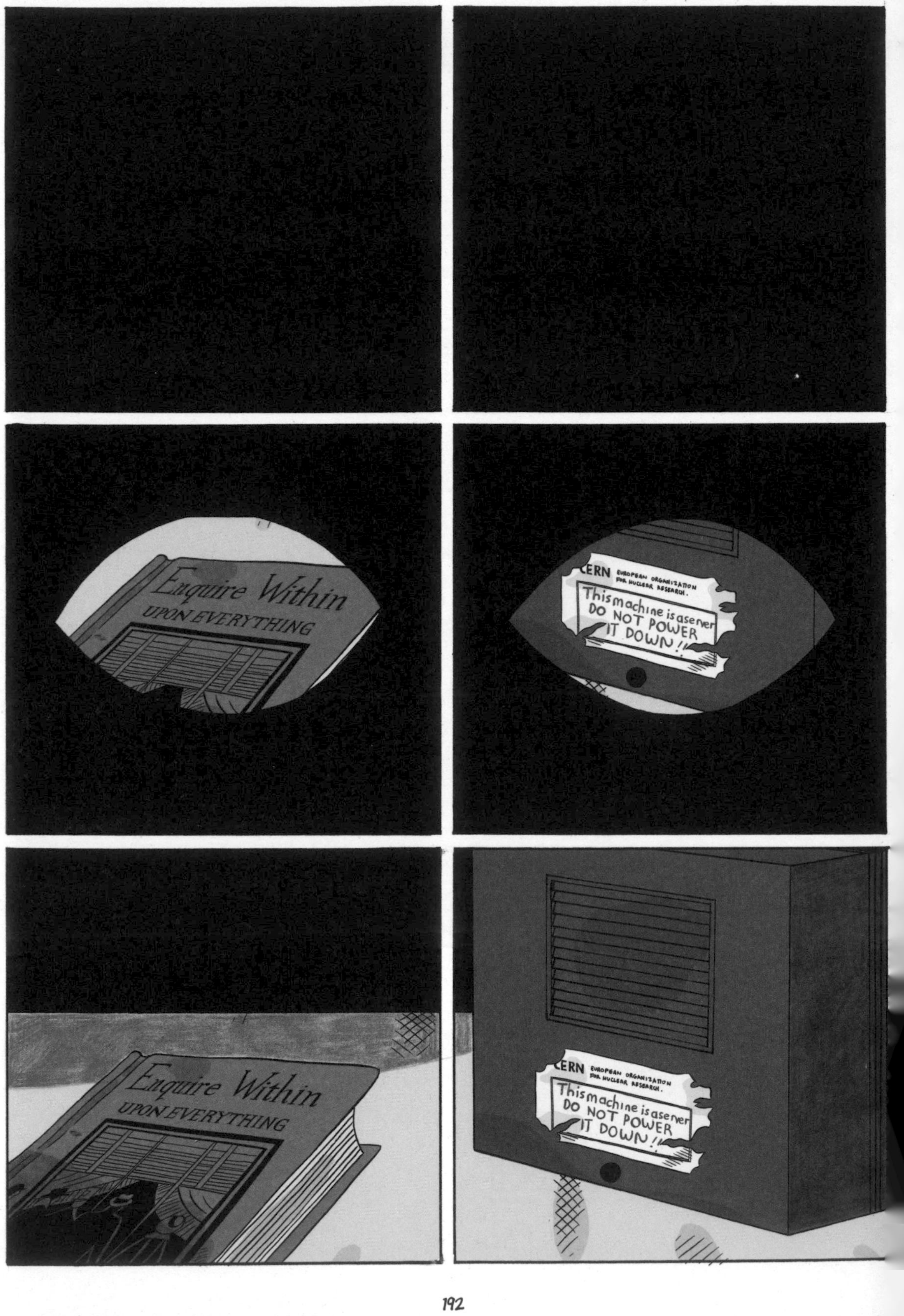
Enquire Within
UPON EVERYTHING
CERN EUROPEAN ORGANIZATION FOR NUCLEAR RESEARCH.
This machine is a server
DO NOT POWER
IT DOWN!!
Enquire Within
UPON EVERYTHING
CERN EUROPEAN ORGANIZATION FOR NUCLEAR RESEARCH.
This machine is a server
DO NOT POWER
IT DOWN!!

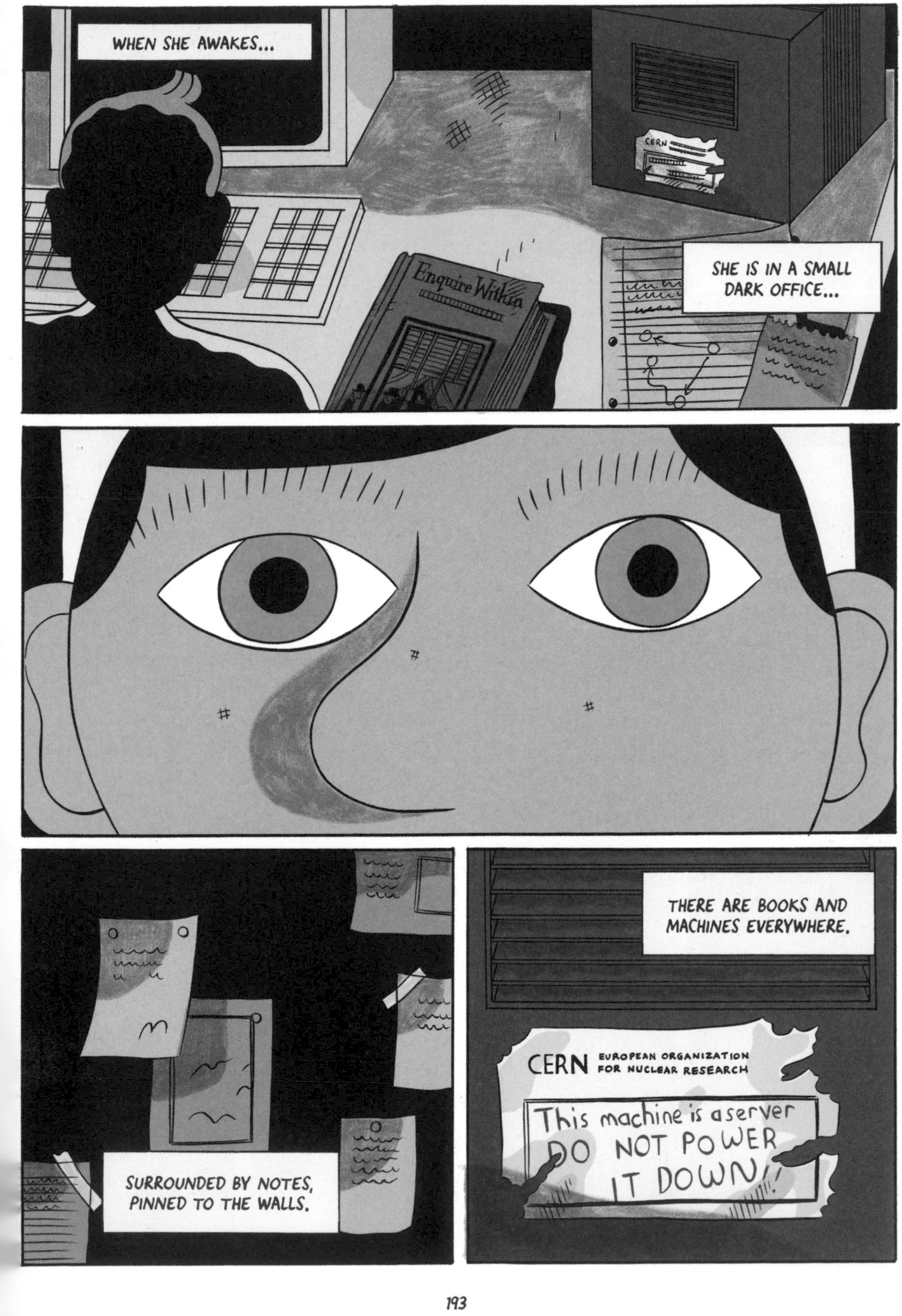
WHEN SHE AWAKES...
CERN
SHE IS IN A SMALL DARK OFFICE...
SURROUNDED BY NOTES, PINNED TO THE WALLS.
THERE ARE BOOKS AND MACHINES EVERYWHERE.
CERN EUROPEAN ORGANIZATION FOR NUCLEAR RESEARCH
This machine is a server
DO NOT POWER IT DOWN!!

A MAN IS TALKING.

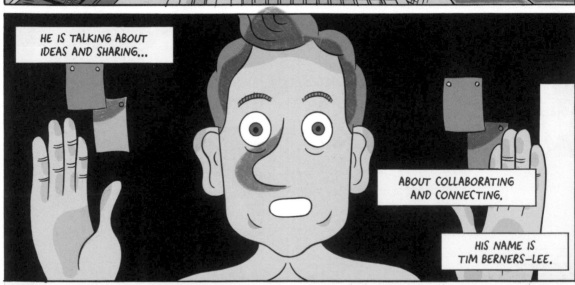

HE IS TALKING ABOUT IDEAS AND SHARING...

ABOUT COLLABORATING AND CONNECTING.

HIS NAME IS TIM BERNERS-LEE.

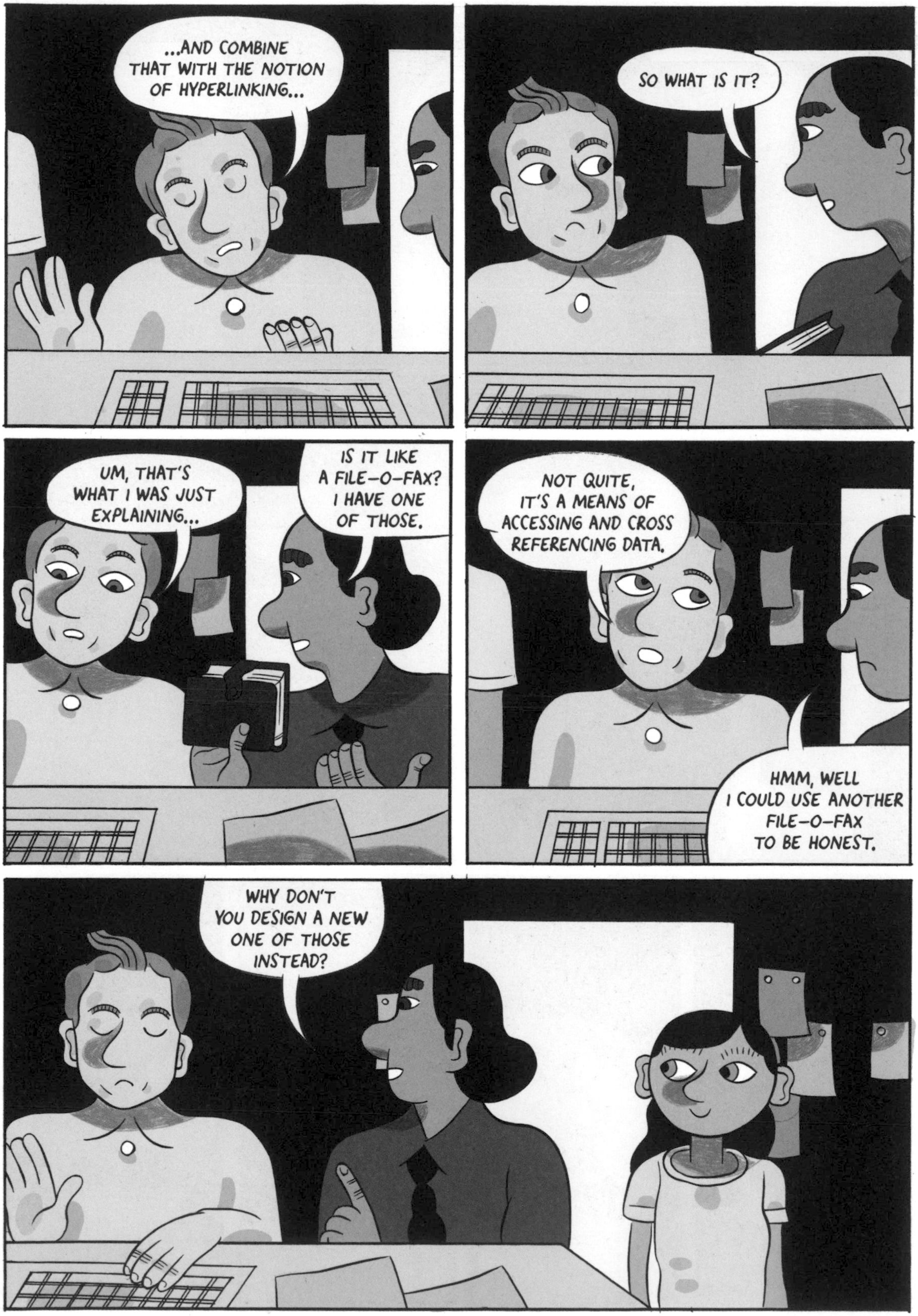
...AND COMBINE THAT WITH THE NOTION OF HYPERLINKING...
SO WHAT IS IT?
UM, THAT'S WHAT I WAS JUST EXPLAINING...
IS IT LIKE A FILE-O-FAX? I HAVE ONE OF THOSE.
NOT QUITE, IT'S A MEANS OF ACCESSING AND CROSS REFERENCING DATA.
HMM, WELL I COULD USE ANOTHER FILE-O-FAX TO BE HONEST.
WHY DON'T YOU DESIGN A NEW ONE OF THOSE INSTEAD?

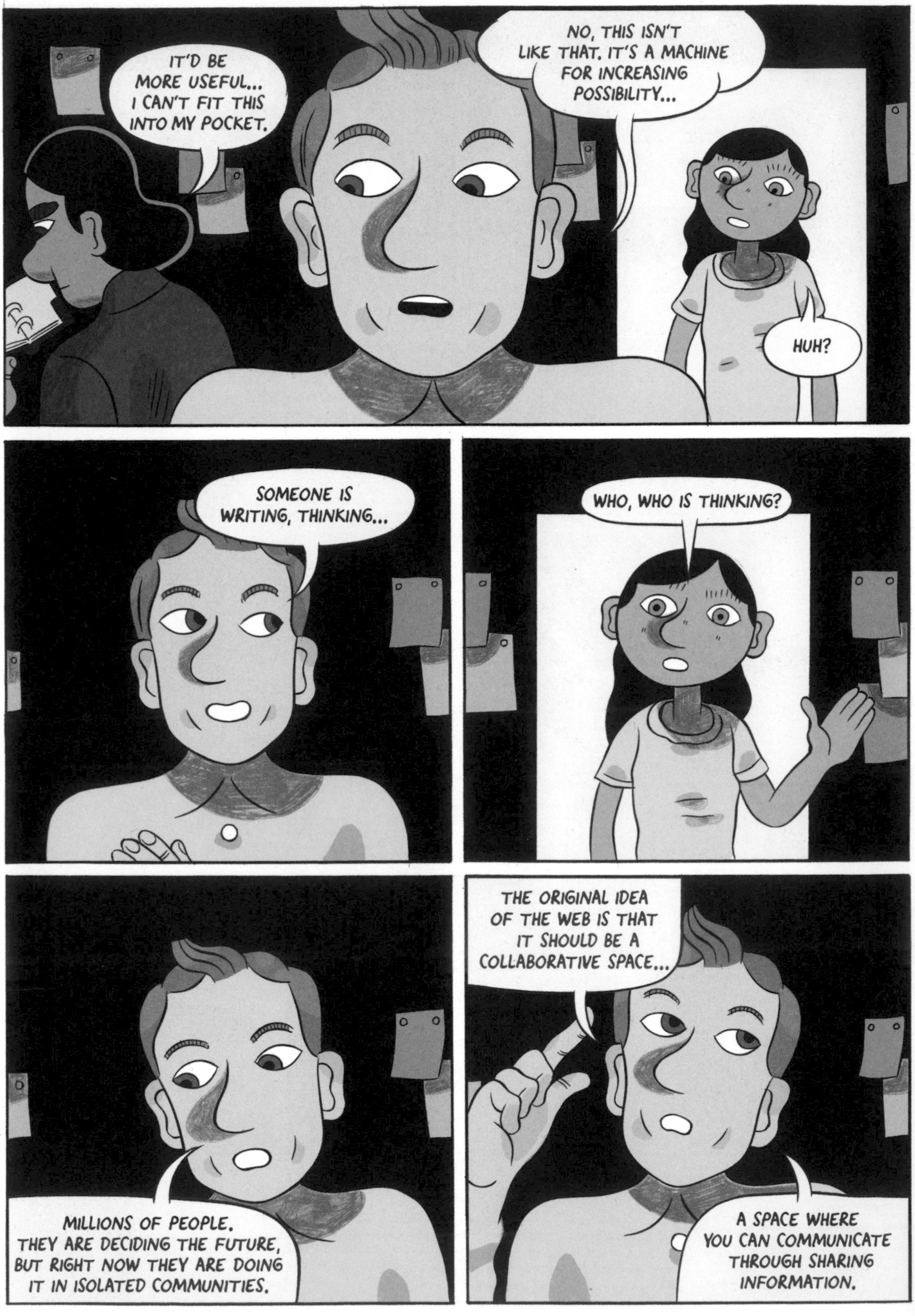
IT'D BE MORE USEFUL... I CAN'T FIT THIS INTO MY POCKET.
NO, THIS ISN'T LIKE THAT. IT'S A MACHINE FOR INCREASING POSSIBILITY...
HUH?
SOMEONE IS WRITING, THINKING...
WHO, WHO IS THINKING?
MILLIONS OF PEOPLE. THEY ARE DECIDING THE FUTURE, BUT RIGHT NOW THEY ARE DOING IT IN ISOLATED COMMUNITIES.
THE ORIGINAL IDEA OF THE WEB IS THAT IT SHOULD BE A COLLABORATIVE SPACE...
A SPACE WHERE YOU CAN COMMUNICATE THROUGH SHARING INFORMATION.

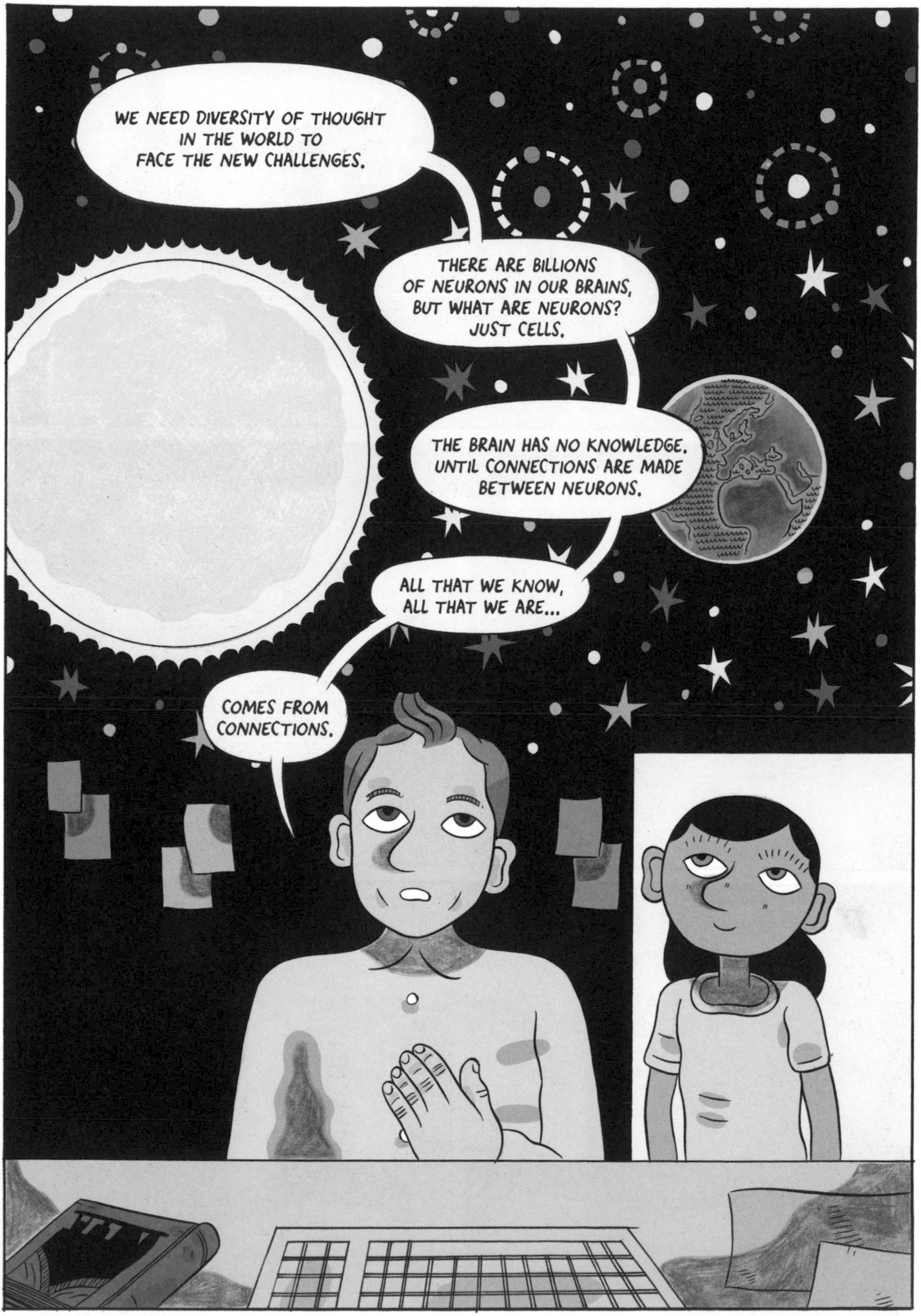
WE NEED DIVERSITY OF THOUGHT IN THE WORLD TO FACE THE NEW CHALLENGES.
THERE ARE BILLIONS OF NEURONS IN OUR BRAINS, BUT WHAT ARE NEURONS? JUST CELLS.
THE BRAIN HAS NO KNOWLEDGE. UNTIL CONNECTIONS ARE MADE BETWEEN NEURONS.
ALL THAT WE KNOW, ALL THAT WE ARE...
COMES FROM CONNECTIONS.

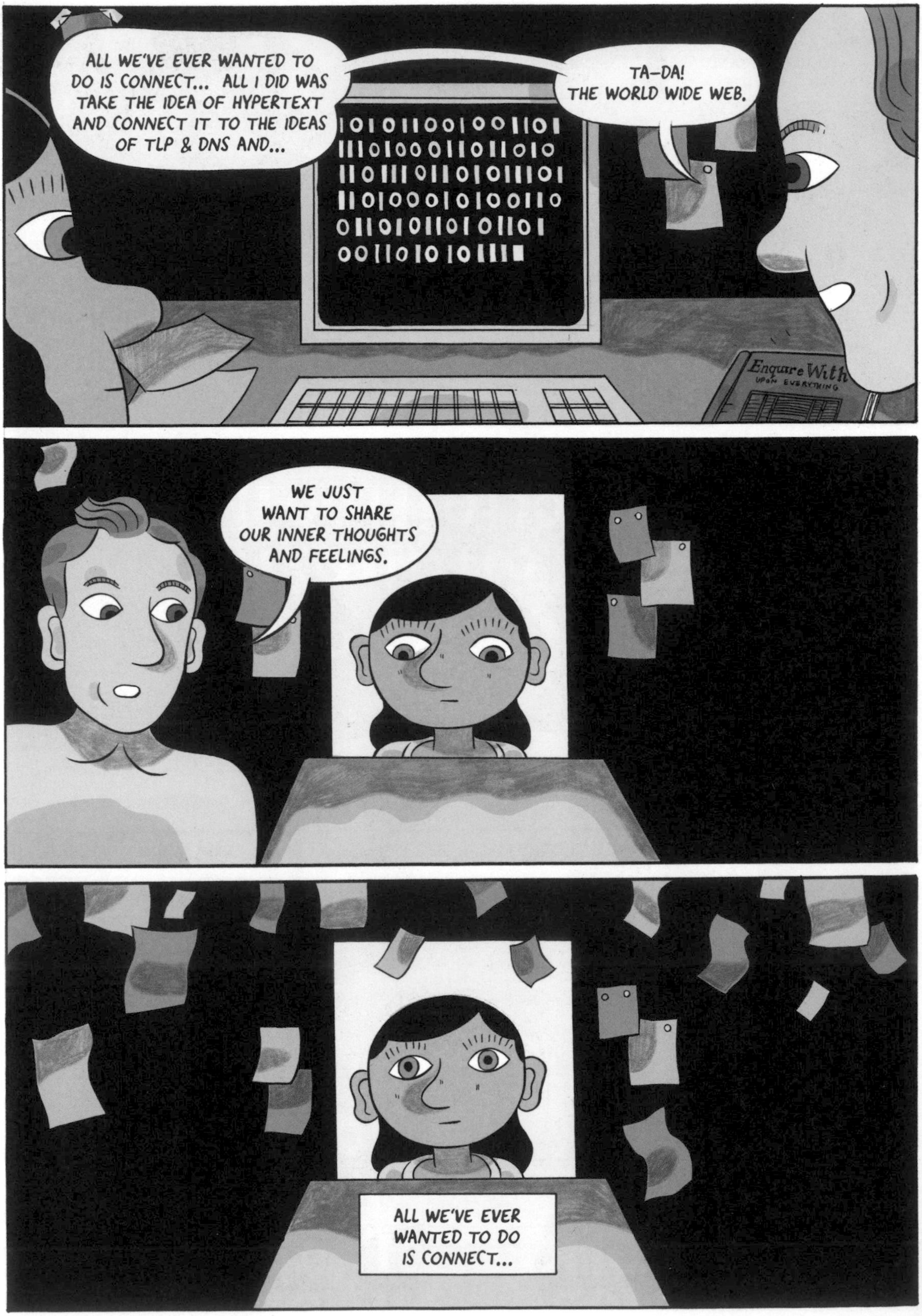
ALL WE'VE EVER WANTED TO DO IS CONNECT... ALL I DID WAS TAKE THE IDEA OF HYPERTEXT AND CONNECT IT TO THE IDEAS OF TLP & DNS AND...
TA-DA! THE WORLD WIDE WEB.
Enquire Within UPON EVERYTHING
WE JUST WANT TO SHARE OUR INNER THOUGHTS AND FEELINGS.
ALL WE'VE EVER WANTED TO DO IS CONNECT...

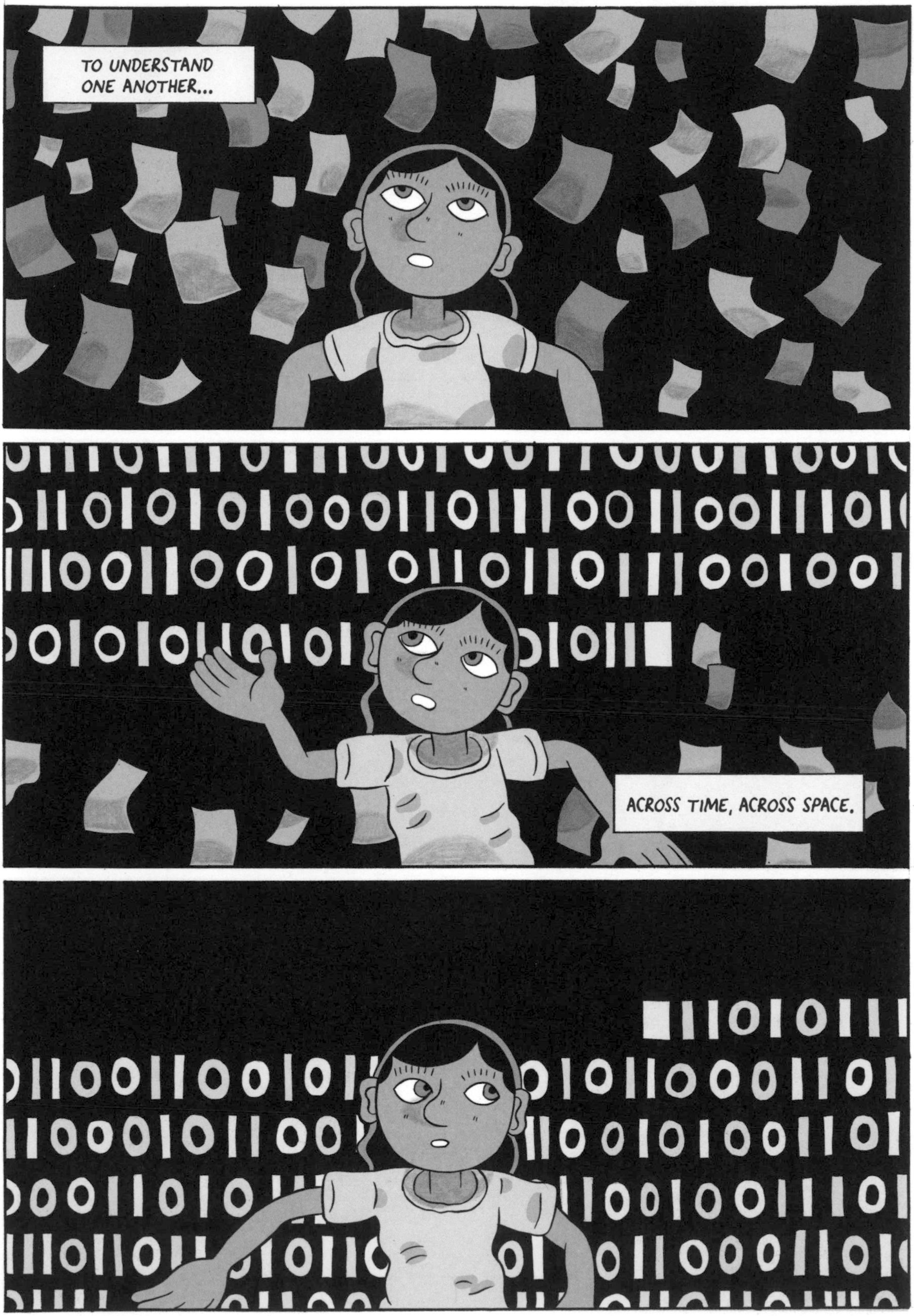
TO UNDERSTAND ONE ANOTHER...
ACROSS TIME, ACROSS SPACE.

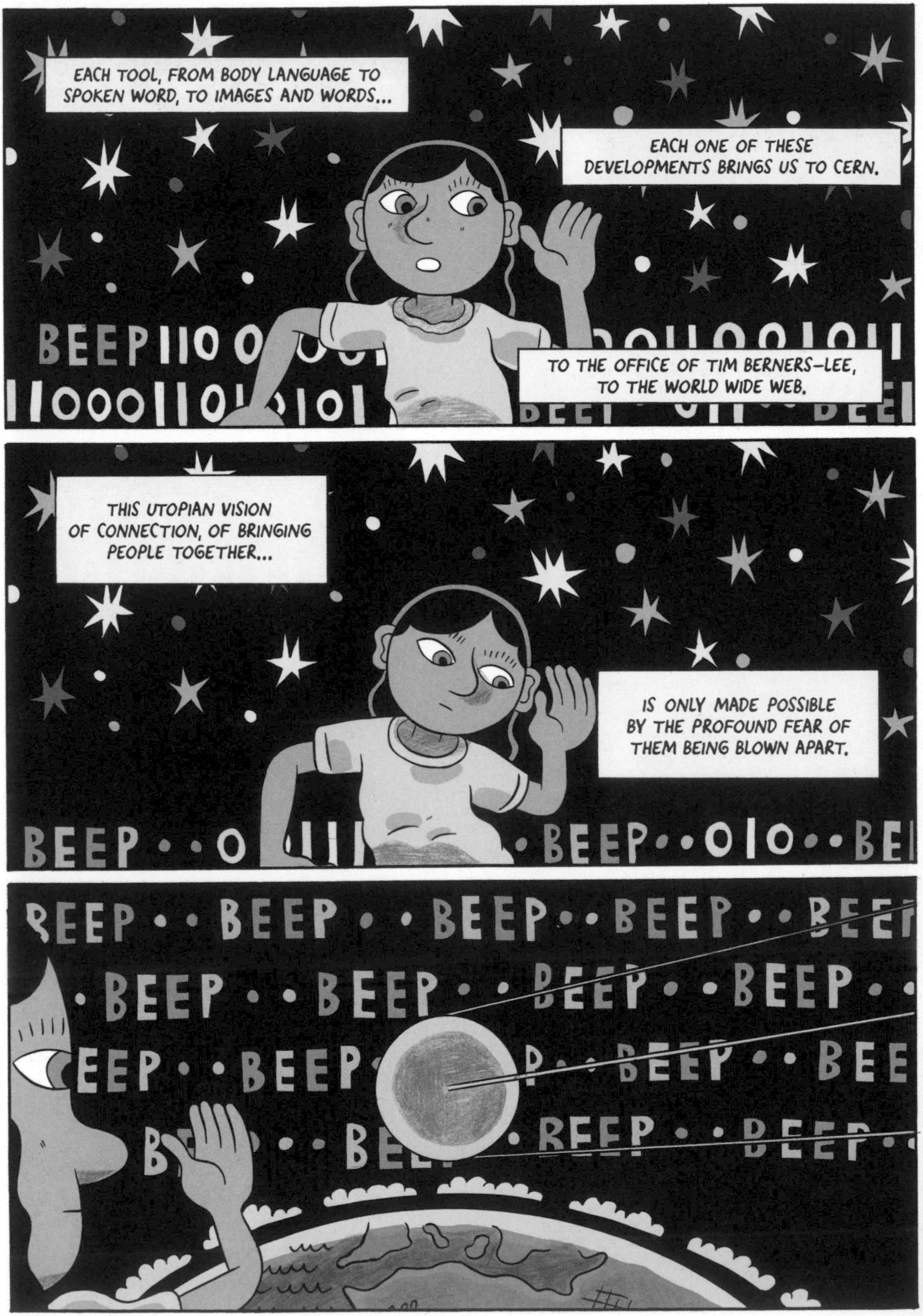
EACH TOOL, FROM BODY LANGUAGE TO SPOKEN WORD, TO IMAGES AND WORDS...
EACH ONE OF THESE DEVELOPMENTS BRINGS US TO CERN.
TO THE OFFICE OF TIM BERNERS-LEE, TO THE WORLD WIDE WEB.
THIS UTOPIAN VISION OF CONNECTION, OF BRINGING PEOPLE TOGETHER...
IS ONLY MADE POSSIBLE BY THE PROFOUND FEAR OF THEM BEING BLOWN APART.
BEEP • • BEEP • • BEEP • • BEEP

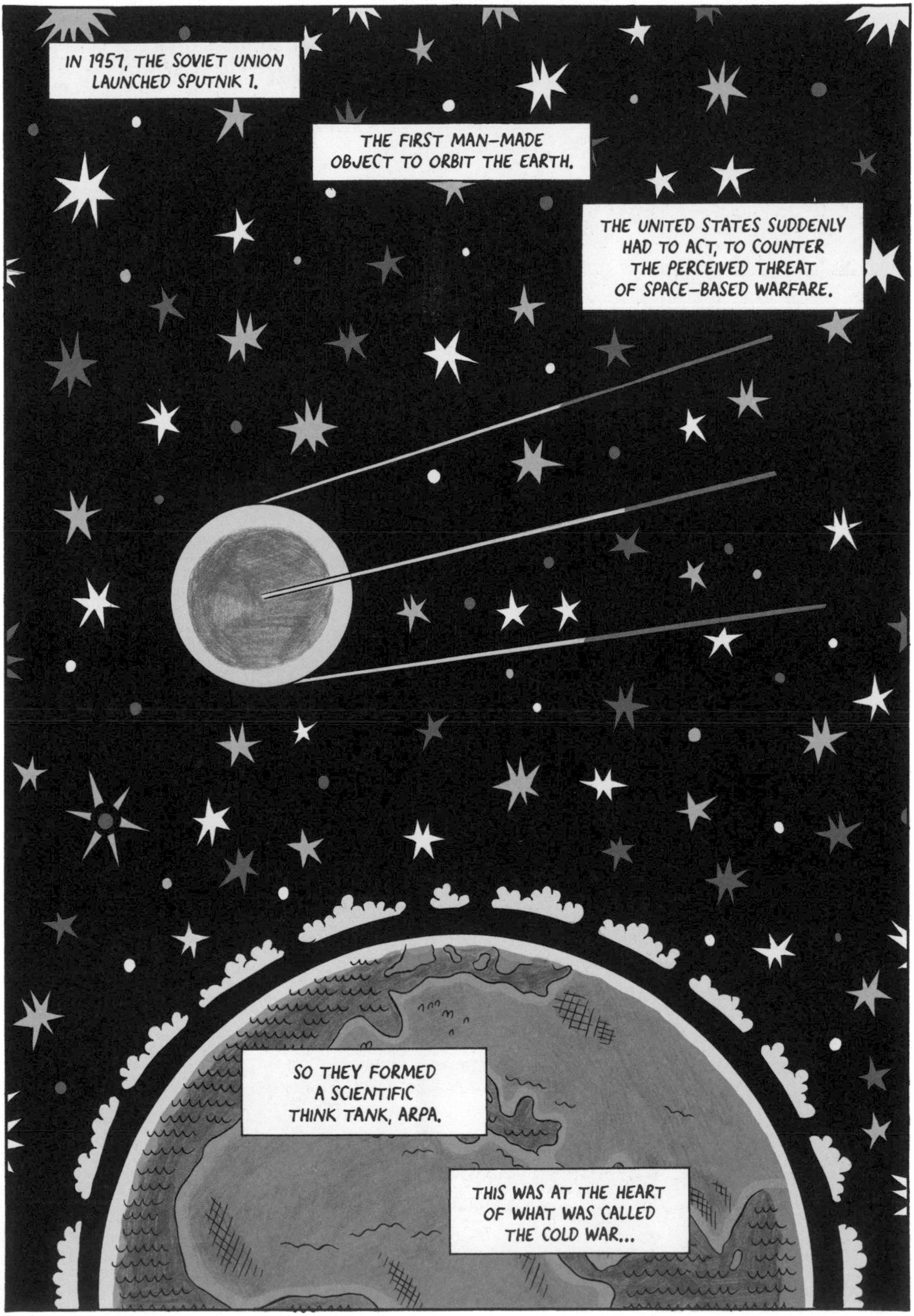
IN 1957, THE SOVIET UNION LAUNCHED SPUTNIK 1.
THE FIRST MAN-MADE OBJECT TO ORBIT THE EARTH.
THE UNITED STATES SUDDENLY HAD TO ACT, TO COUNTER THE PERCEIVED THREAT OF SPACE-BASED WARFARE.
SO THEY FORMED A SCIENTIFIC THINK TANK, ARPA.
THIS WAS AT THE HEART OF WHAT WAS CALLED THE COLD WAR...

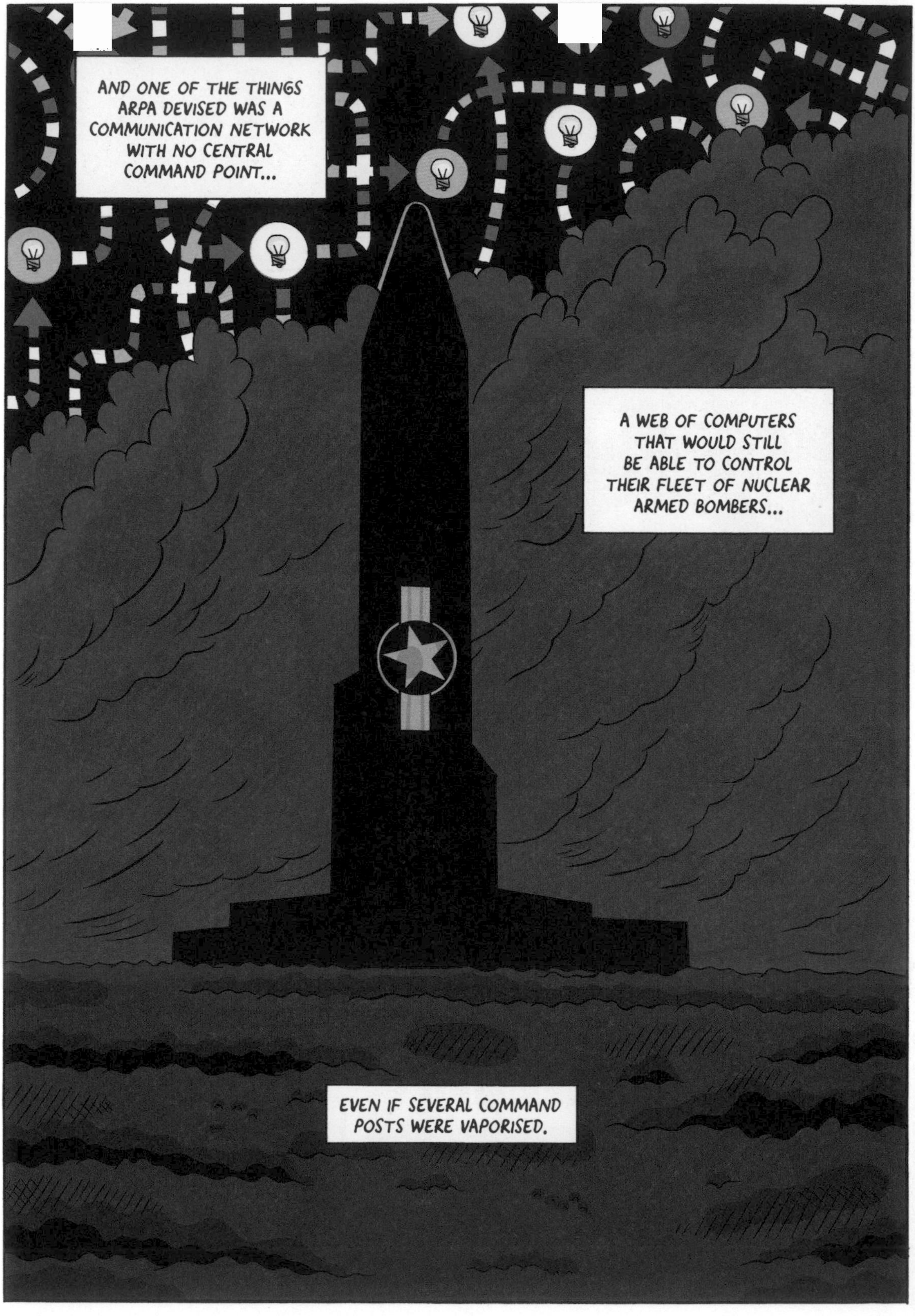
AND ONE OF THE THINGS ARPA DEVISED WAS A COMMUNICATION NETWORK WITH NO CENTRAL COMMAND POINT...
A WEB OF COMPUTERS THAT WOULD STILL BE ABLE TO CONTROL THEIR FLEET OF NUCLEAR ARMED BOMBERS...
EVEN IF SEVERAL COMMAND POSTS WERE VAPORISED.

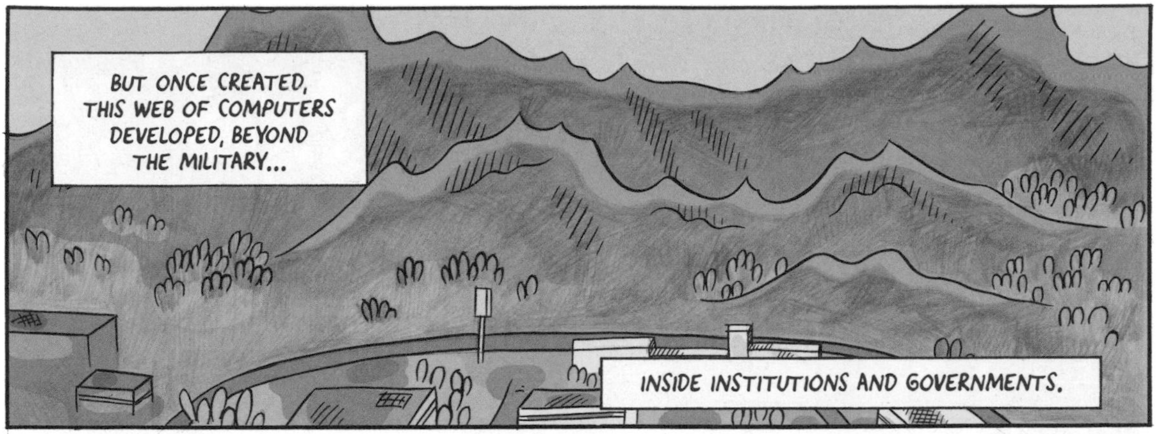
BUT ONCE CREATED, THIS WEB OF COMPUTERS DEVELOPED, BEYOND THE MILITARY...
INSIDE INSTITUTIONS AND GOVERNMENTS.

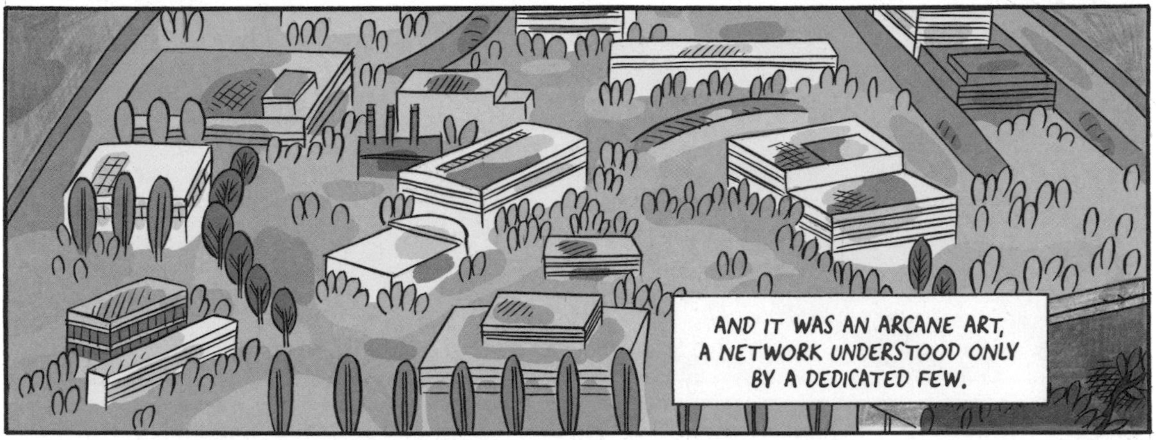
AND IT WAS AN ARCANE ART, A NETWORK UNDERSTOOD ONLY BY A DEDICATED FEW.

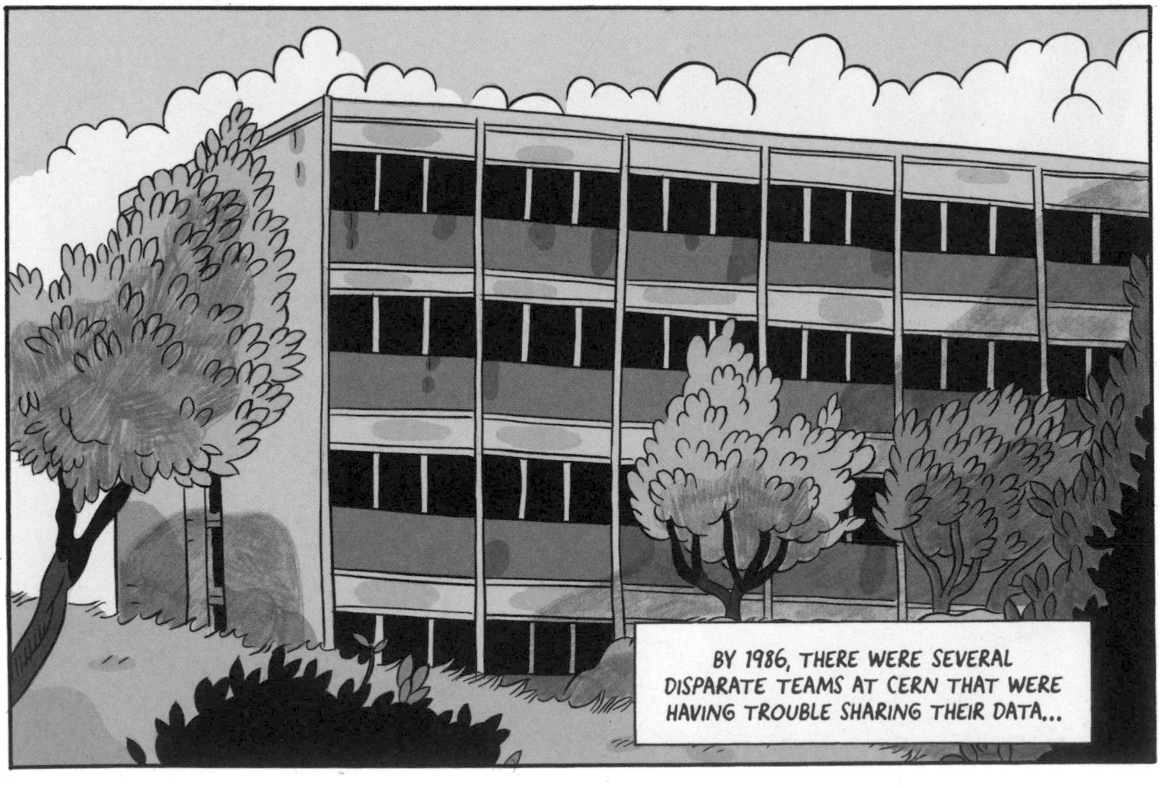
BY 1986, THERE WERE SEVERAL DISPARATE TEAMS AT CERN THAT WERE HAVING TROUBLE SHARING THEIR DATA...

...THEIR THOUGHTS AND IDEAS.

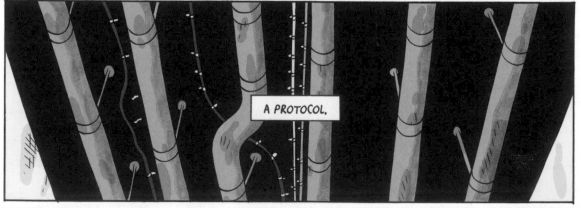

SO BERNERS-LEE DEVISED A COMMON COMPUTER LANGUAGE...

A PROTOCOL.

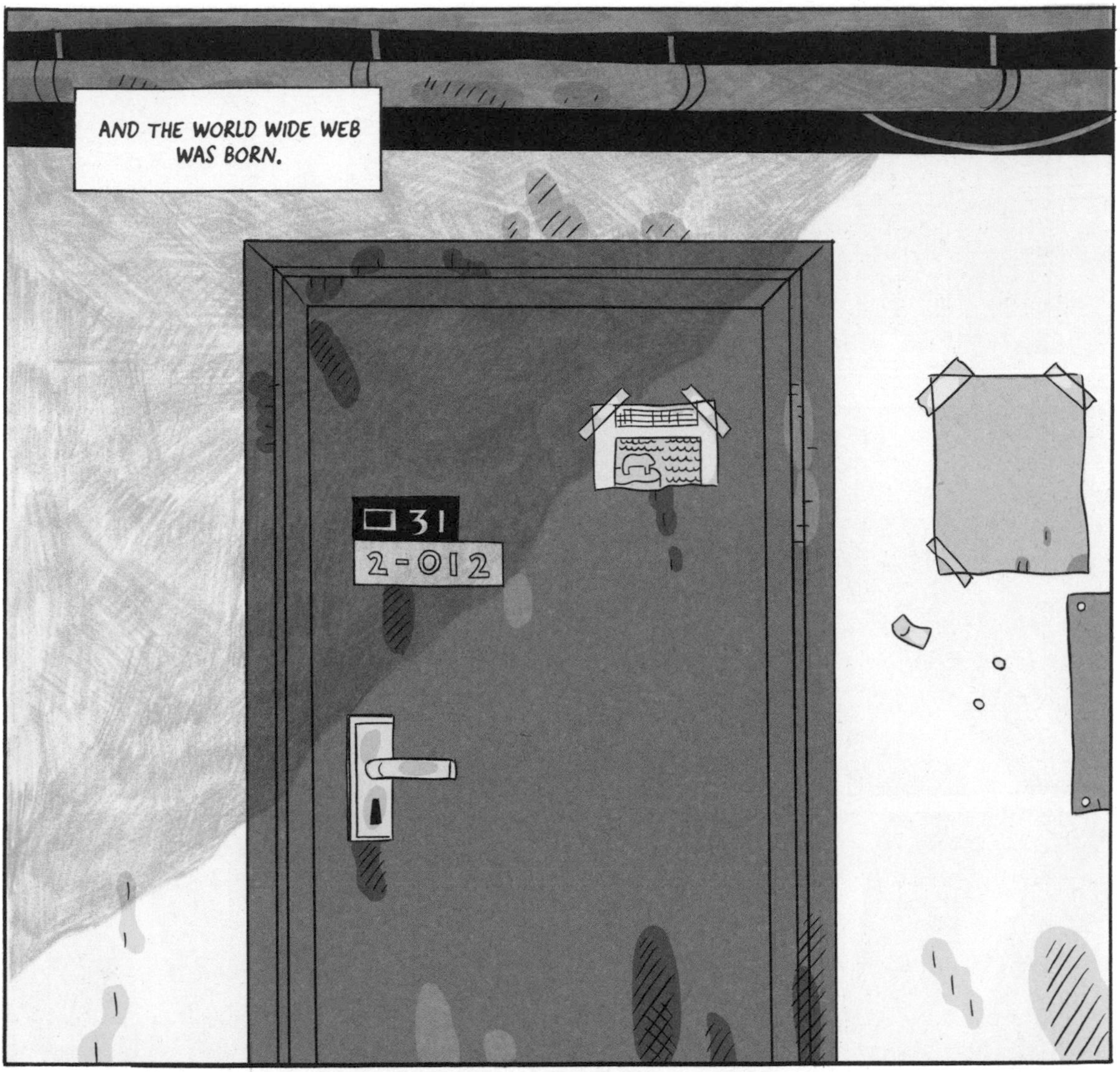
AND THE WORLD WIDE WEB WAS BORN.
31
2-012

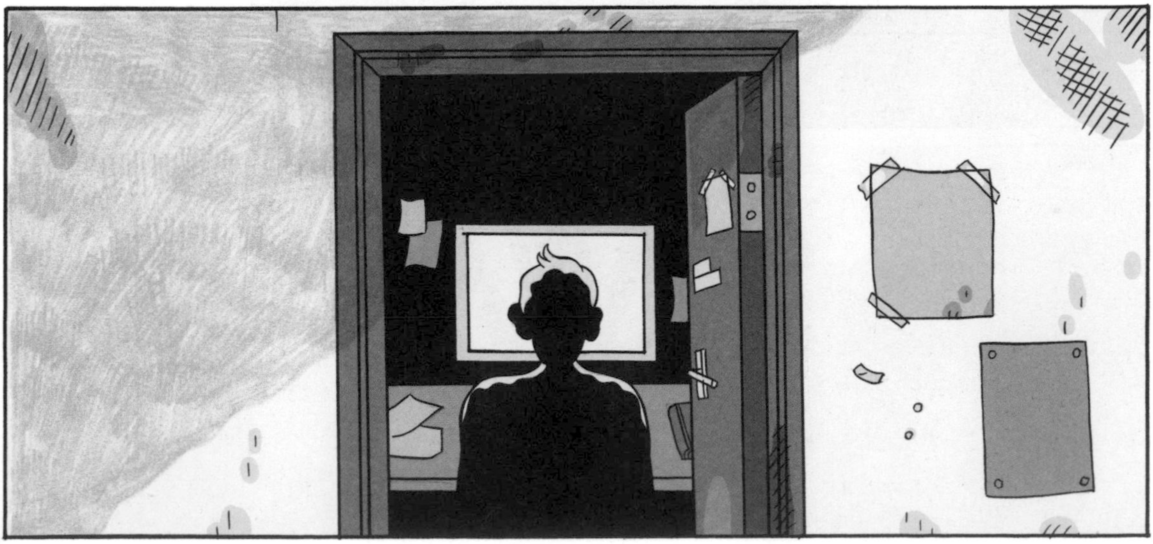

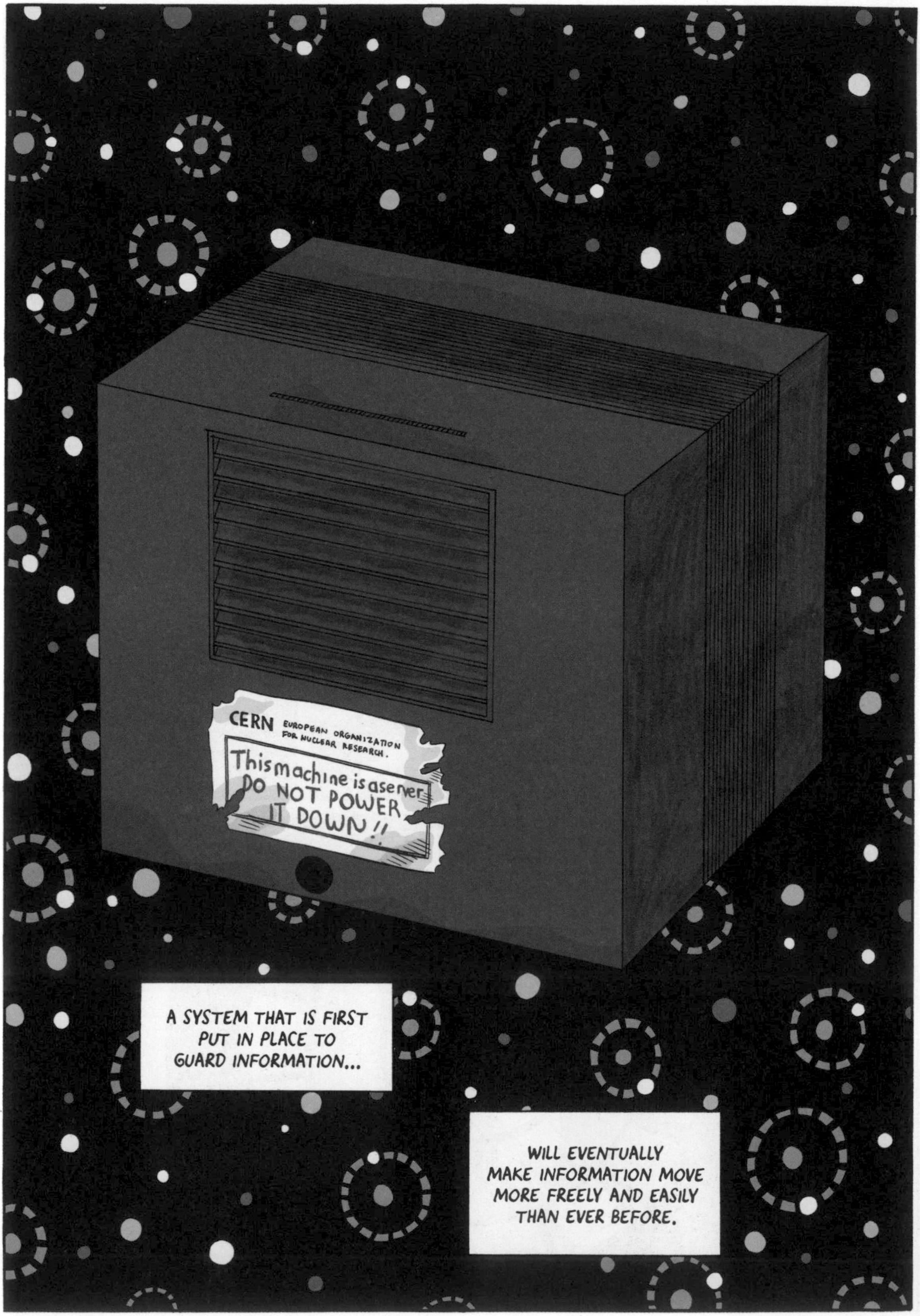
CERN EUROPEAN ORGANIZATION FOR NUCLEAR RESEARCH.
This machine is a server DO NOT POWER IT DOWN!!
A SYSTEM THAT IS FIRST PUT IN PLACE TO GUARD INFORMATION...
WILL EVENTUALLY MAKE INFORMATION MOVE MORE FREELY AND EASILY THAN EVER BEFORE.

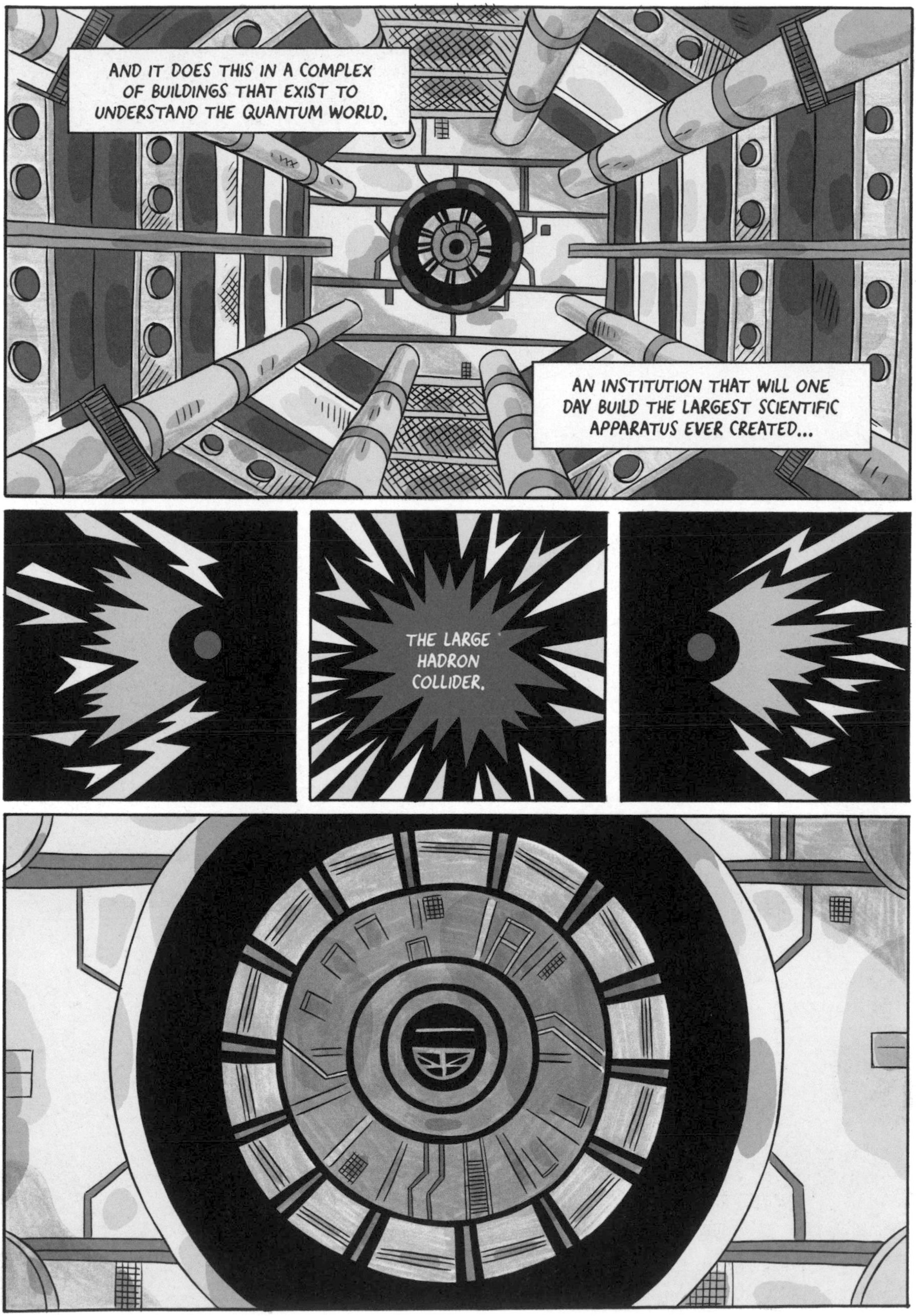
AND IT DOES THIS IN A COMPLEX OF BUILDINGS THAT EXIST TO UNDERSTAND THE QUANTUM WORLD.
AN INSTITUTION THAT WILL ONE DAY BUILD THE LARGEST SCIENTIFIC APPARATUS EVER CREATED...
THE LARGE HADRON COLLIDER.

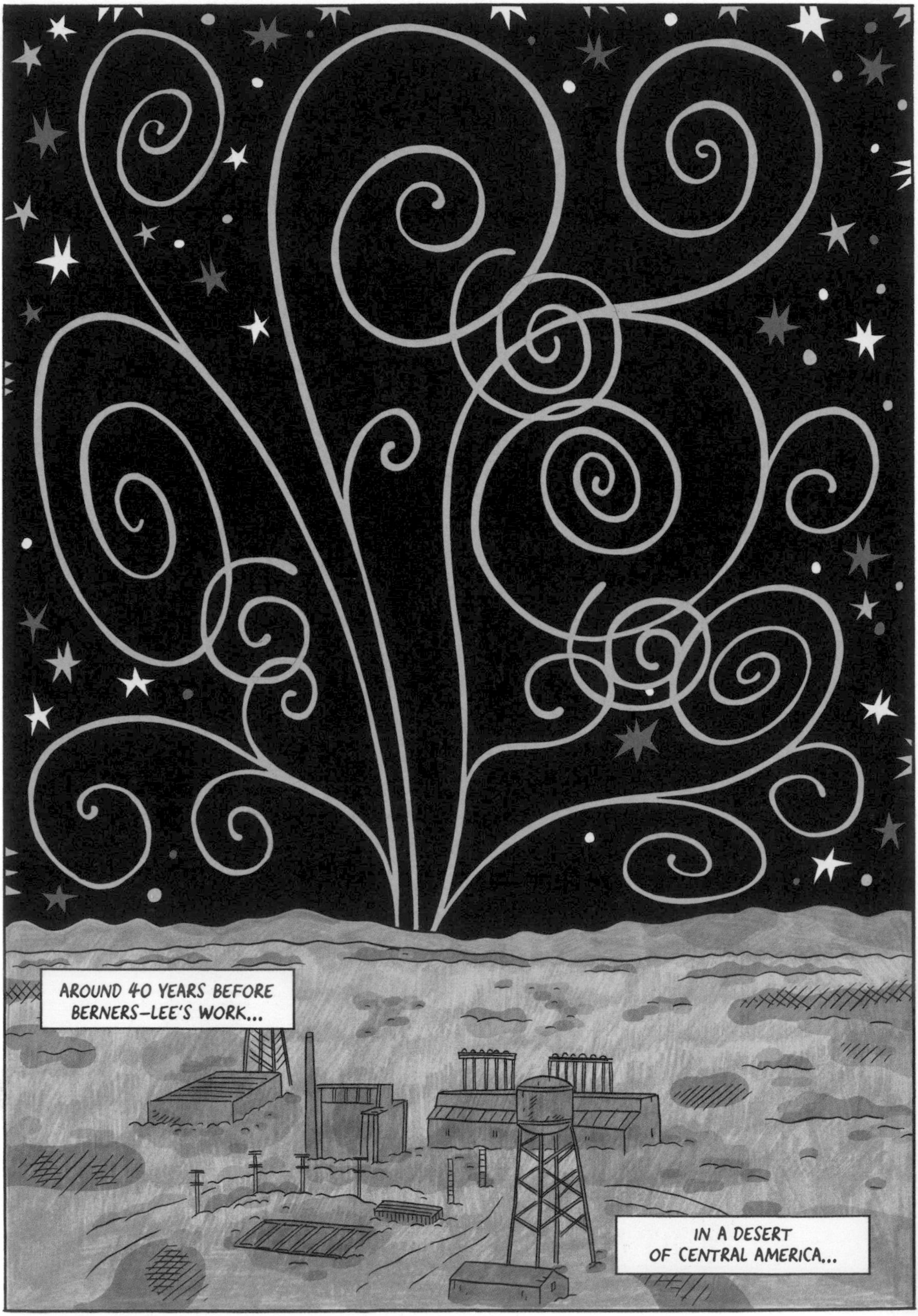
AROUND 40 YEARS BEFORE BERNERS-LEE'S WORK...
IN A DESERT OF CENTRAL AMERICA...

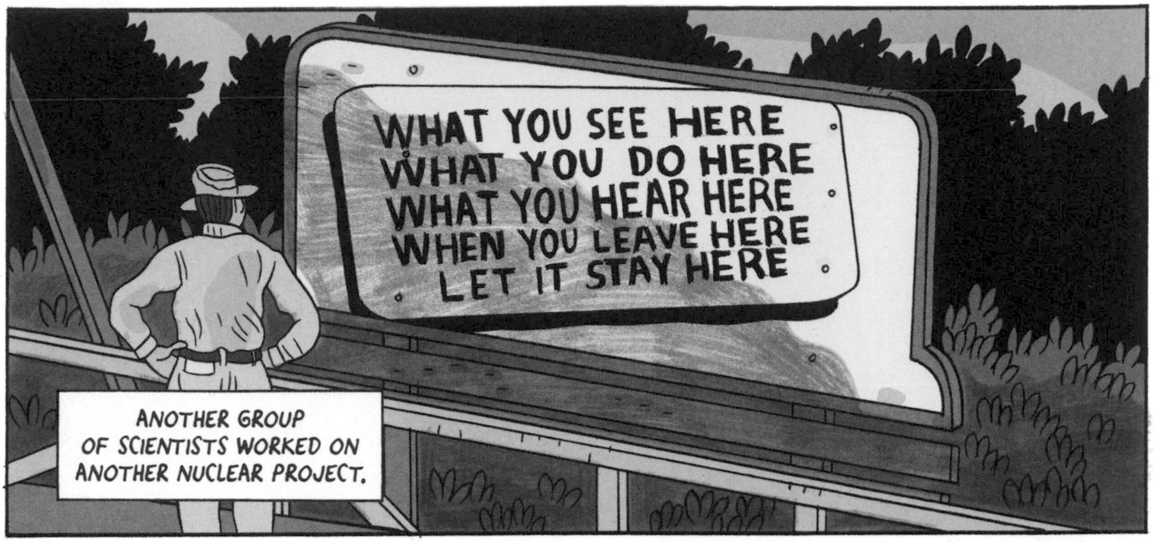
WHAT YOU SEE HERE
WHAT YOU DO HERE
WHAT YOU HEAR HERE
WHEN YOU LEAVE HERE
LET IT STAY HERE
ANOTHER GROUP OF SCIENTISTS WORKED ON ANOTHER NUCLEAR PROJECT.

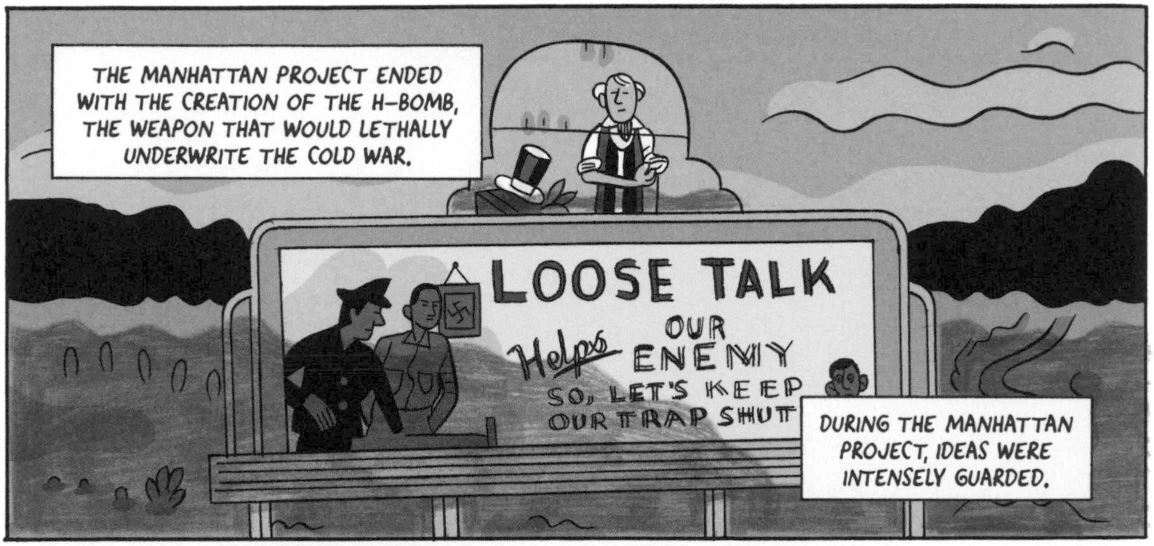
THE MANHATTAN PROJECT ENDED WITH THE CREATION OF THE H-BOMB, THE WEAPON THAT WOULD LETHALLY UNDERWRITE THE COLD WAR.
LOOSE TALK
Helps OUR ENEMY
SO, LET'S KEEP OUR TRAP SHUT
DURING THE MANHATTAN PROJECT, IDEAS WERE INTENSELY GUARDED.

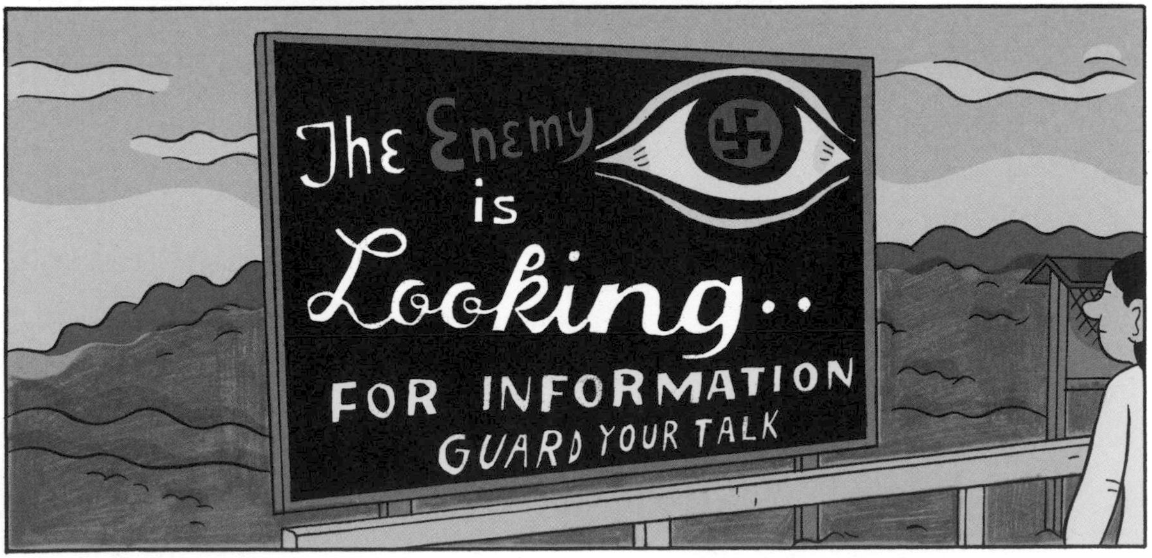
The Enemy is Looking..
FOR INFORMATION
GUARD YOUR TALK

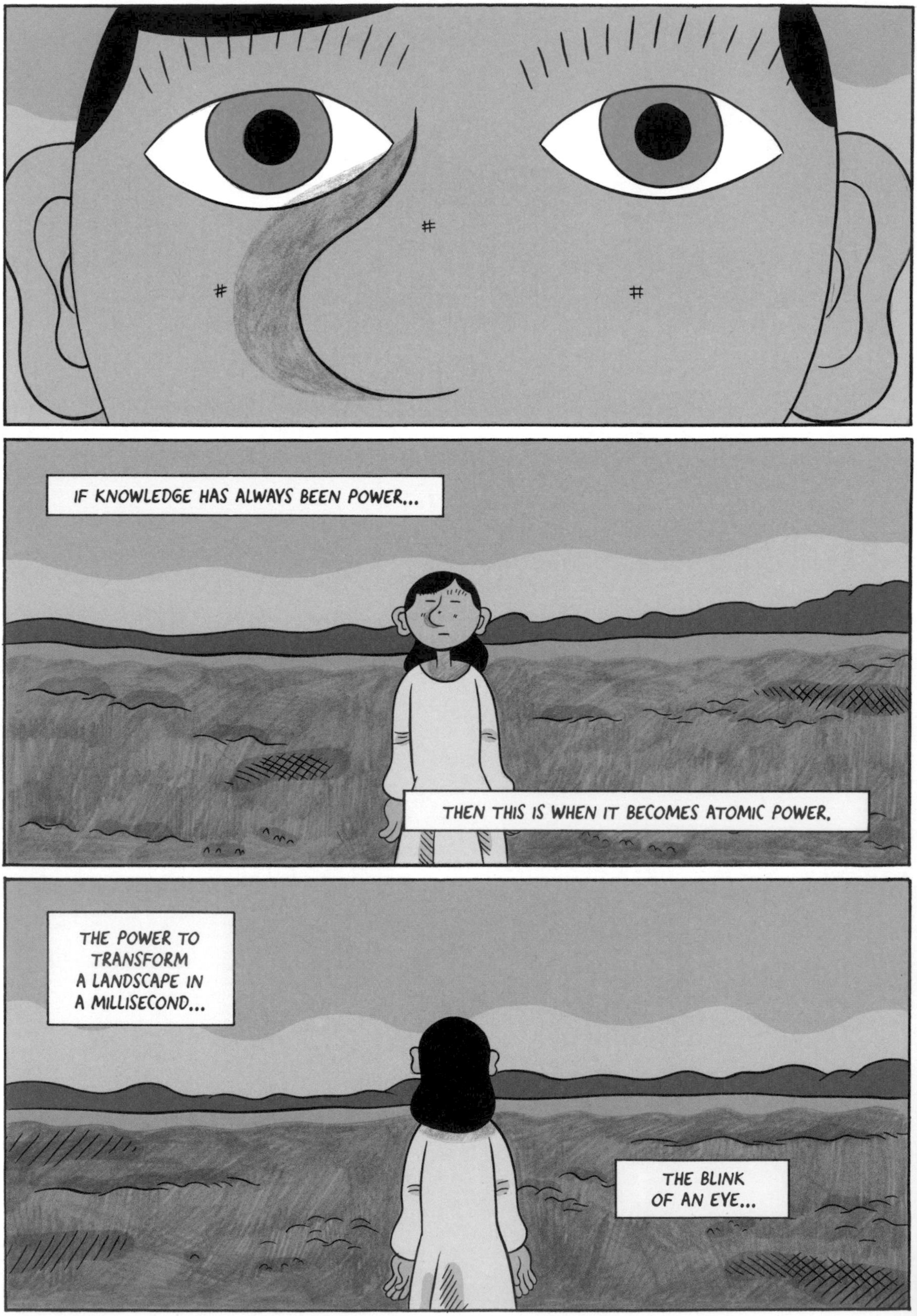
IF KNOWLEDGE HAS ALWAYS BEEN POWER...
THEN THIS IS WHEN IT BECOMES ATOMIC POWER.
THE POWER TO TRANSFORM A LANDSCAPE IN A MILLISECOND...
THE BLINK OF AN EYE...

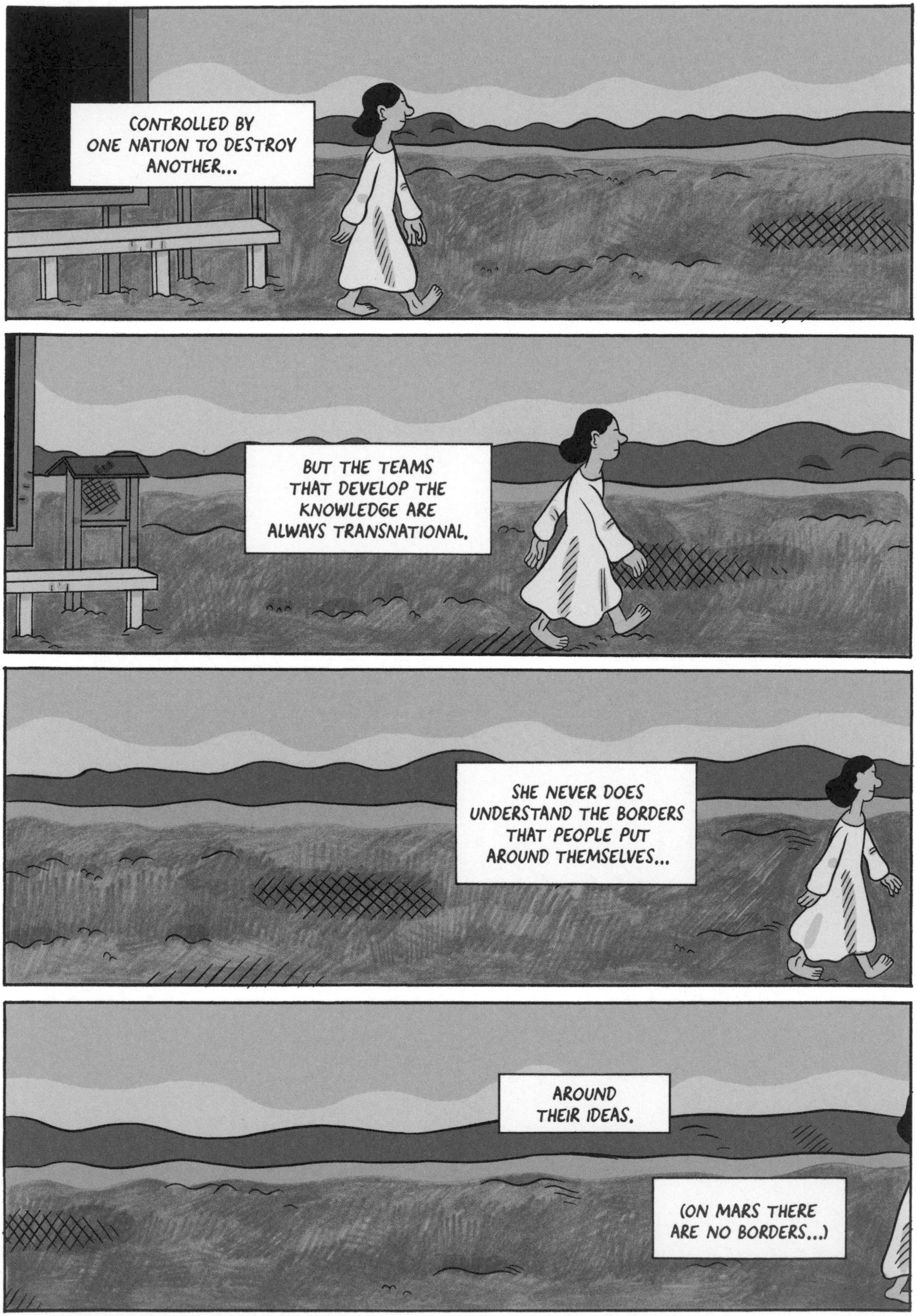
CONTROLLED BY ONE NATION TO DESTROY ANOTHER...
BUT THE TEAMS THAT DEVELOP THE KNOWLEDGE ARE ALWAYS TRANSNATIONAL.
SHE NEVER DOES UNDERSTAND THE BORDERS THAT PEOPLE PUT AROUND THEMSELVES...
AROUND THEIR IDEAS.
(ON MARS THERE ARE NO BORDERS...)

the future...

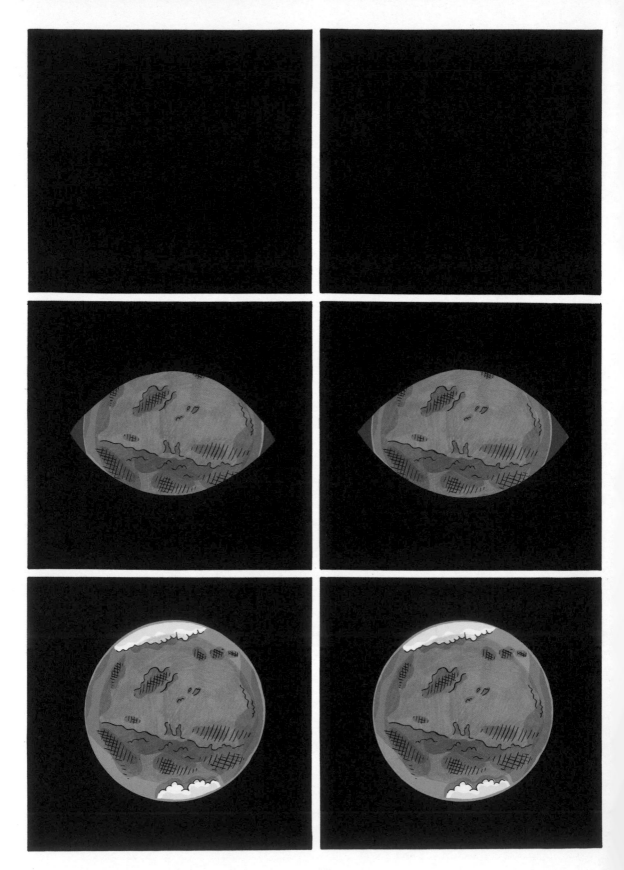

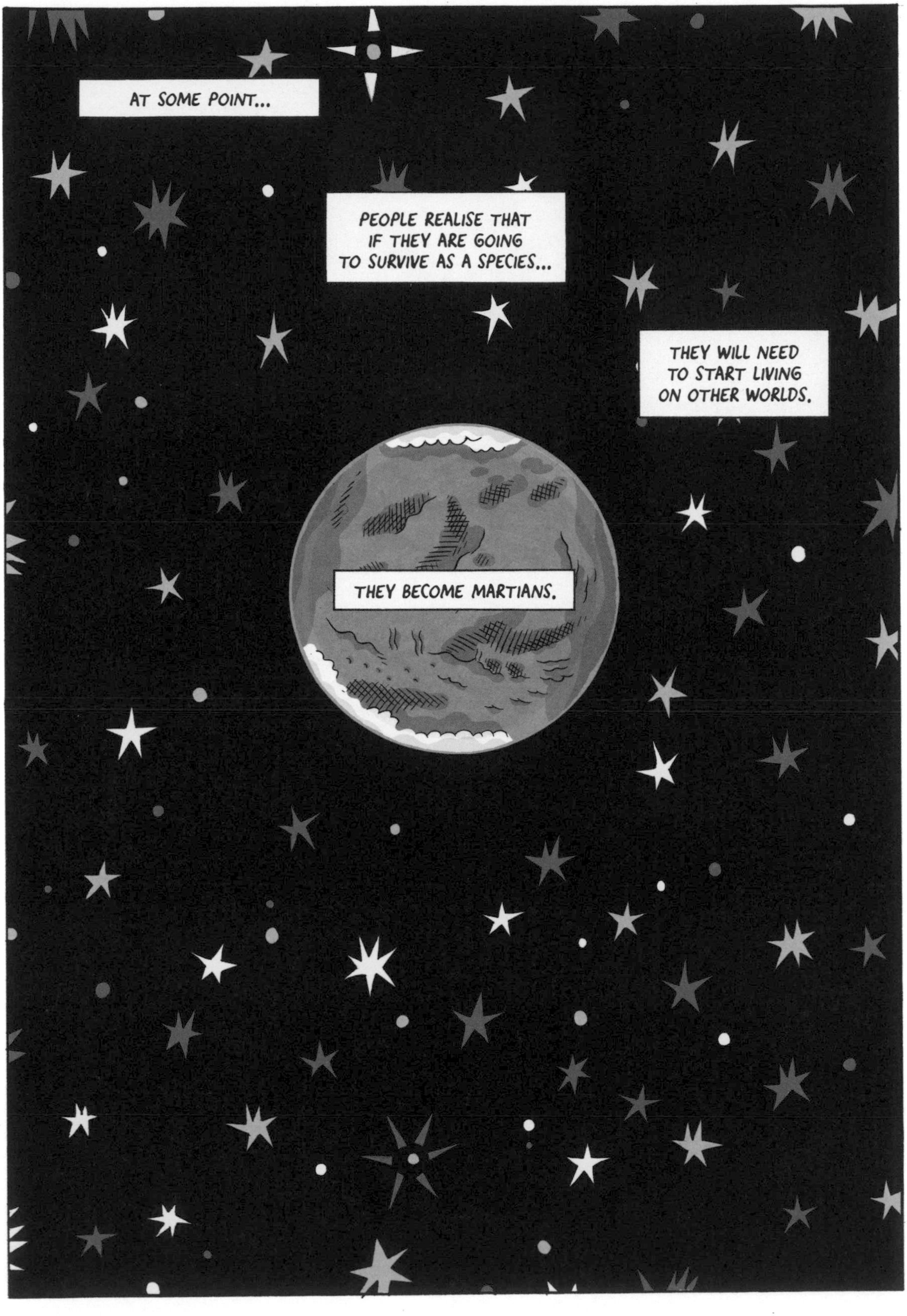
AT SOME POINT...
PEOPLE REALISE THAT IF THEY ARE GOING TO SURVIVE AS A SPECIES...
THEY WILL NEED TO START LIVING ON OTHER WORLDS.
THEY BECOME MARTIANS.

MARINER 9.

A PIECE OF THE EARTH SENT TO ORBIT ANOTHER WORLD...

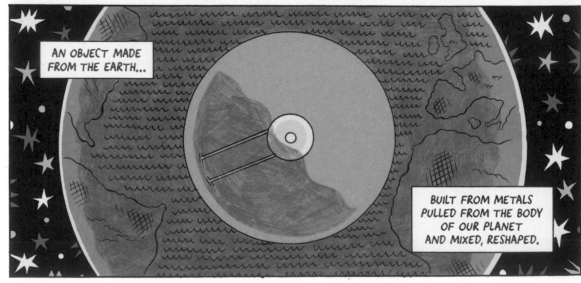

AN OBJECT MADE FROM THE EARTH...

BUILT FROM METALS PULLED FROM THE BODY OF OUR PLANET AND MIXED, RESHAPED.

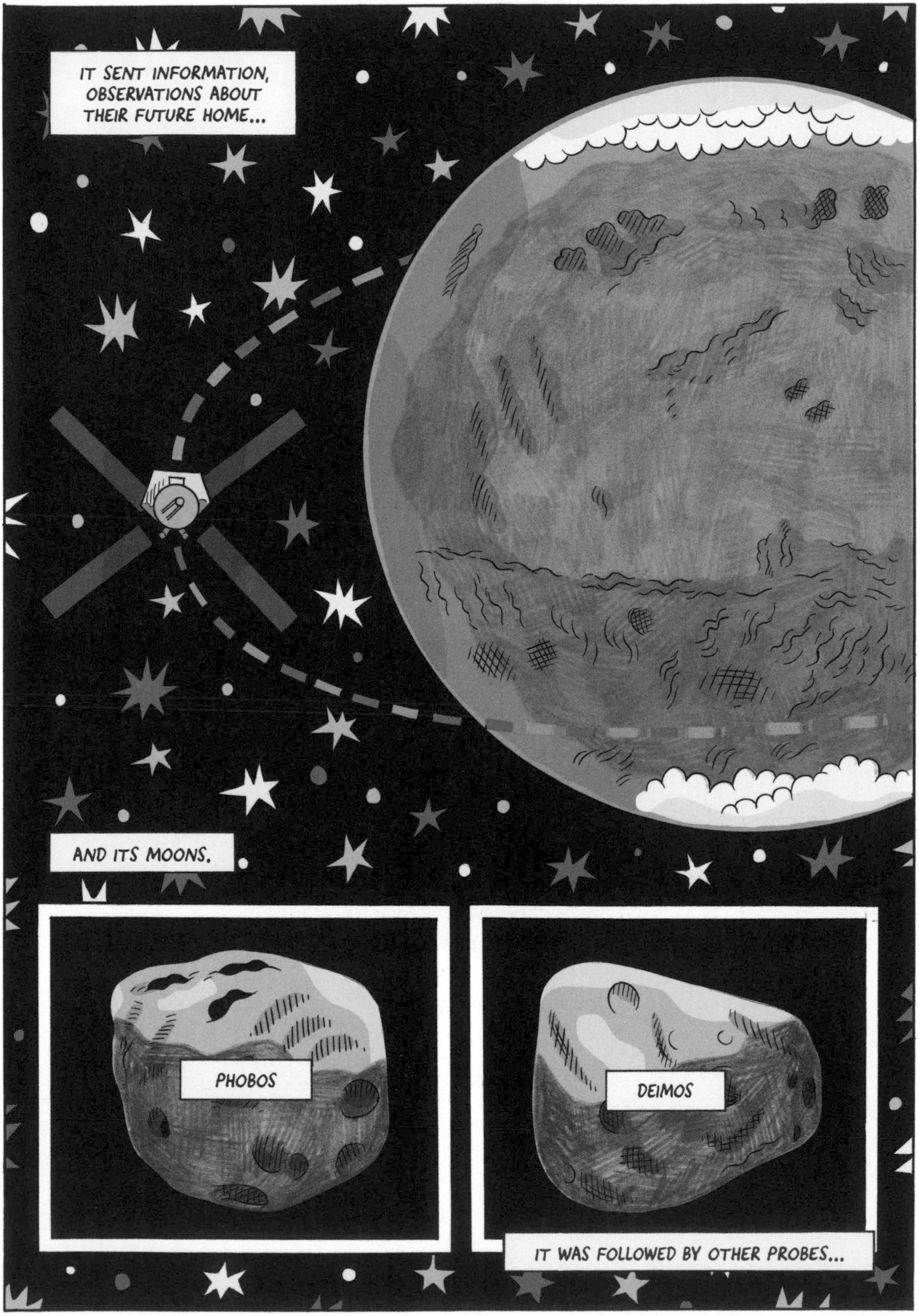
IT SENT INFORMATION, OBSERVATIONS ABOUT THEIR FUTURE HOME...
AND ITS MOONS.
PHOBOS
DEIMOS
IT WAS FOLLOWED BY OTHER PROBES...

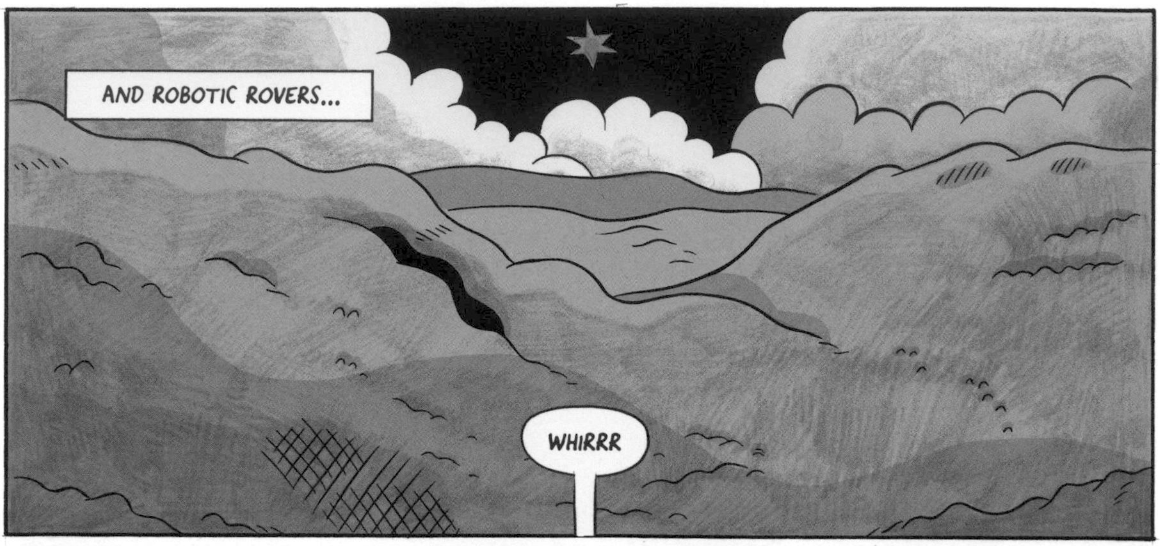
AND ROBOTIC ROVERS...
WHIRRR

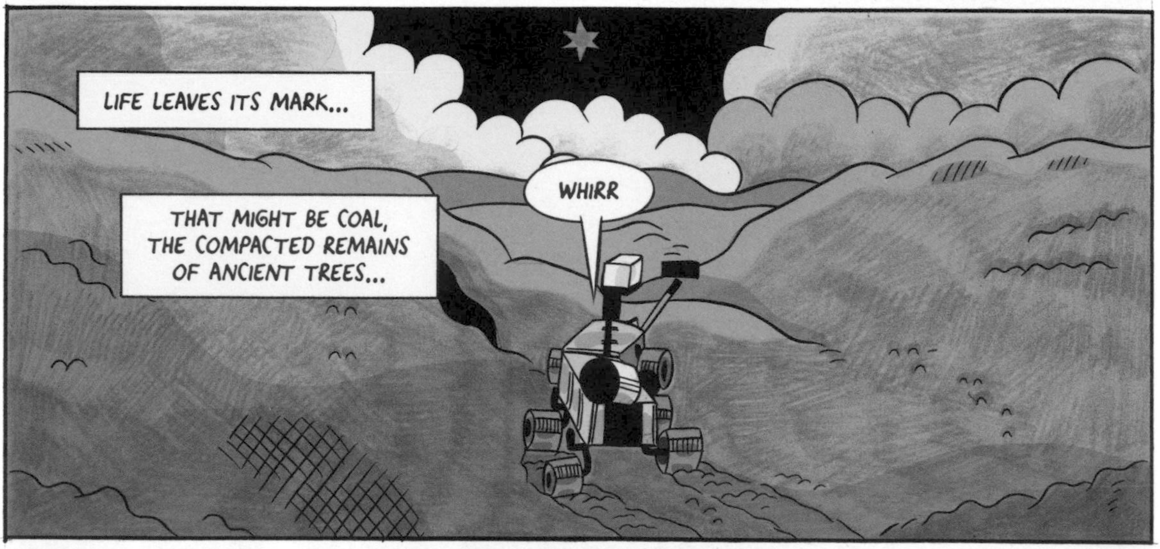
LIFE LEAVES ITS MARK...
WHIRR
THAT MIGHT BE COAL, THE COMPACTED REMAINS OF ANCIENT TREES...

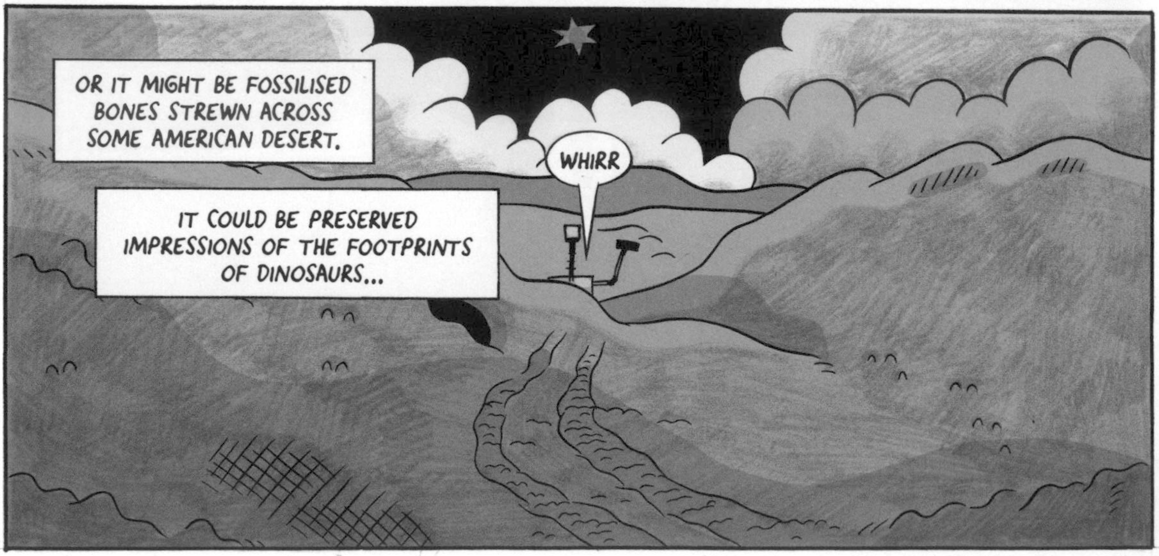
OR IT MIGHT BE FOSSILISED BONES STREWN ACROSS SOME AMERICAN DESERT.
WHIRR
IT COULD BE PRESERVED IMPRESSIONS OF THE FOOTPRINTS OF DINOSAURS...

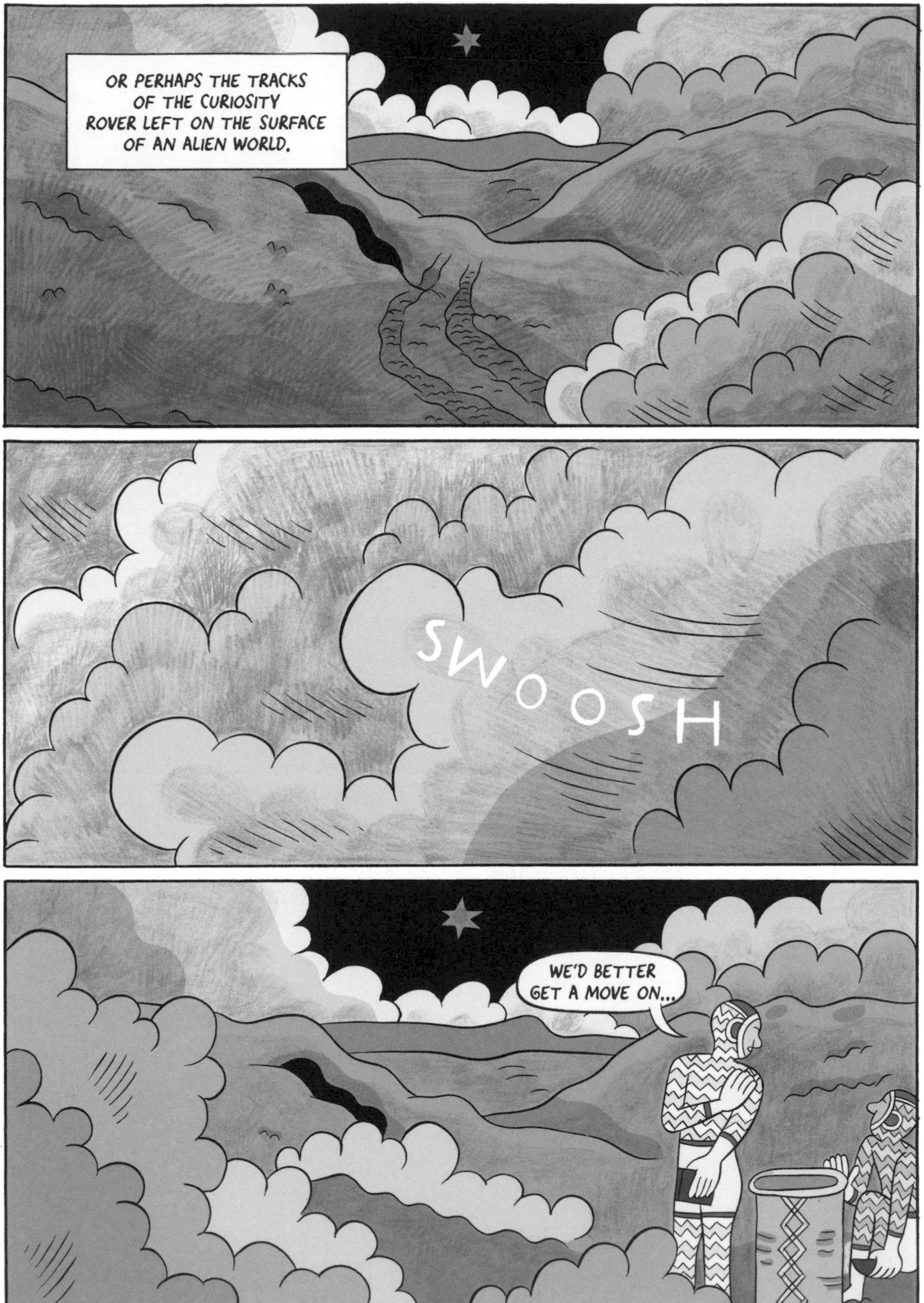
OR PERHAPS THE TRACKS OF THE CURIOSITY ROVER LEFT ON THE SURFACE OF AN ALIEN WORLD.
SWOOSH
WE'D BETTER GET A MOVE ON...

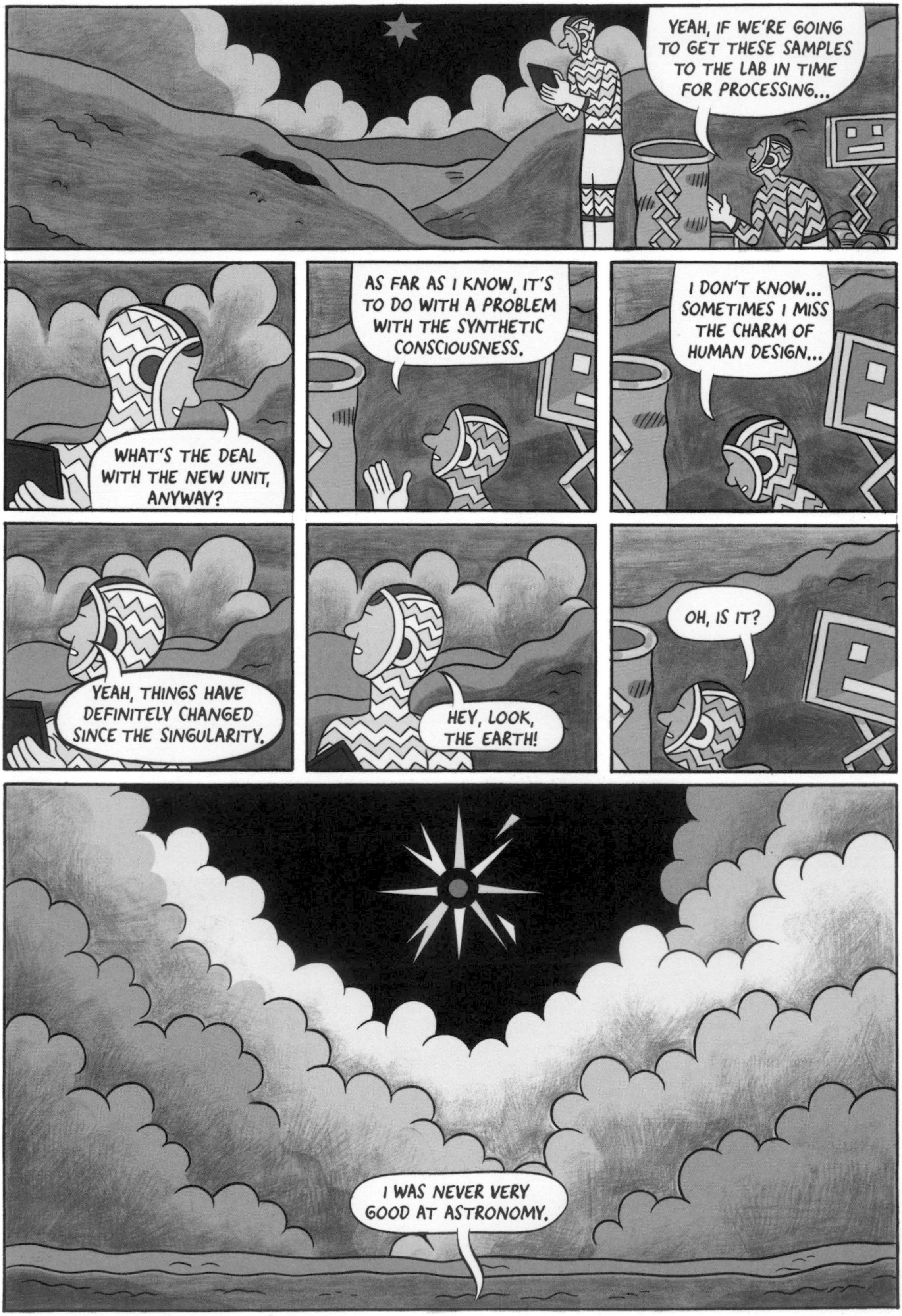
YEAH, IF WE'RE GOING TO GET THESE SAMPLES TO THE LAB IN TIME FOR PROCESSING...
WHAT'S THE DEAL WITH THE NEW UNIT, ANYWAY?
AS FAR AS I KNOW, IT'S TO DO WITH A PROBLEM WITH THE SYNTHETIC CONSCIOUSNESS.
I DON'T KNOW... SOMETIMES I MISS THE CHARM OF HUMAN DESIGN...
YEAH, THINGS HAVE DEFINITELY CHANGED SINCE THE SINGULARITY.
HEY, LOOK, THE EARTH!
OH, IS IT?
I WAS NEVER VERY GOOD AT ASTRONOMY.

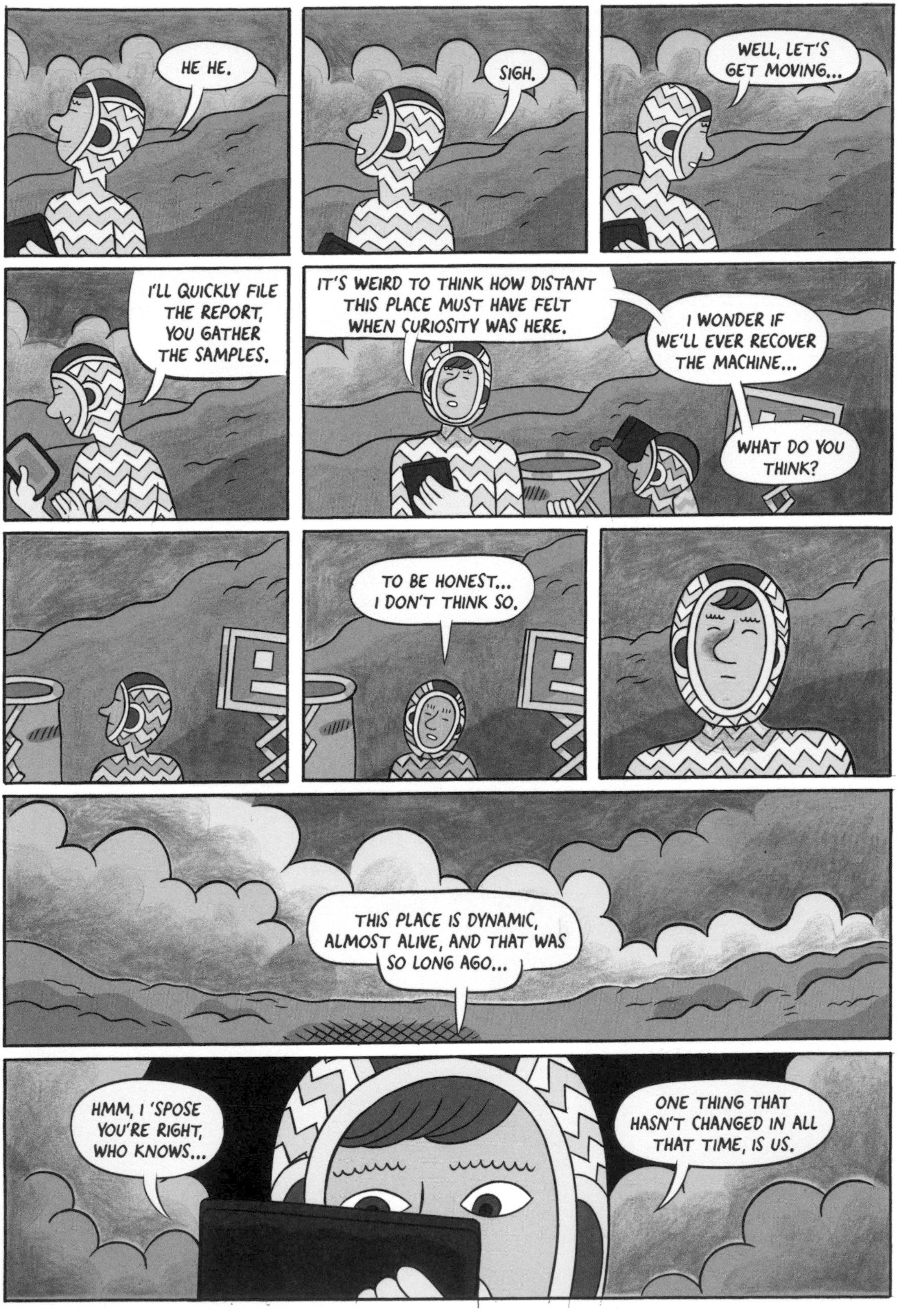

HE HE.
SIGH.
WELL, LET'S GET MOVING...
I'LL QUICKLY FILE THE REPORT, YOU GATHER THE SAMPLES.
IT'S WEIRD TO THINK HOW DISTANT THIS PLACE MUST HAVE FELT WHEN CURIOSITY WAS HERE.
I WONDER IF WE'LL EVER RECOVER THE MACHINE...
WHAT DO YOU THINK?
TO BE HONEST... I DON'T THINK SO.
THIS PLACE IS DYNAMIC, ALMOST ALIVE, AND THAT WAS SO LONG AGO...
HMM, I 'SPOSE YOU'RE RIGHT, WHO KNOWS...
ONE THING THAT HASN'T CHANGED IN ALL THAT TIME, IS US.

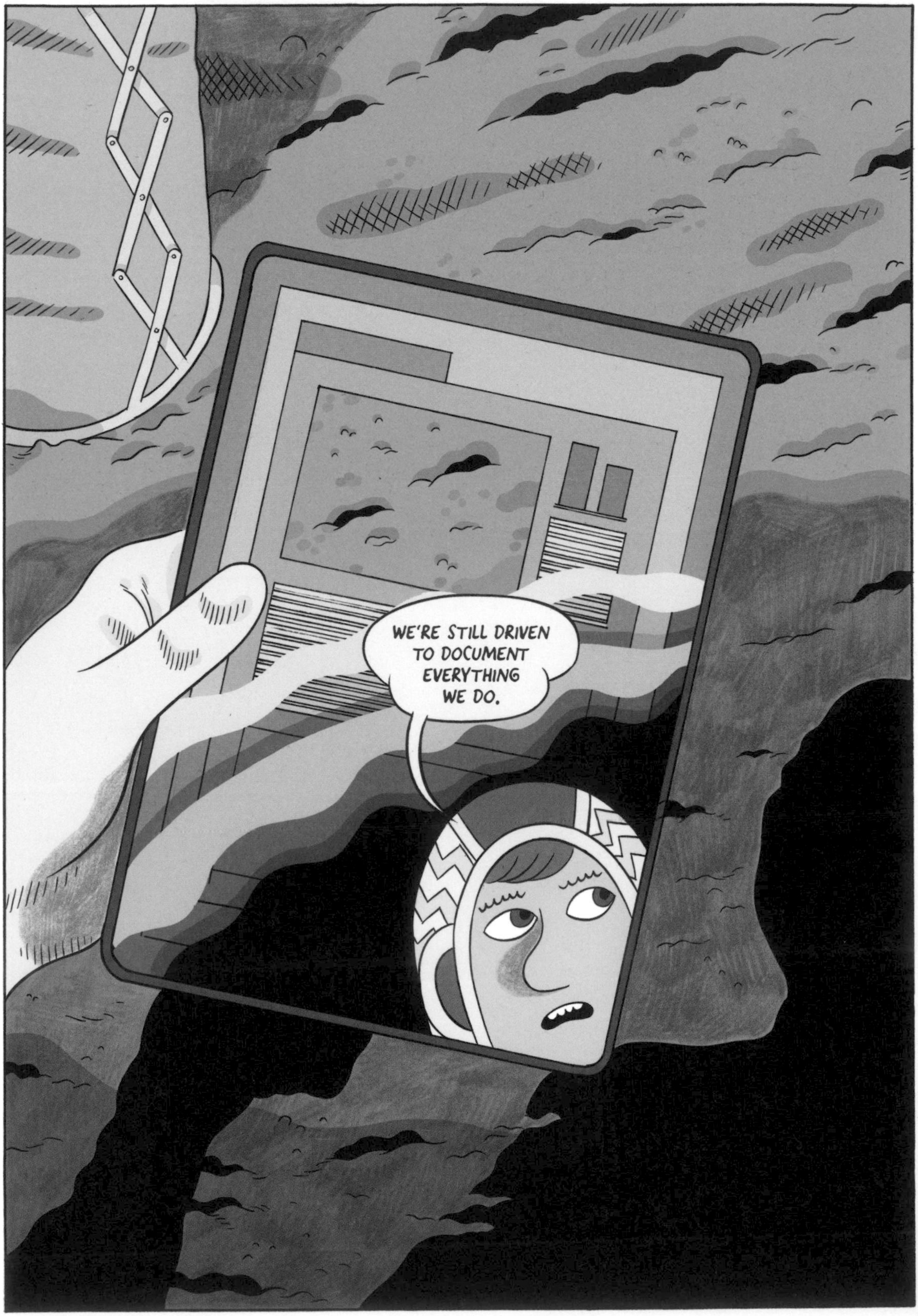
WE'RE STILL DRIVEN TO DOCUMENT EVERYTHING WE DO.

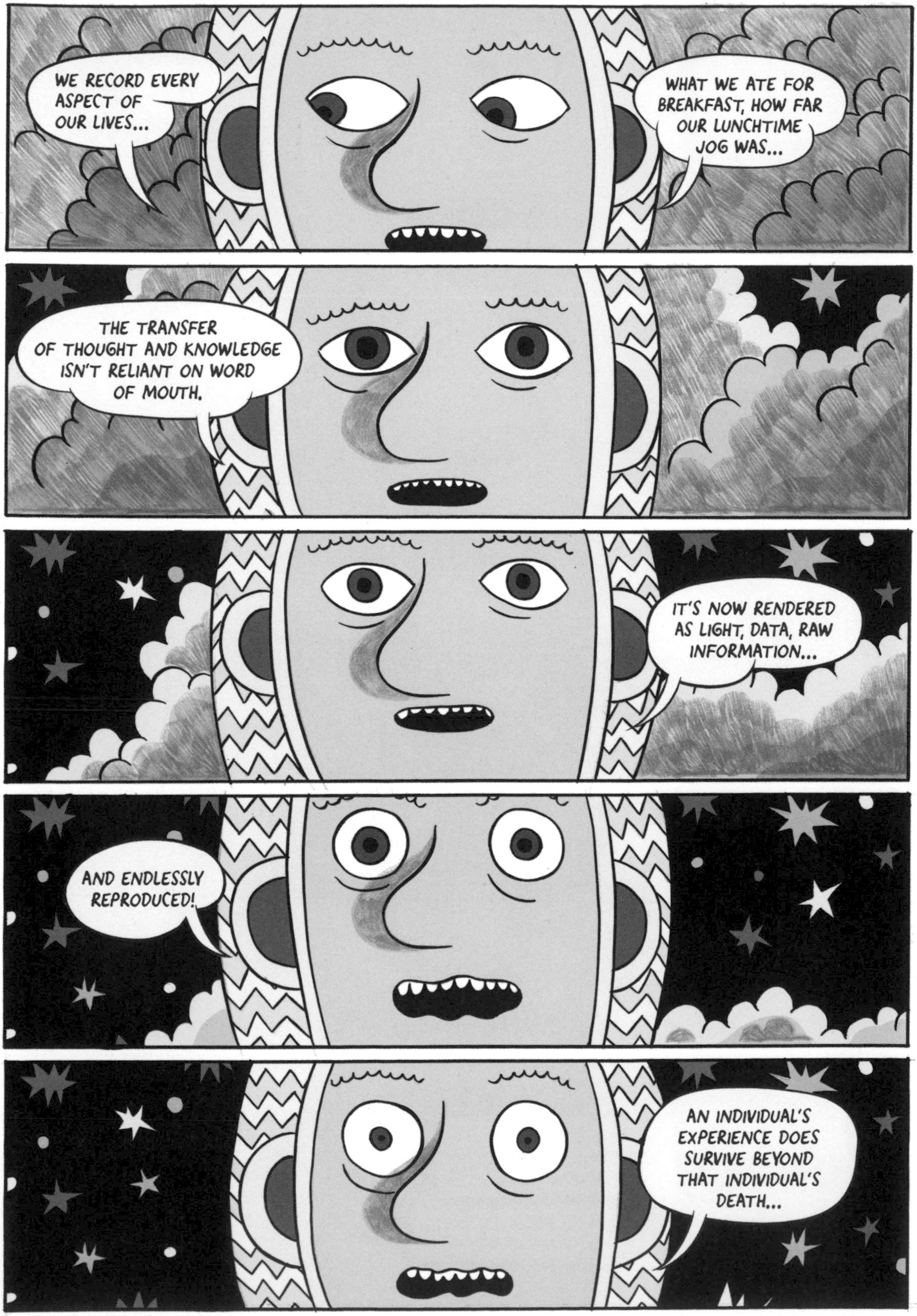
WE RECORD EVERY ASPECT OF OUR LIVES...
WHAT WE ATE FOR BREAKFAST, HOW FAR OUR LUNCHTIME JOG WAS...
THE TRANSFER OF THOUGHT AND KNOWLEDGE ISN'T RELIANT ON WORD OF MOUTH.
IT'S NOW RENDERED AS LIGHT, DATA, RAW INFORMATION...
AND ENDLESSLY REPRODUCED!
AN INDIVIDUAL'S EXPERIENCE DOES SURVIVE BEYOND THAT INDIVIDUAL'S DEATH...

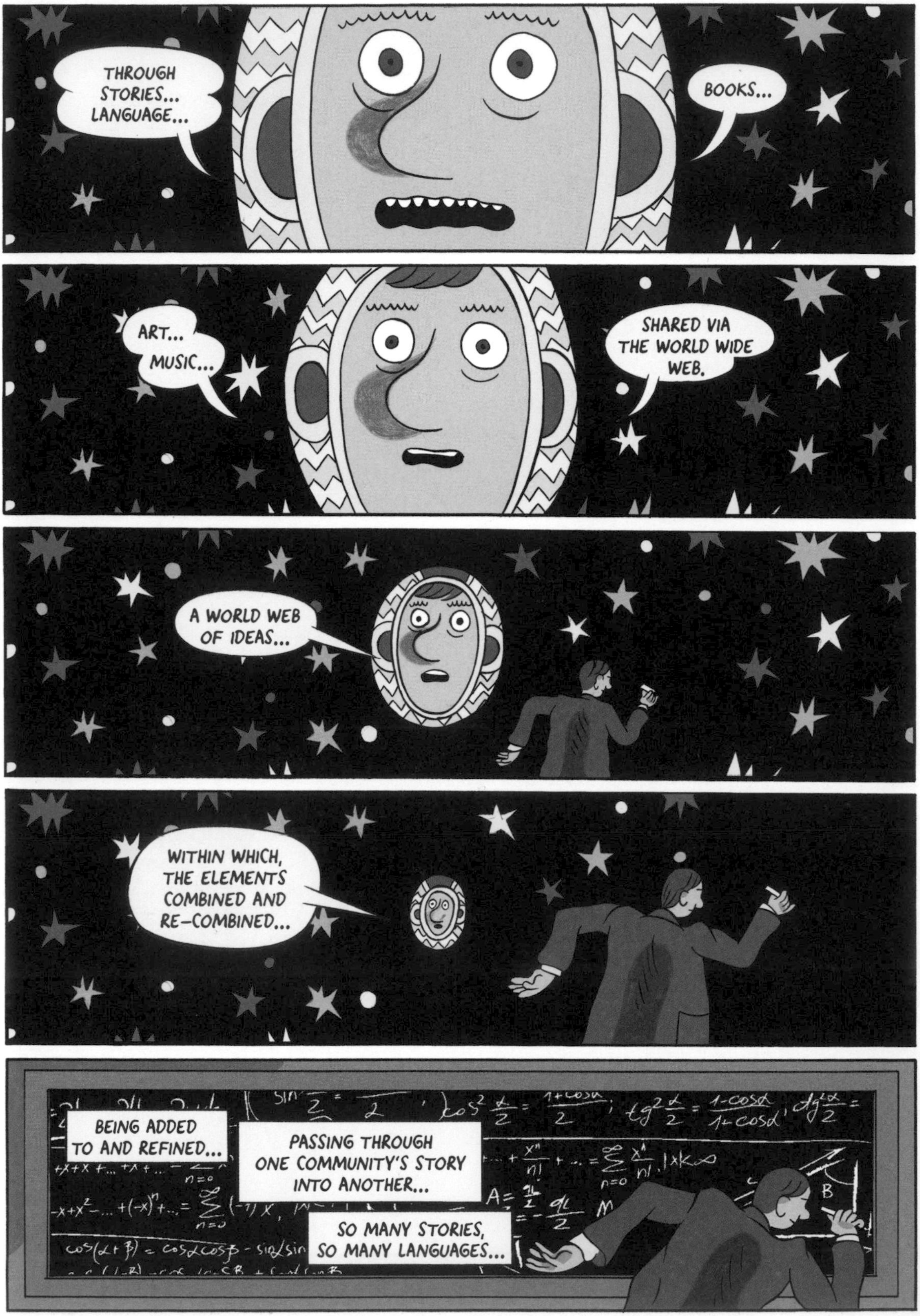
THROUGH STORIES... LANGUAGE...
BOOKS...
ART... MUSIC...
SHARED VIA THE WORLD WIDE WEB.
A WORLD WEB OF IDEAS...
WITHIN WHICH, THE ELEMENTS COMBINED AND RE-COMBINED...
BEING ADDED TO AND REFINED...
PASSING THROUGH ONE COMMUNITY'S STORY INTO ANOTHER...
SO MANY STORIES, SO MANY LANGUAGES...

EACH ITS OWN MODEL OF THE WORLD.

ONE POWERFUL LANGUAGE, MATHEMATICS, LOGIC AND THE SCIENTIFIC METHOD...

HOLDS THE POTENTIAL TO UNLOCK THE FUNDAMENTAL STORY OF THE UNIVERSE.

ENOLA GAY

IDEAS BUILT UPON, DEVELOPED,

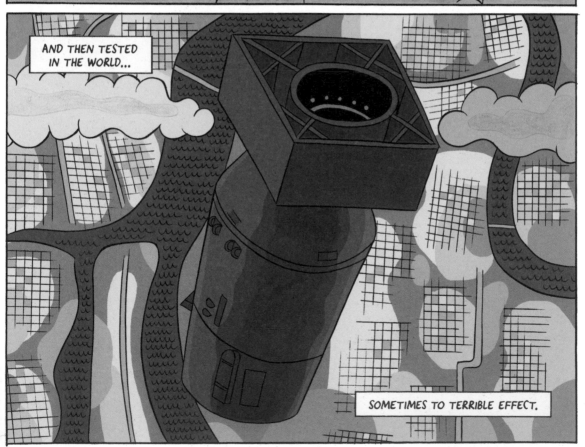

AND THEN TESTED IN THE WORLD...

SOMETIMES TO TERRIBLE EFFECT.

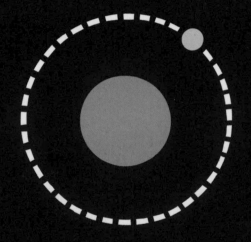

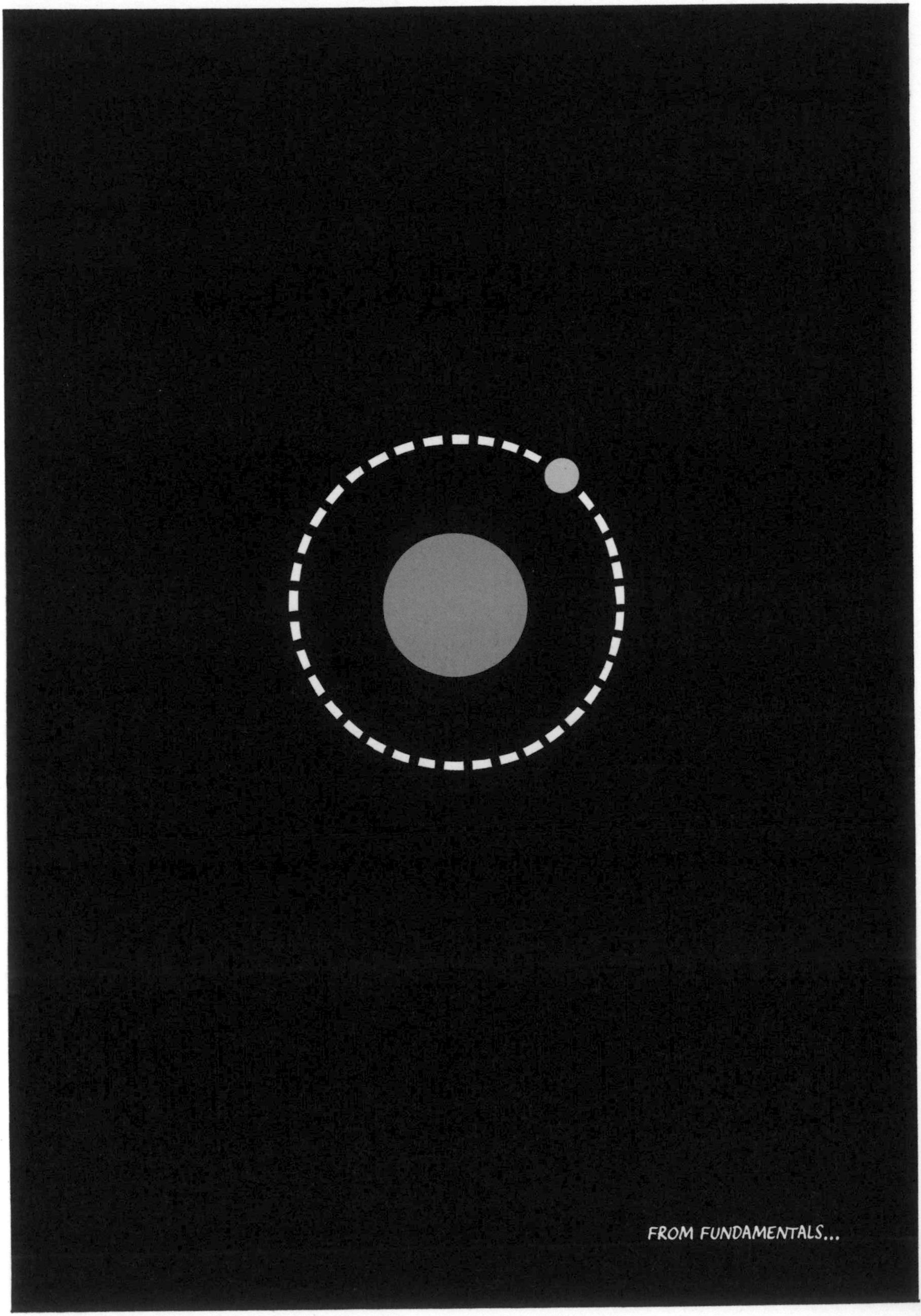

FROM FUNDAMENTALS...

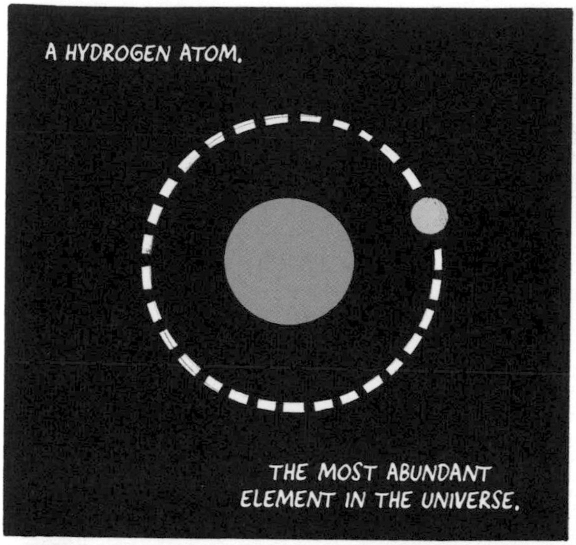
A HYDROGEN ATOM.
THE MOST ABUNDANT ELEMENT IN THE UNIVERSE.

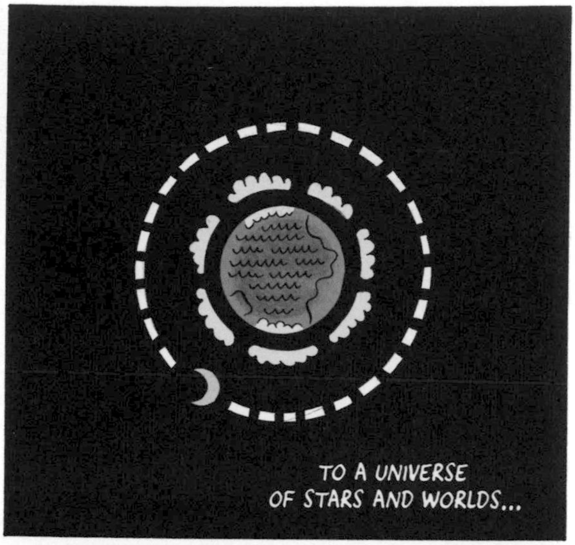
TO A UNIVERSE OF STARS AND WORLDS...

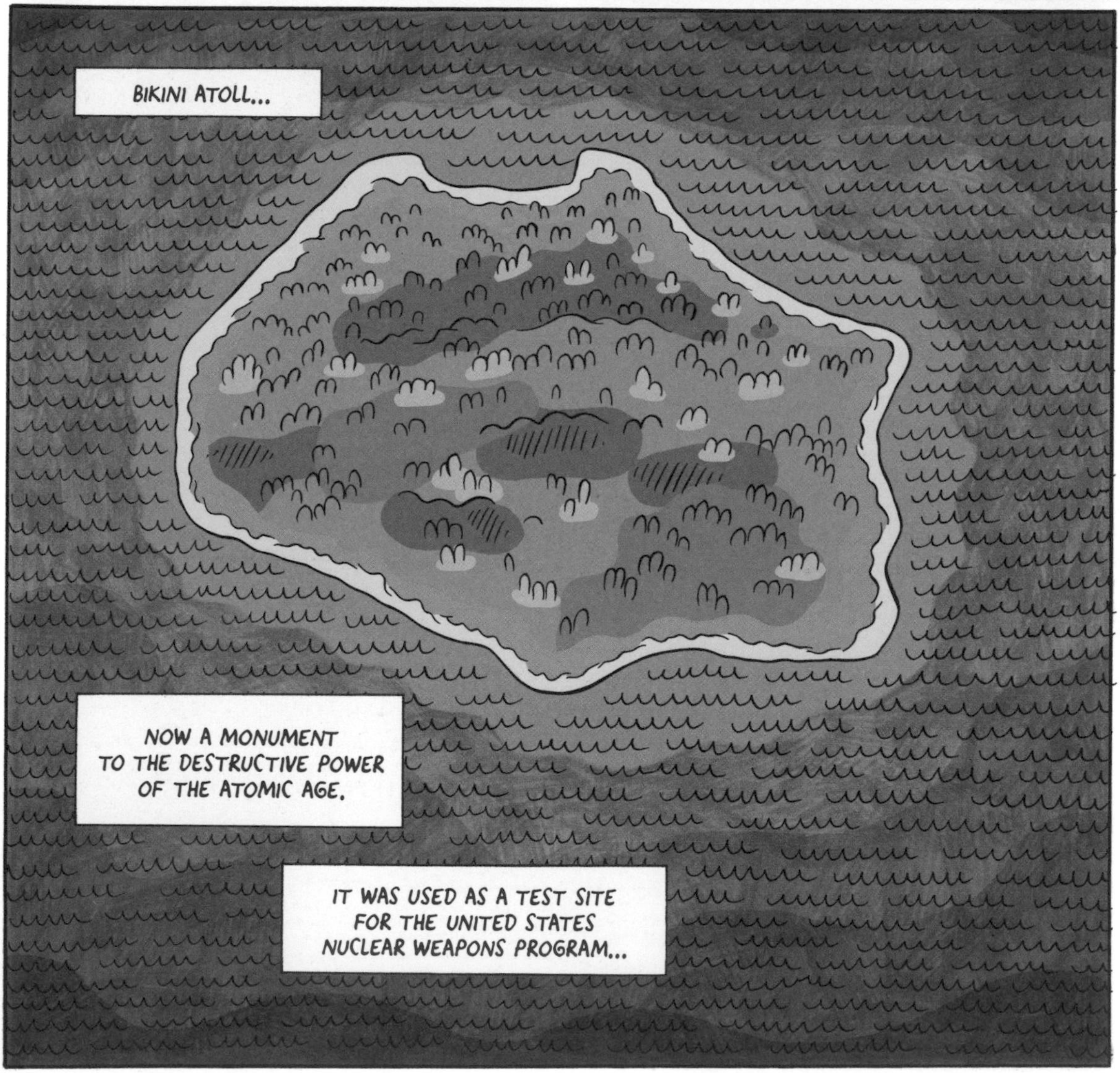
BIKINI ATOLL...
NOW A MONUMENT TO THE DESTRUCTIVE POWER OF THE ATOMIC AGE.
IT WAS USED AS A TEST SITE FOR THE UNITED STATES NUCLEAR WEAPONS PROGRAM...

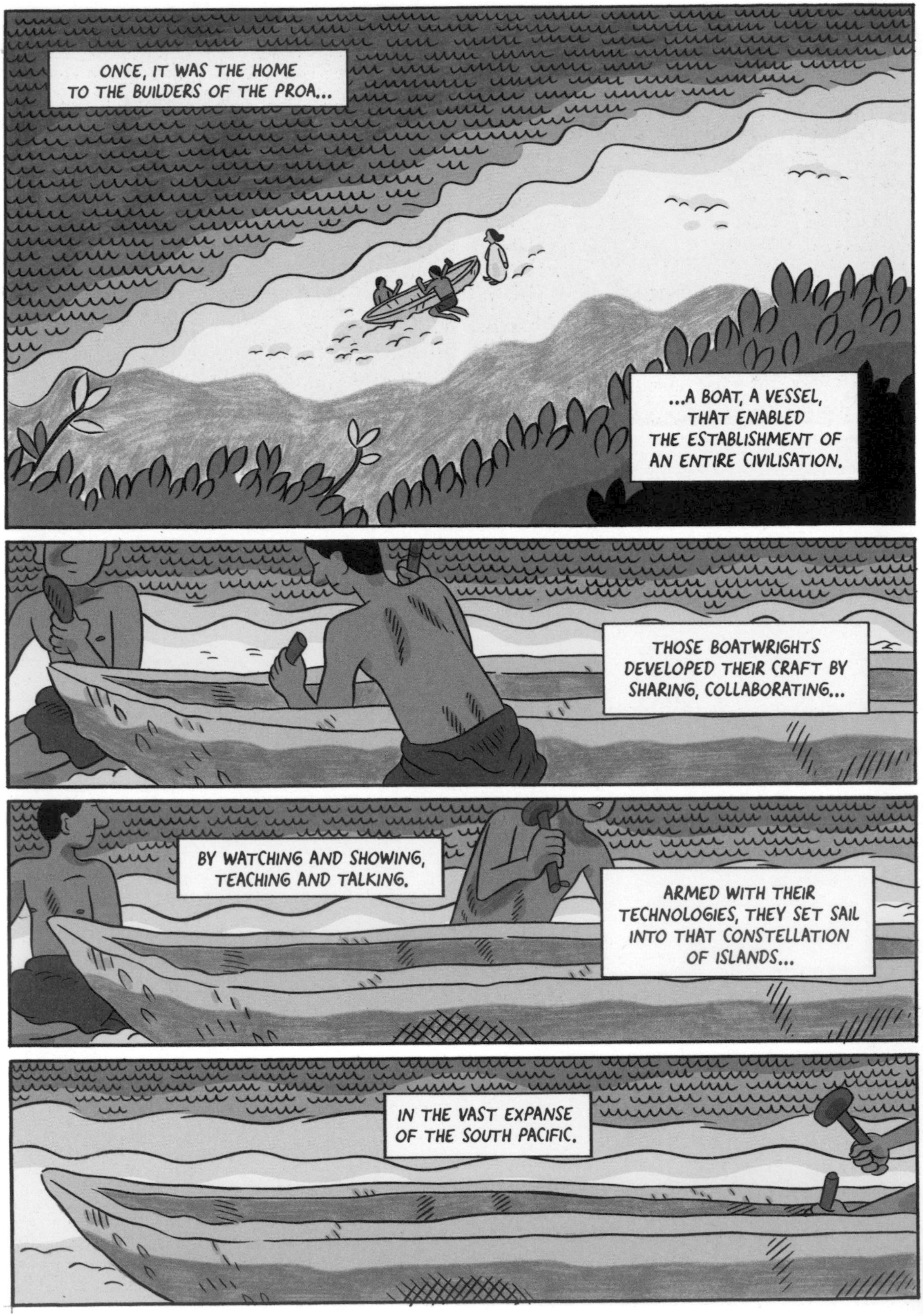
ONCE, IT WAS THE HOME TO THE BUILDERS OF THE PROA...
...A BOAT, A VESSEL, THAT ENABLED THE ESTABLISHMENT OF AN ENTIRE CIVILISATION.
THOSE BOATWRIGHTS DEVELOPED THEIR CRAFT BY SHARING, COLLABORATING...
BY WATCHING AND SHOWING, TEACHING AND TALKING.
ARMED WITH THEIR TECHNOLOGIES, THEY SET SAIL INTO THAT CONSTELLATION OF ISLANDS...
IN THE VAST EXPANSE OF THE SOUTH PACIFIC.

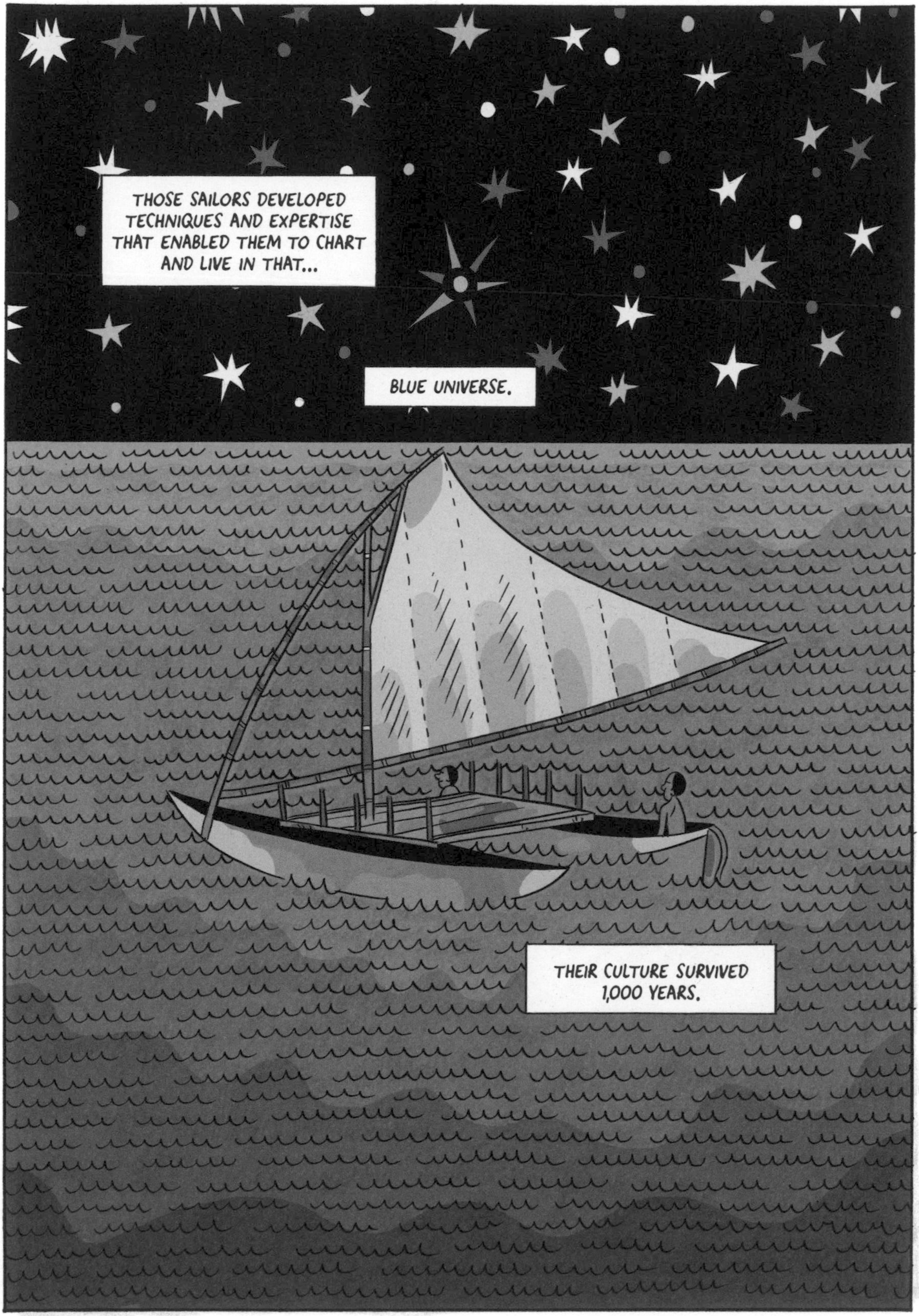
THOSE SAILORS DEVELOPED TECHNIQUES AND EXPERTISE THAT ENABLED THEM TO CHART AND LIVE IN THAT...
BLUE UNIVERSE.
THEIR CULTURE SURVIVED 1,000 YEARS.

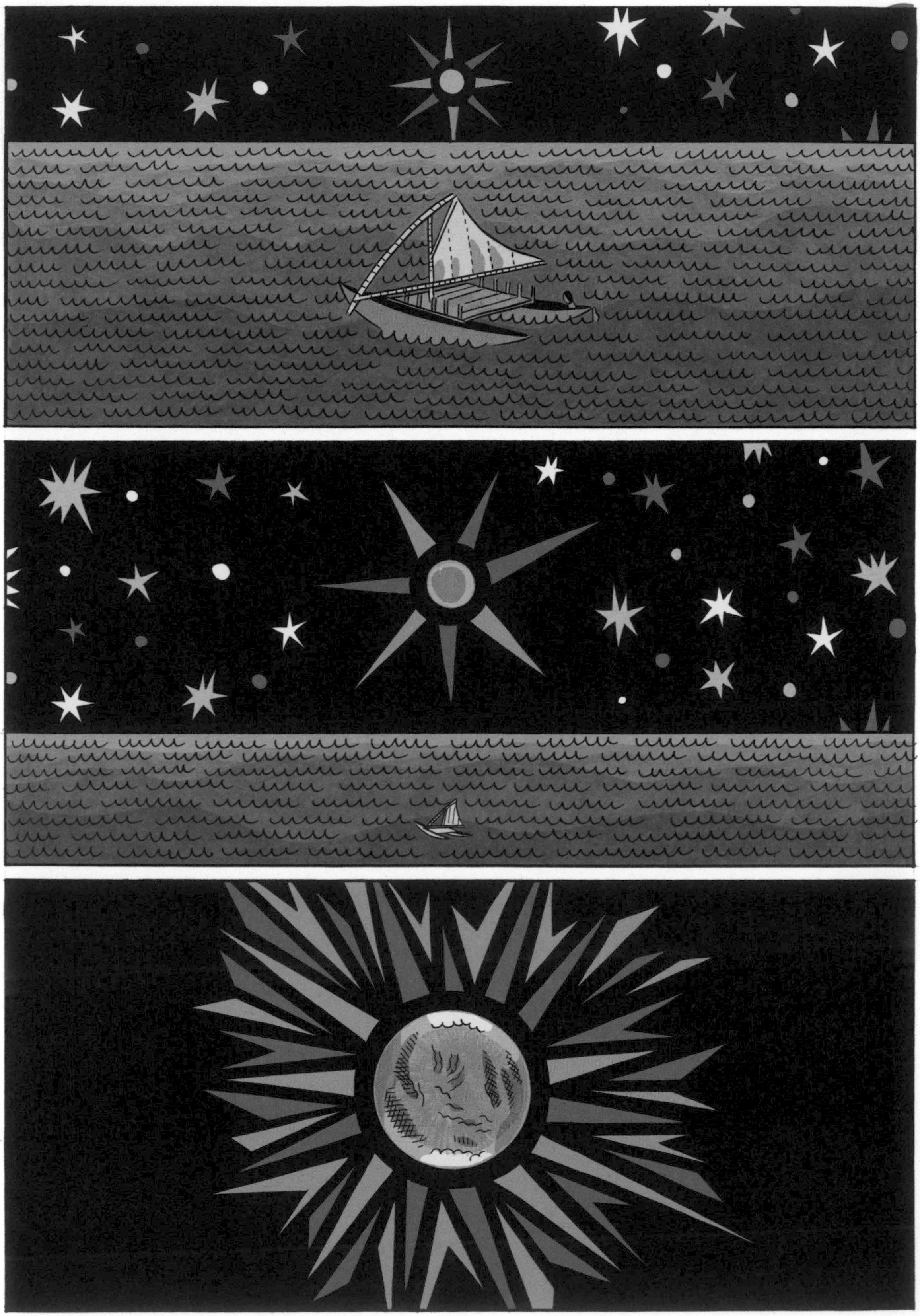

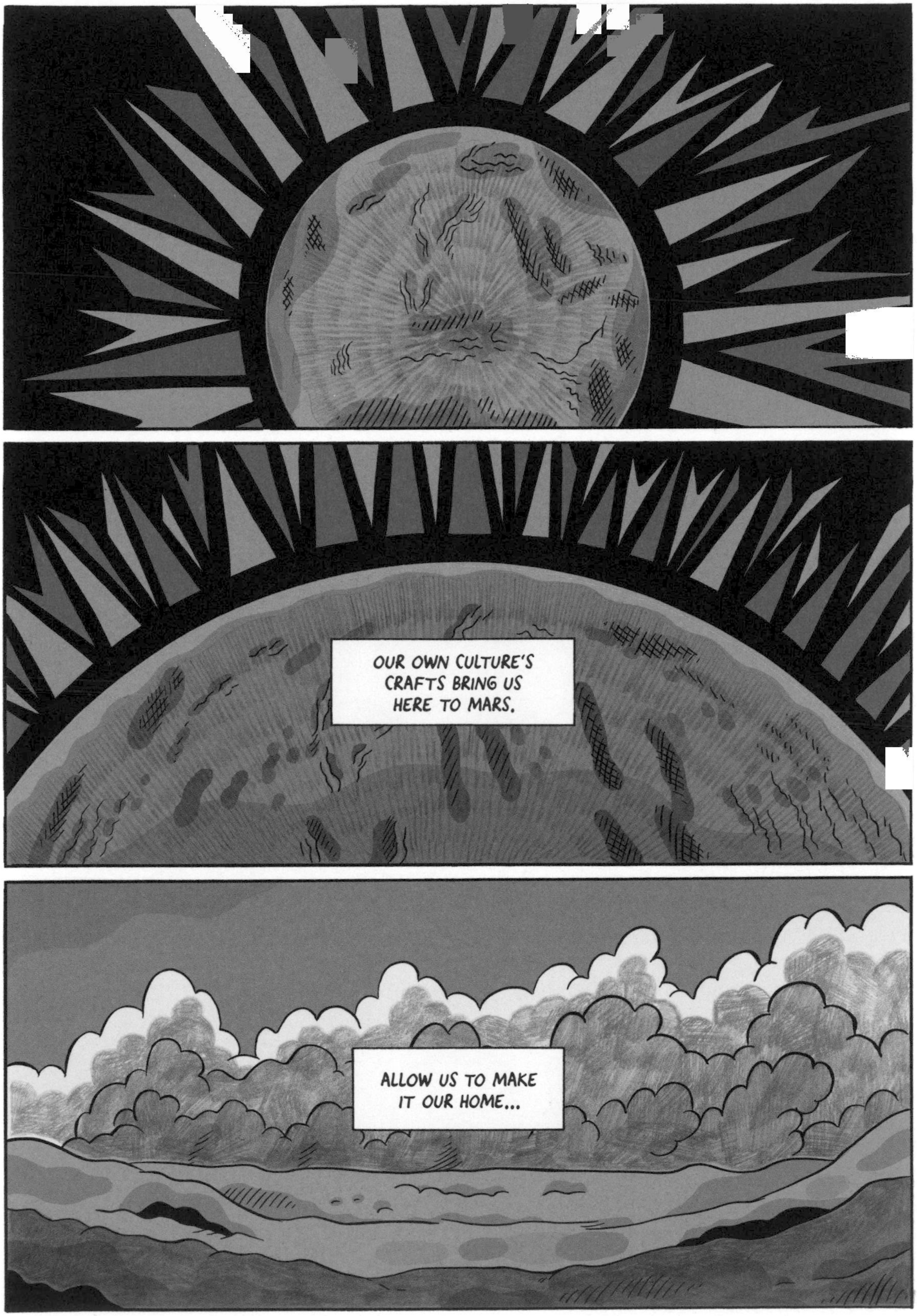

OUR OWN CULTURE'S CRAFTS BRING US HERE TO MARS.
ALLOW US TO MAKE IT OUR HOME...

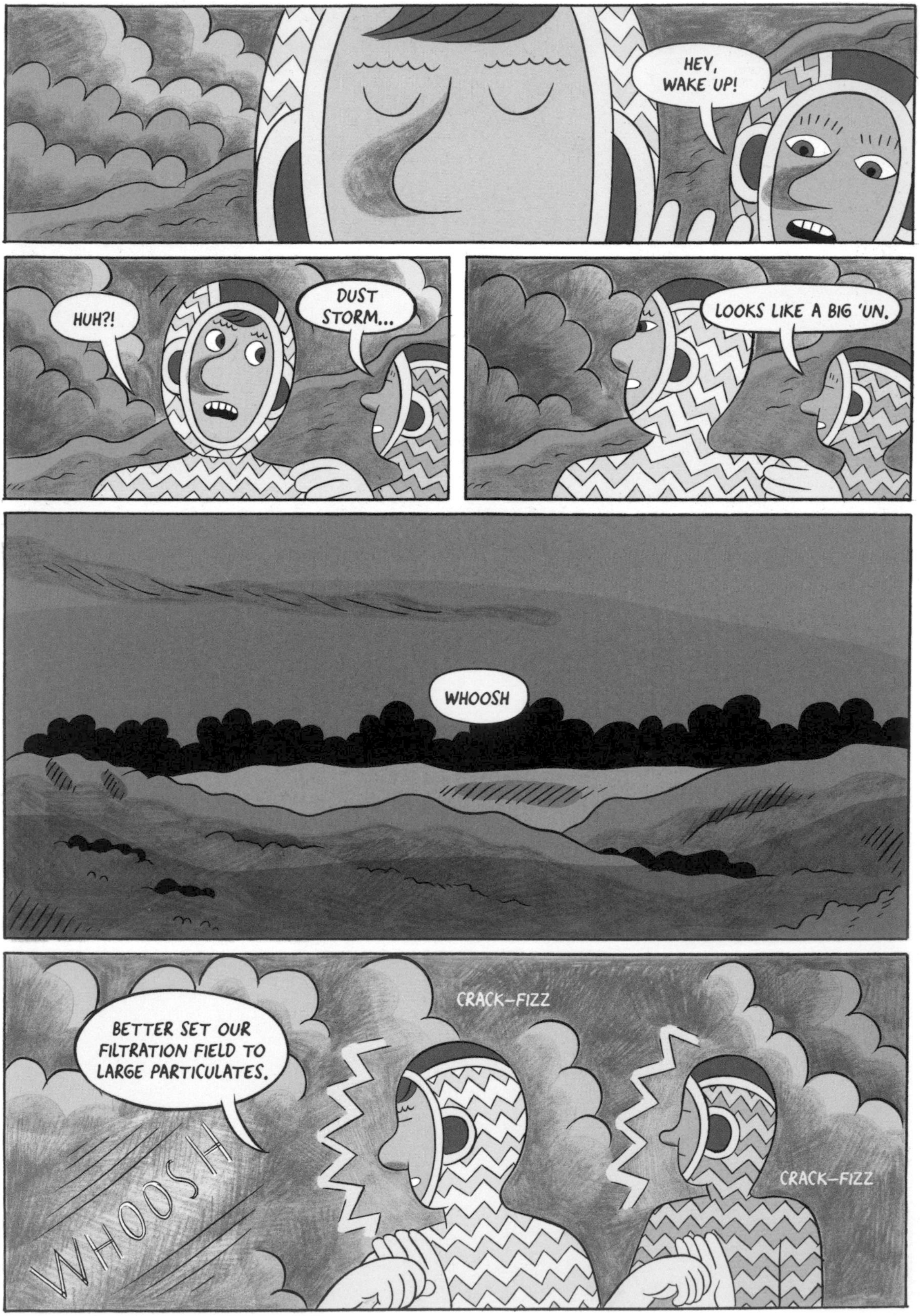
HEY, WAKE UP!
HUH?!
DUST STORM...
LOOKS LIKE A BIG 'UN.
WHOOSH
BETTER SET OUR FILTRATION FIELD TO LARGE PARTICULATES.
CRACK–FIZZ
WHOOSH
CRACK–FIZZ

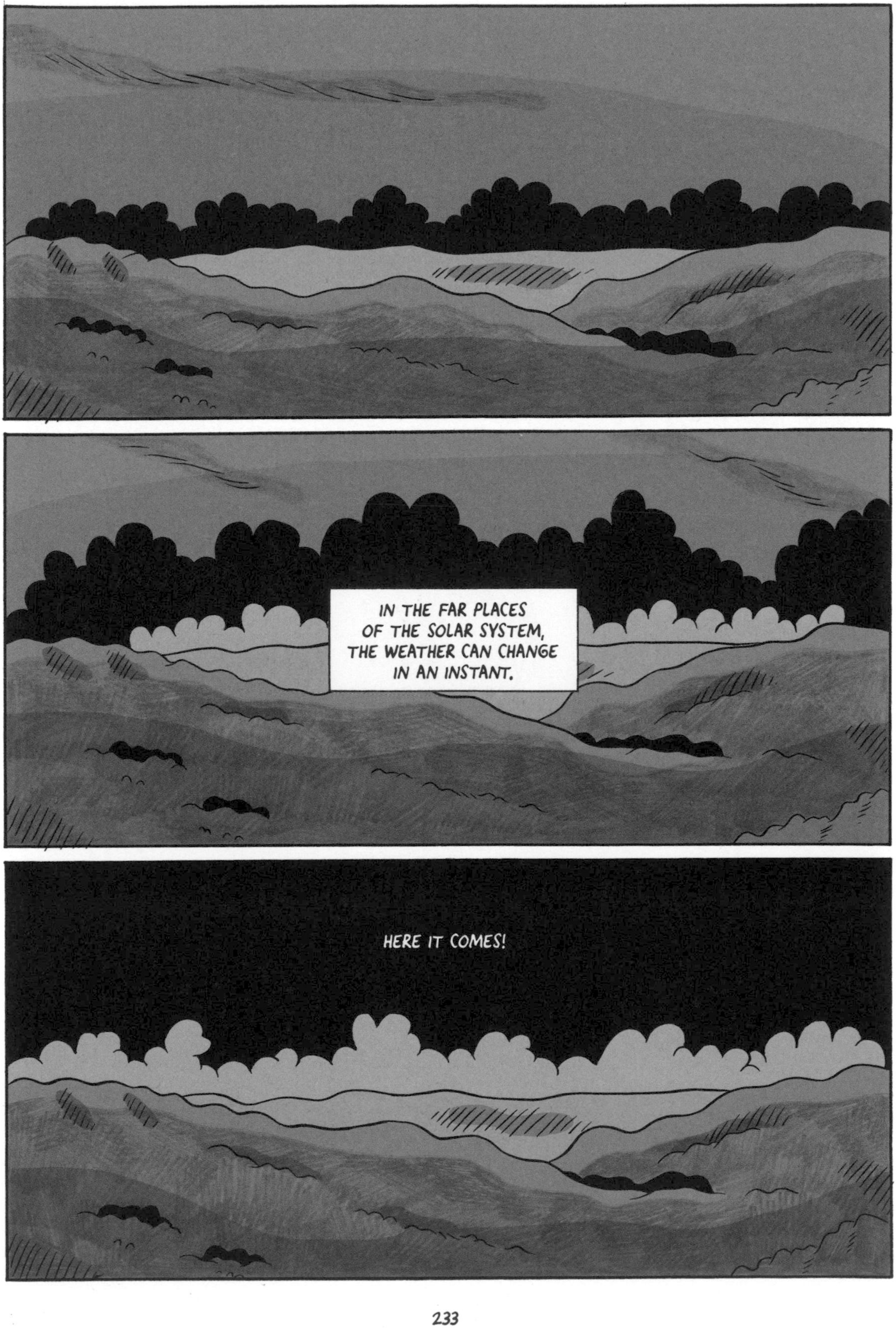
IN THE FAR PLACES OF THE SOLAR SYSTEM, THE WEATHER CAN CHANGE IN AN INSTANT.
HERE IT COMES!

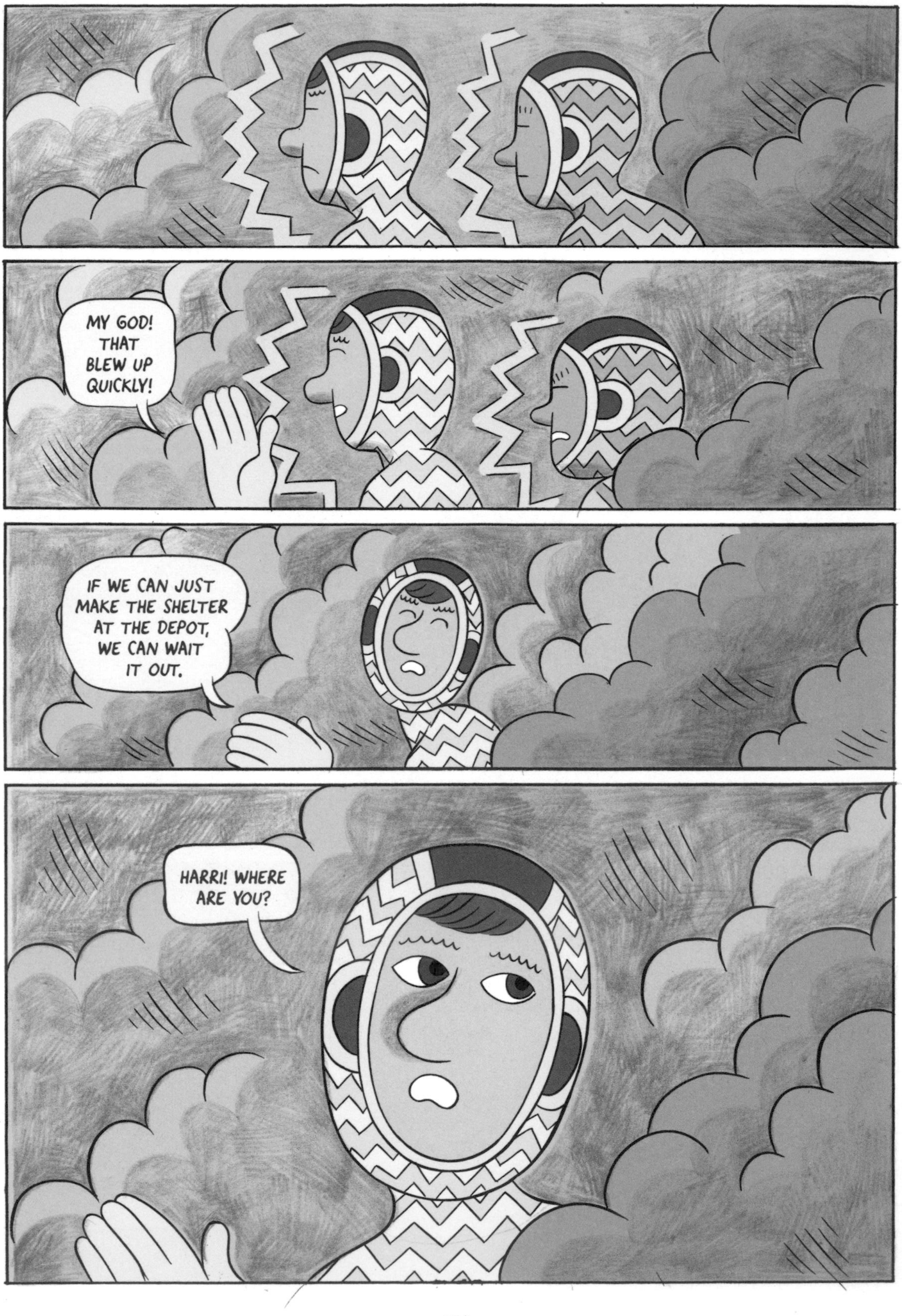
MY GOD! THAT BLEW UP QUICKLY!
IF WE CAN JUST MAKE THE SHELTER AT THE DEPOT, WE CAN WAIT IT OUT.
HARRI! WHERE ARE YOU?

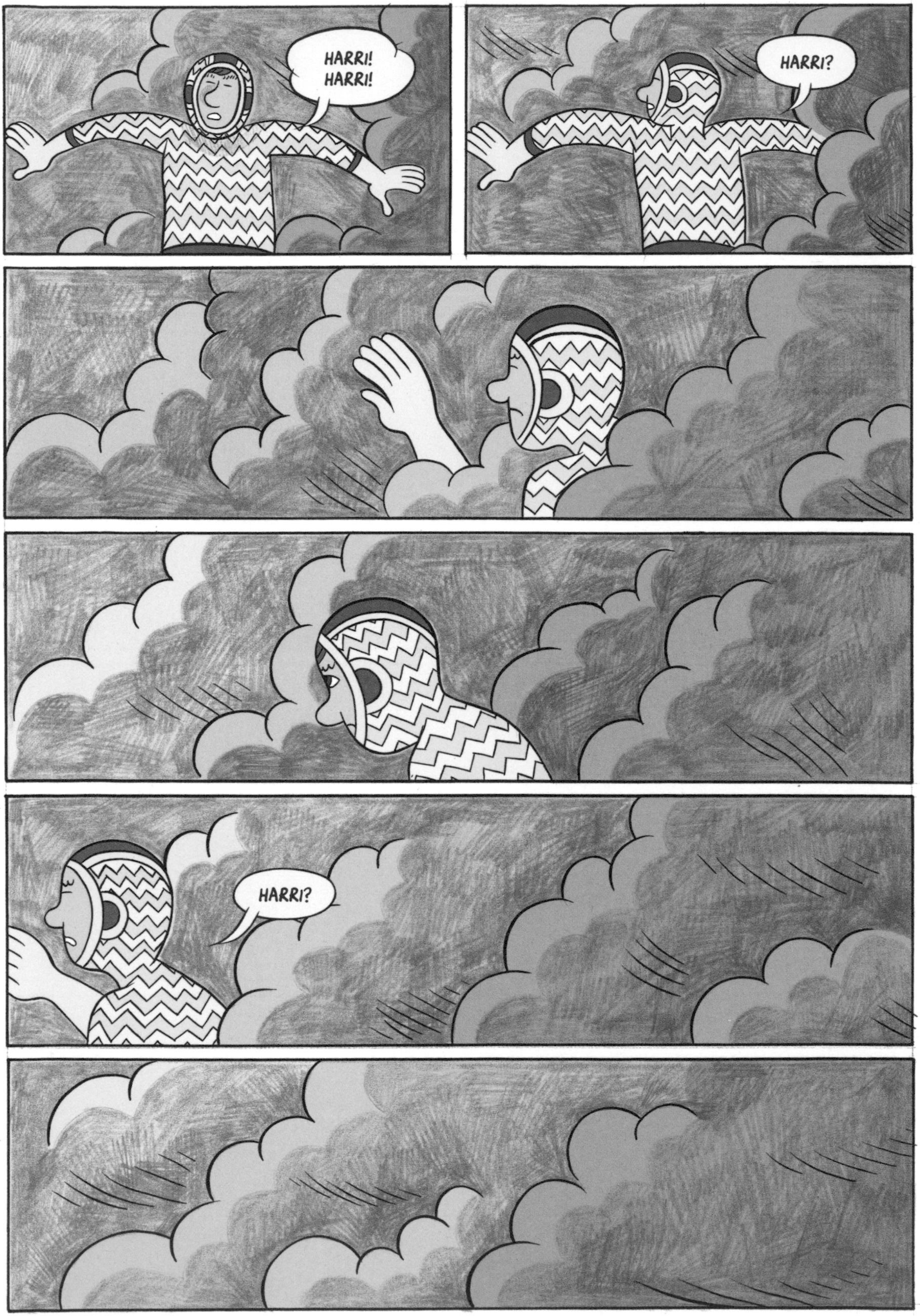
HARRI!
HARRI!
HARRI?
HARRI?

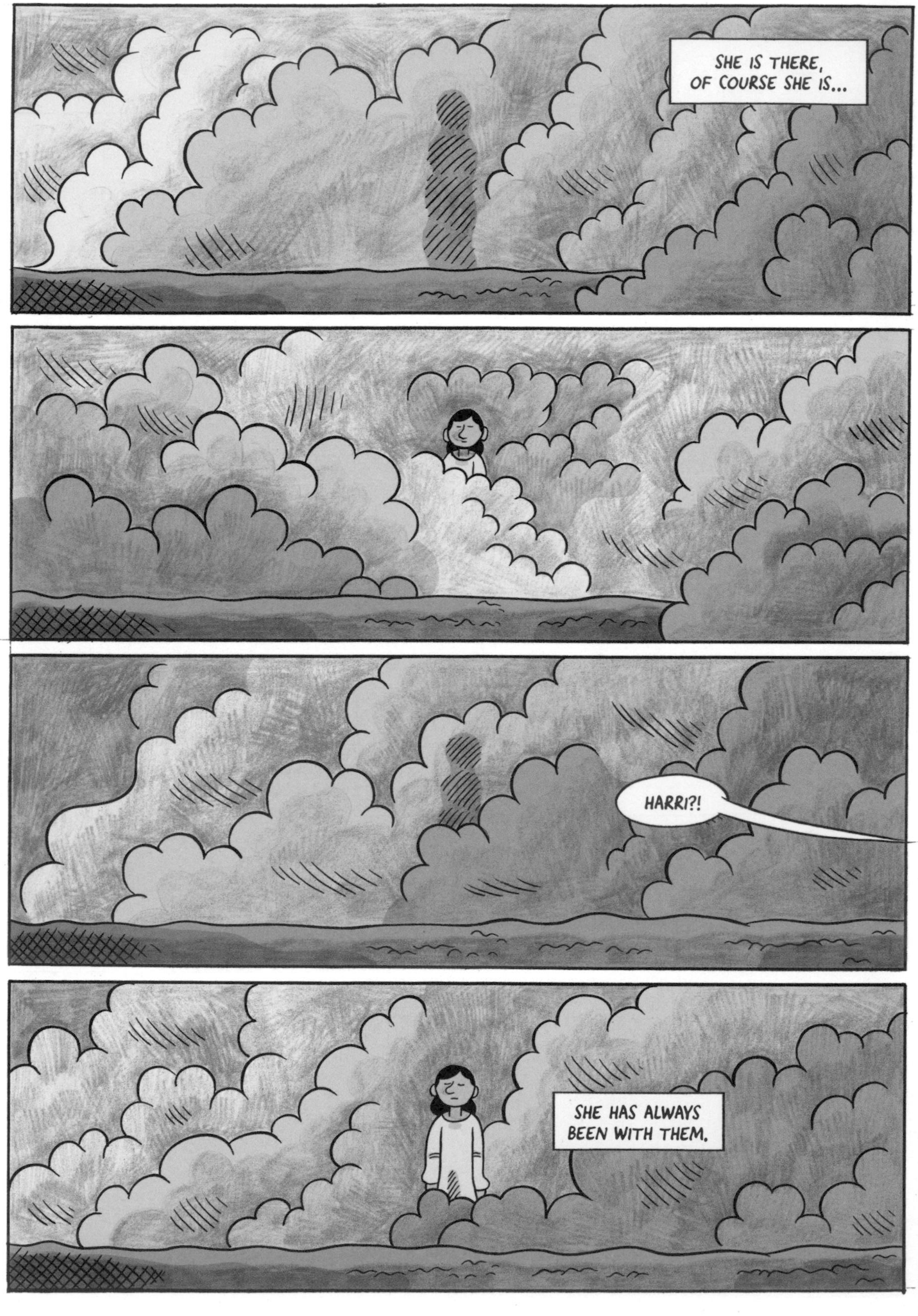
SHE IS THERE, OF COURSE SHE IS...
HARRI?!
SHE HAS ALWAYS BEEN WITH THEM.

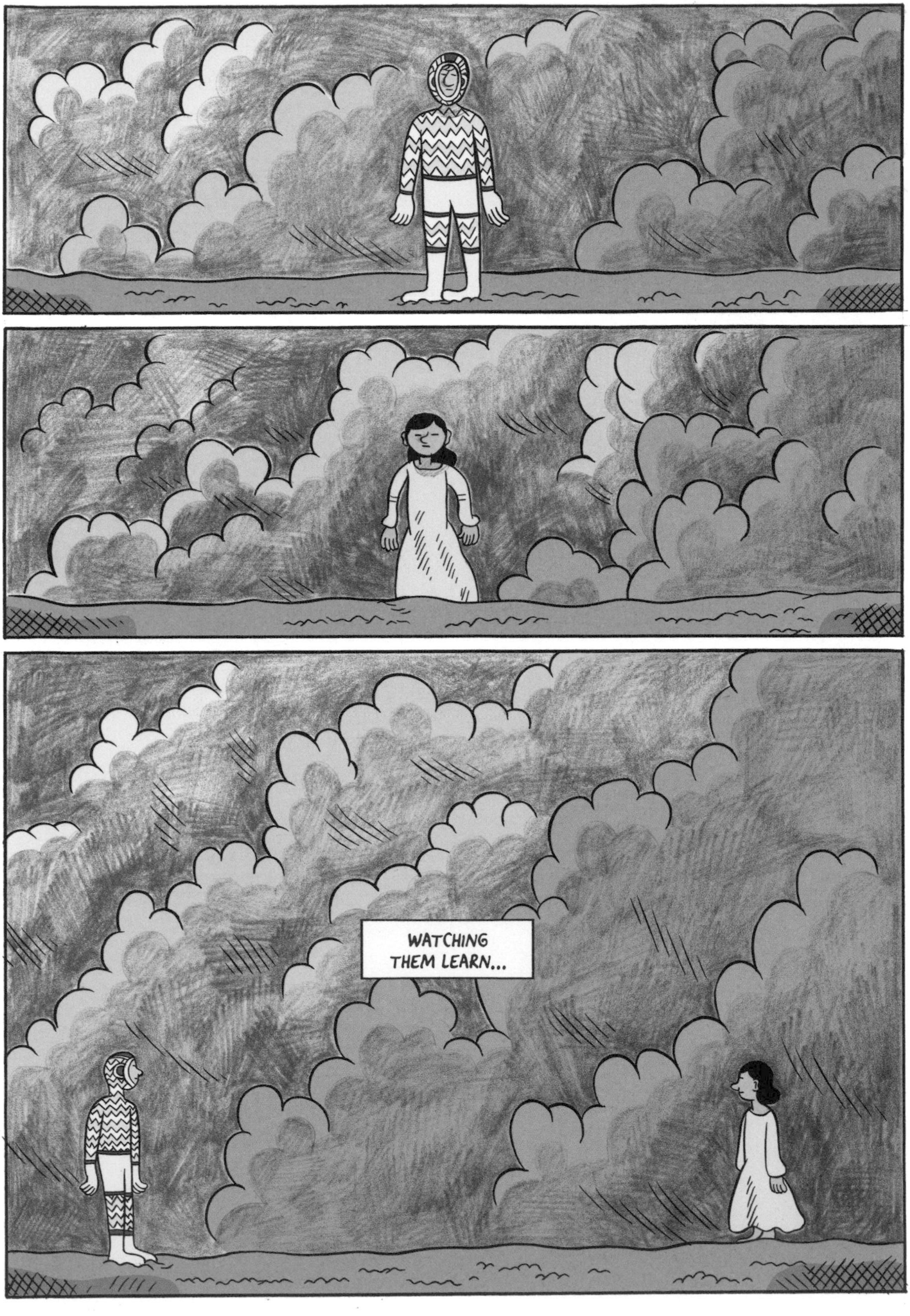
WATCHING THEM LEARN...

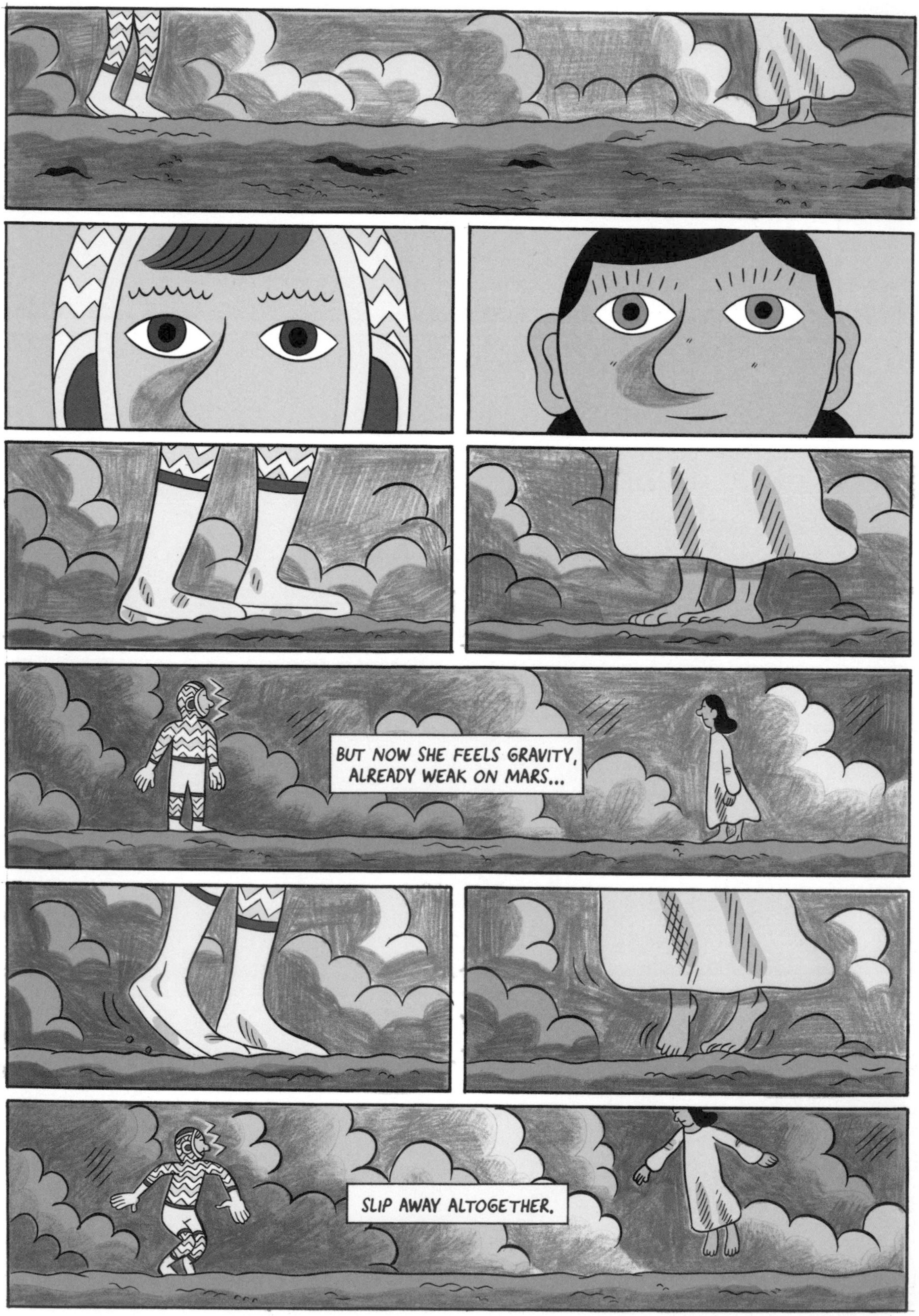
BUT NOW SHE FEELS GRAVITY, ALREADY WEAK ON MARS...
SLIP AWAY ALTOGETHER.

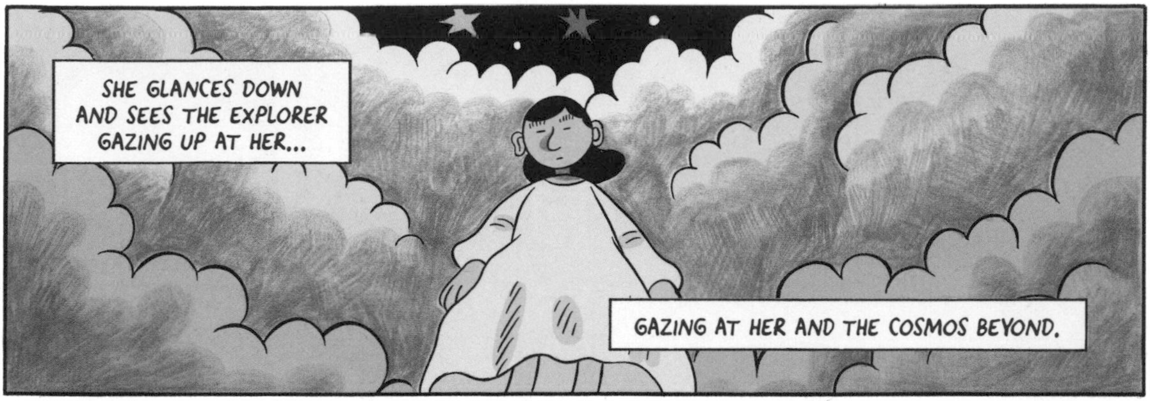
SHE GLANCES DOWN AND SEES THE EXPLORER GAZING UP AT HER...
GAZING AT HER AND THE COSMOS BEYOND.

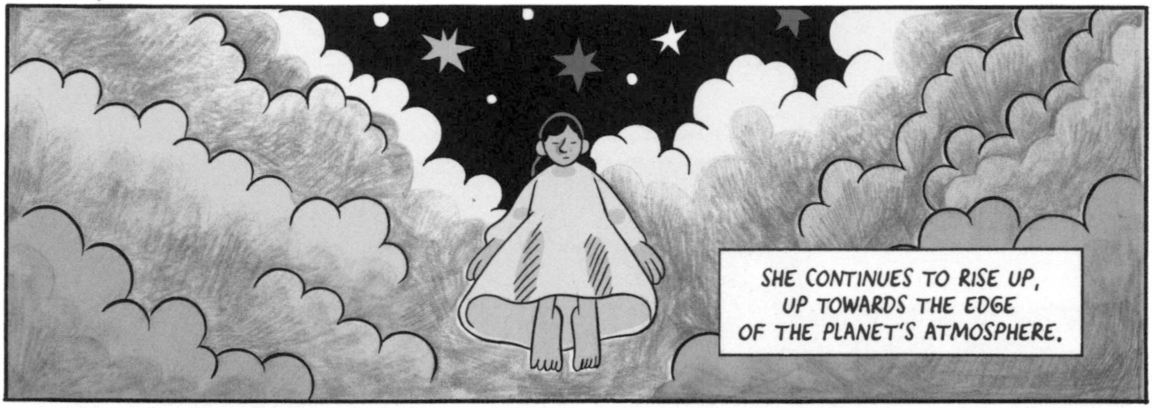
SHE CONTINUES TO RISE UP, UP TOWARDS THE EDGE OF THE PLANET'S ATMOSPHERE.

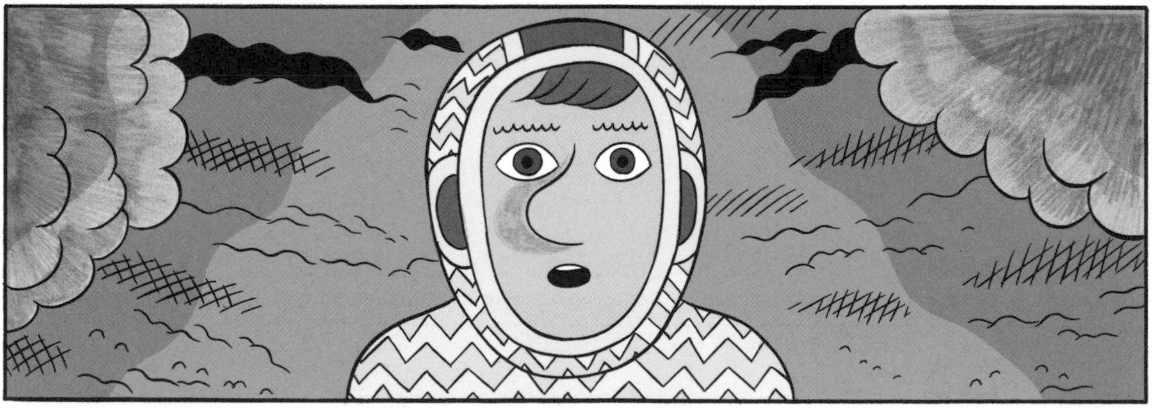

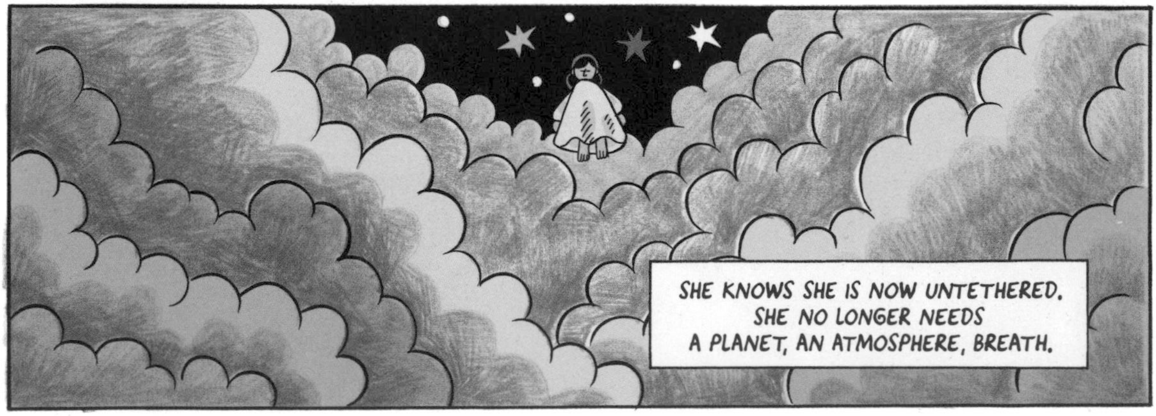
SHE KNOWS SHE IS NOW UNTETHERED. SHE NO LONGER NEEDS A PLANET, AN ATMOSPHERE, BREATH.

SHE TAKES HER LAST BREATH.

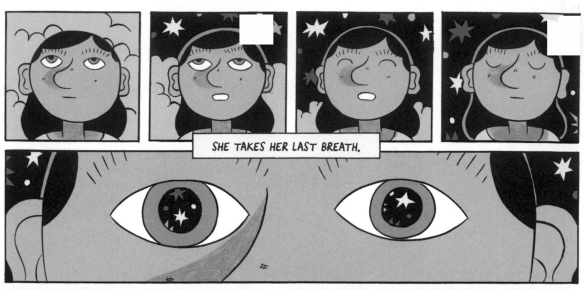

STARS SHINING MORE BRIGHTLY...

CALLING HER...

ONCE AGAIN SHE IS JUST WATCHING.

AND JUST AS GRAVITY AND THE PLANET FALLS BENEATH HER, SO TOO DOES TIME.

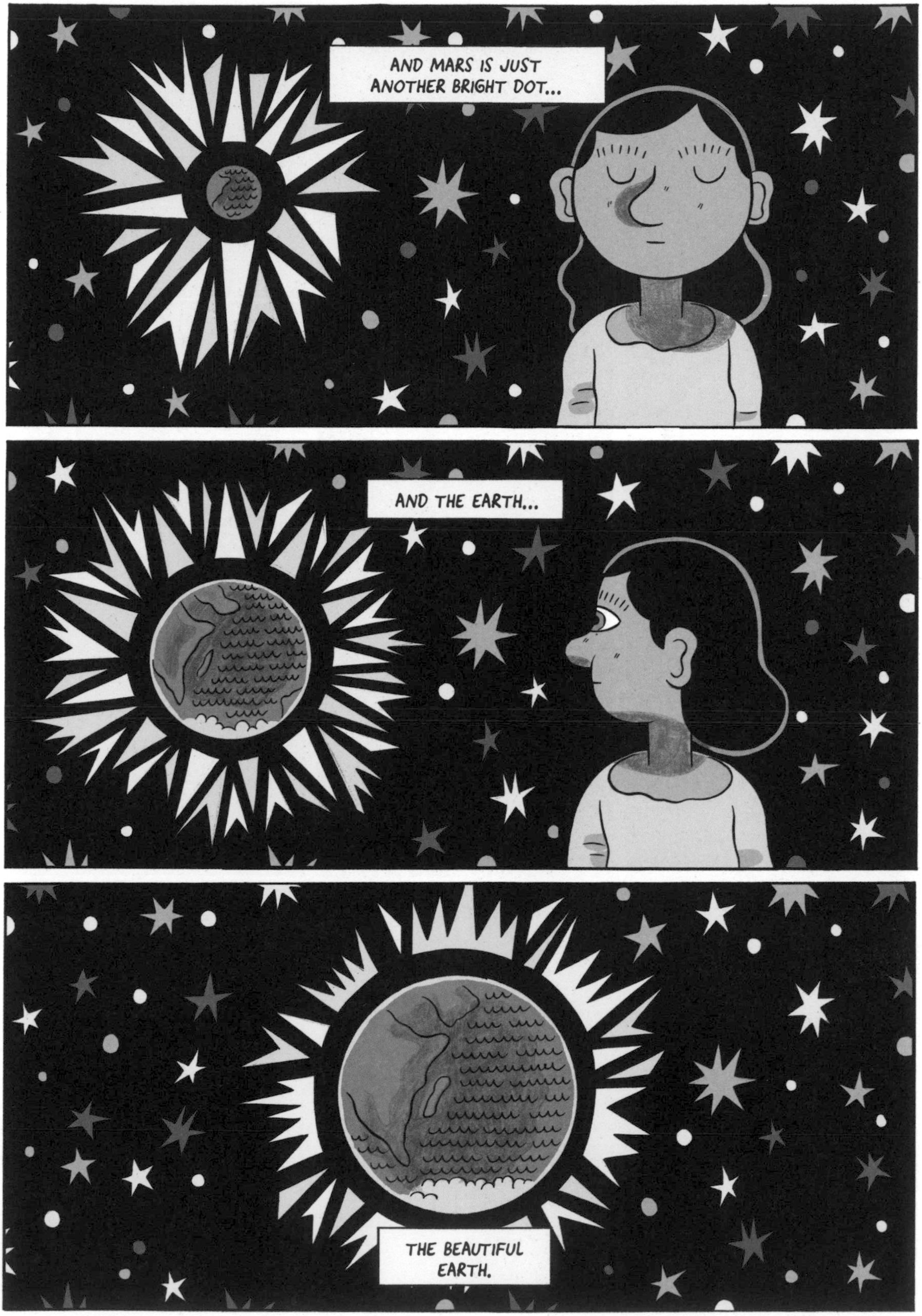
AND MARS IS JUST ANOTHER BRIGHT DOT...
AND THE EARTH...
THE BEAUTIFUL EARTH.

MARS AND EARTH!

TWINS!

BOTH ISLANDS OF LIFE.

AND THERE ARE OTHER WORLDS OUT THERE, SHE KNOWS...

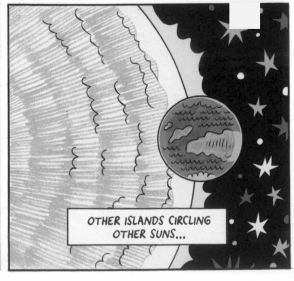

OTHER ISLANDS CIRCLING OTHER SUNS...

SHE HAS TAKEN ON THE SELF-CONSCIOUSNESS OF ALL THOSE HUMAN BEINGS SHE LIVED WITH...

WALKING AMONGST THEM.

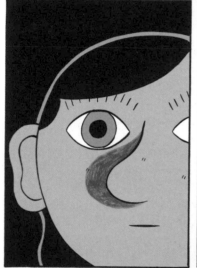

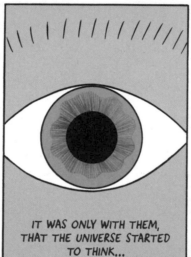

IT WAS ONLY WITH THEM, THAT THE UNIVERSE STARTED TO THINK...

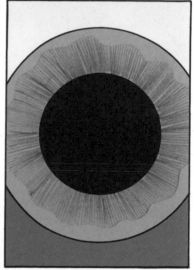

TO REFLECT UPON ITSELF.

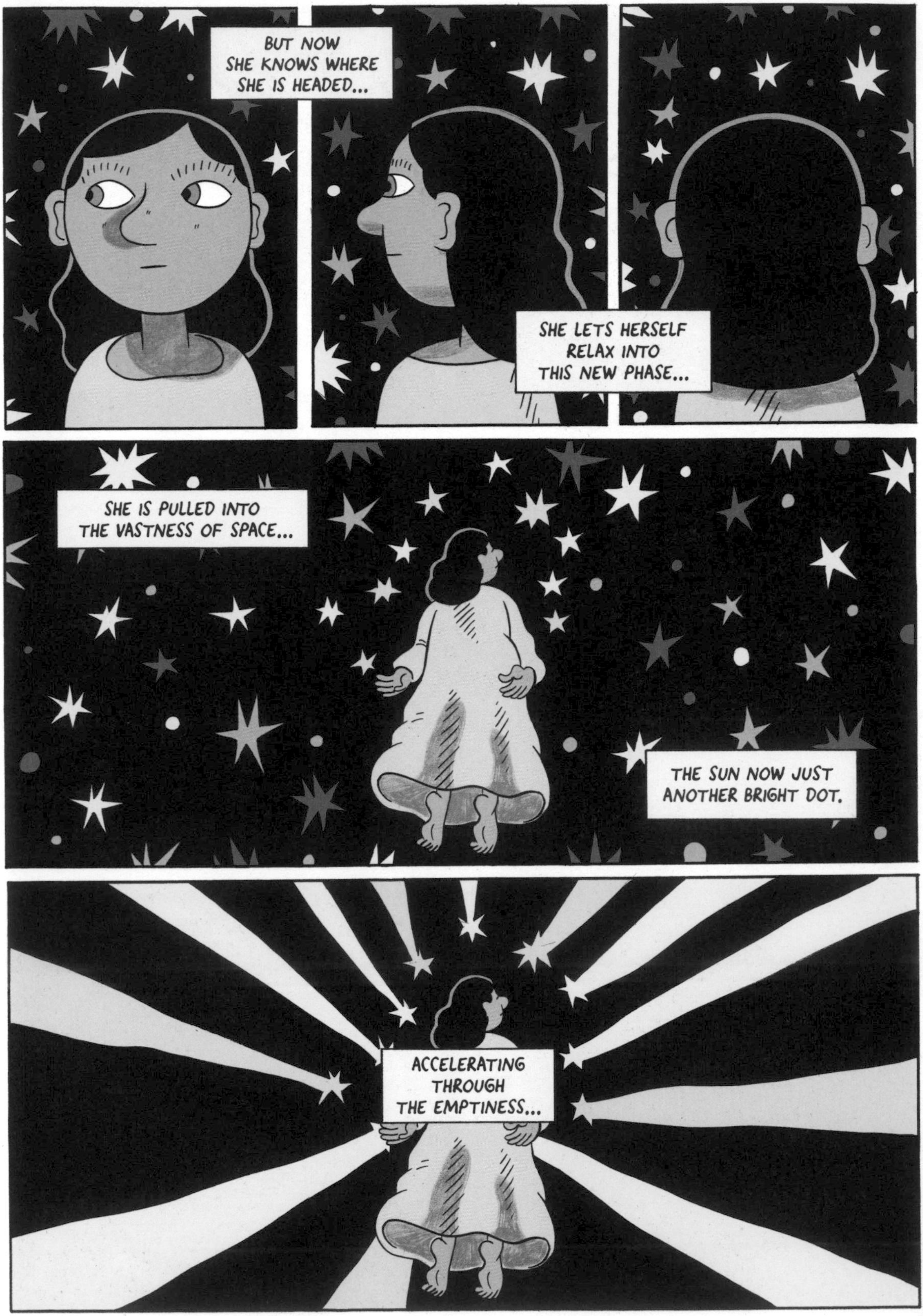
BUT NOW SHE KNOWS WHERE SHE IS HEADED...
SHE LETS HERSELF RELAX INTO THIS NEW PHASE...
SHE IS PULLED INTO THE VASTNESS OF SPACE...
THE SUN NOW JUST ANOTHER BRIGHT DOT.
ACCELERATING THROUGH THE EMPTINESS...

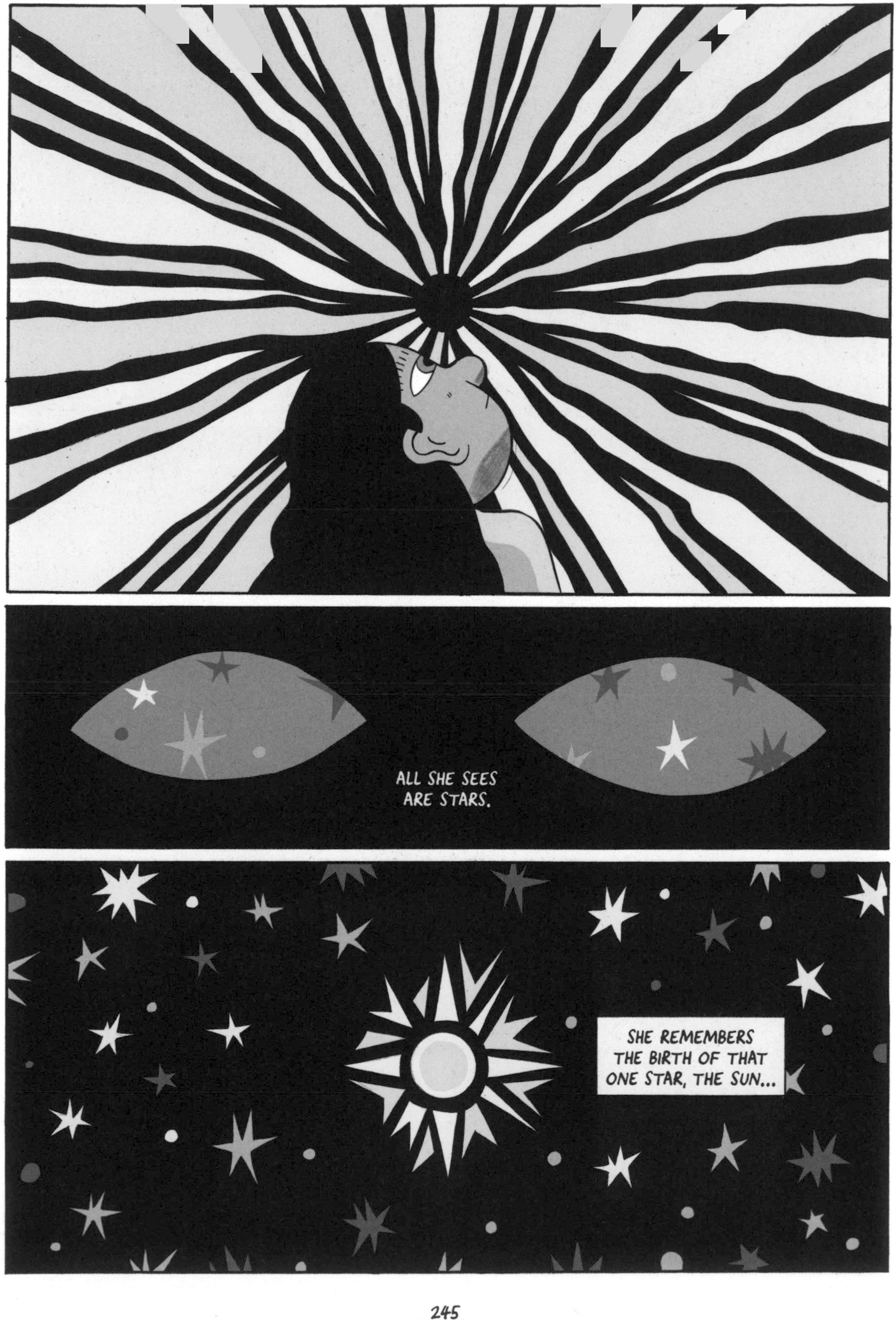
ALL SHE SEES ARE STARS.
SHE REMEMBERS THE BIRTH OF THAT ONE STAR, THE SUN...

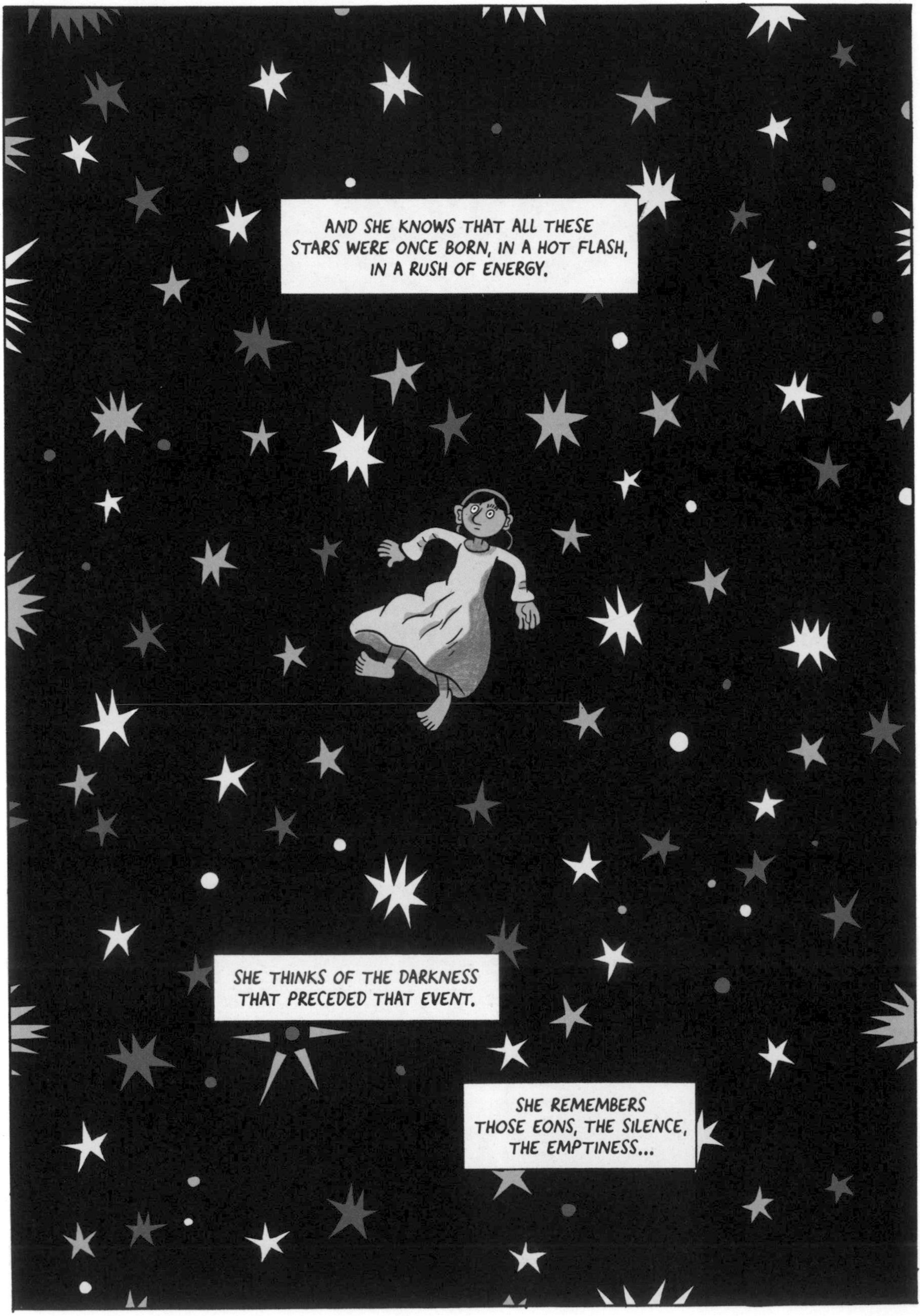
AND SHE KNOWS THAT ALL THESE STARS WERE ONCE BORN, IN A HOT FLASH, IN A RUSH OF ENERGY.
SHE THINKS OF THE DARKNESS THAT PRECEDED THAT EVENT.
SHE REMEMBERS THOSE EONS, THE SILENCE, THE EMPTINESS...

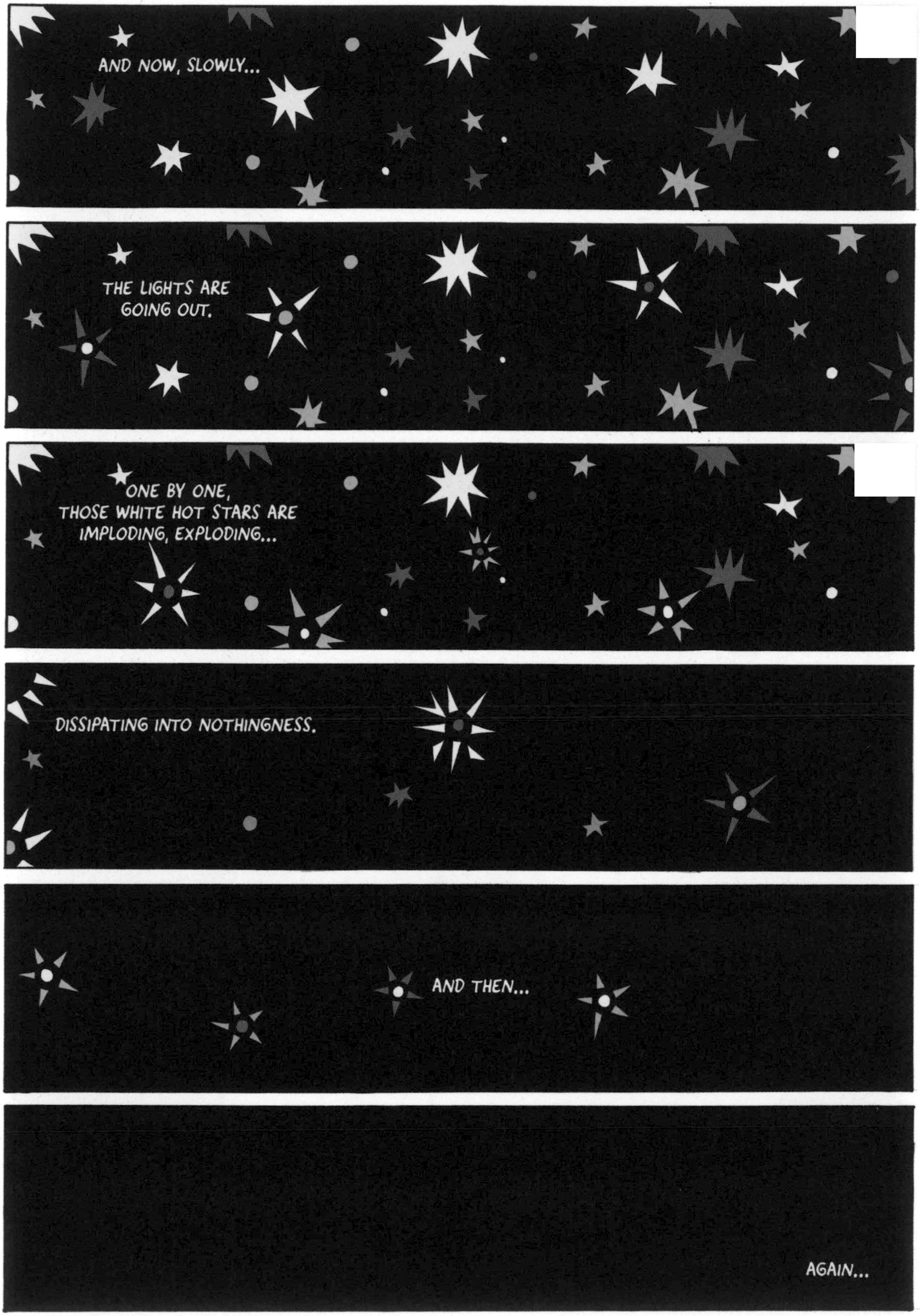
AND NOW, SLOWLY...
THE LIGHTS ARE GOING OUT.
ONE BY ONE, THOSE WHITE HOT STARS ARE IMPLODING, EXPLODING...
DISSIPATING INTO NOTHINGNESS.
AND THEN...
AGAIN...

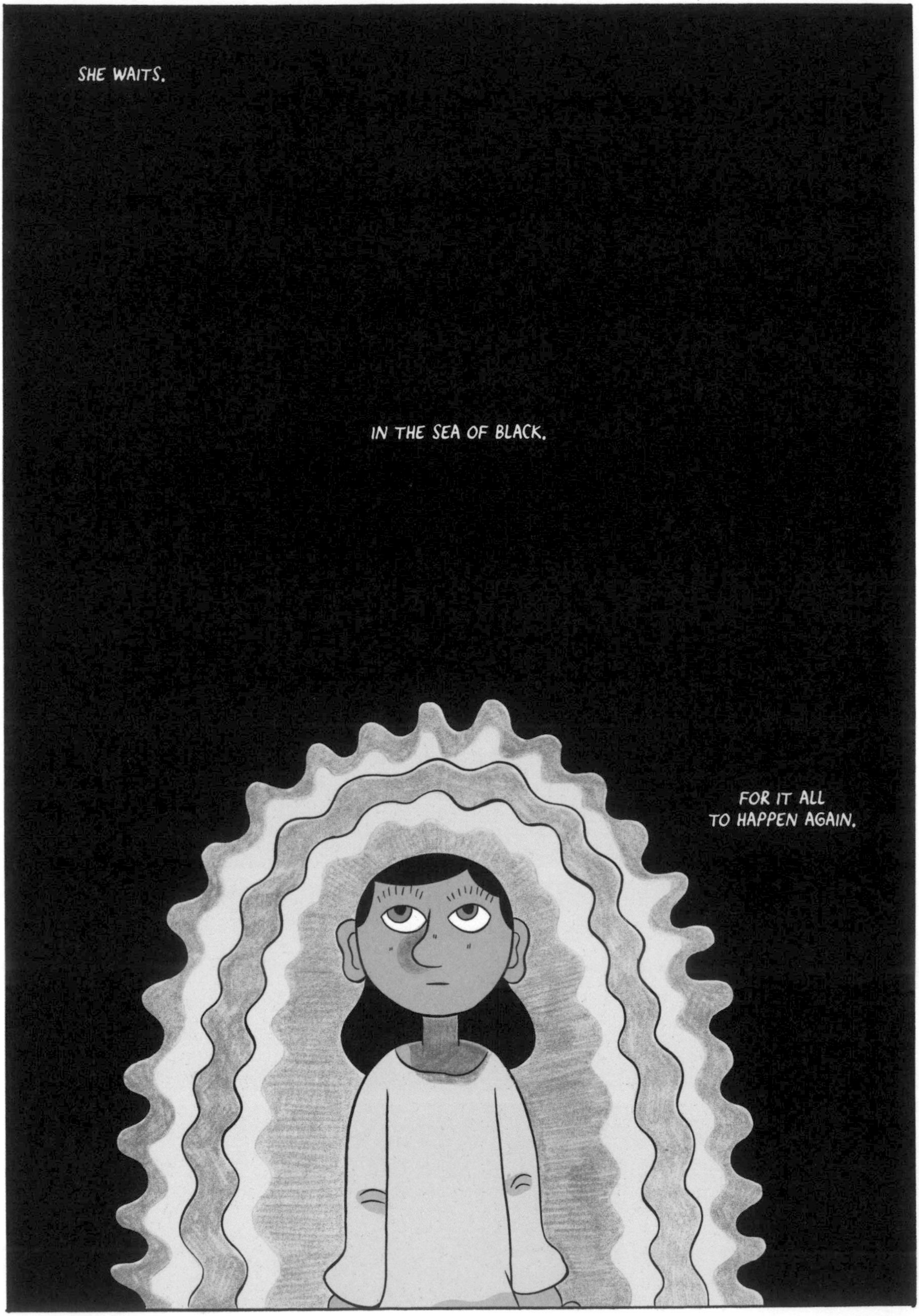
SHE WAITS.
IN THE SEA OF BLACK.
FOR IT ALL TO HAPPEN AGAIN.

DANIEL LOCKE

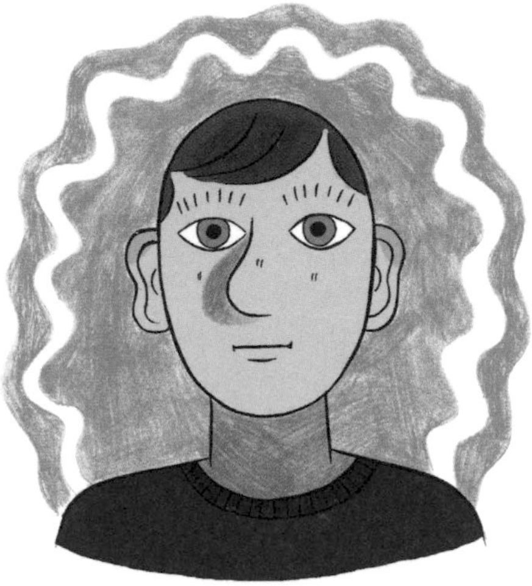

Daniel Locke is an artist and graphic novelist based in the UK. His work has regularly featured in a wide variety of illustration and comics anthologies and publications, and he has created commissions for organisations such as the Wellcome Trust, the Arts Council England and the National Trust. Since 2013 much of his work has been informed and shaped by the discoveries of contemporary science. Alongside producing *Out of Nothing*, Daniel has been illustrating a history of the careers of noted neurologists Profs. Uta and Chris Frith. Daniel is a committed environmentalist and helps to run the Brighton-based organisation Rewilding Sussex. *Out of Nothing* is his first full-length graphic novel.

DAVID BLANDY

David Blandy is a contemporary artist, based in Brighton, who uses video, performance and comics to address identity formation and how it relates to popular culture. His work deals with our problematic relationship with popular culture, highlighting the slippage and tension between fantasy and reality. David has exhibited widely in the UK and internationally, including at Tate Modern in London, The Baltic in Gateshead, Spike Island in Bristol, and Platform China Project Space in Beijing. His work is distributed by LUX and he is represented by Seventeen Gallery in London.

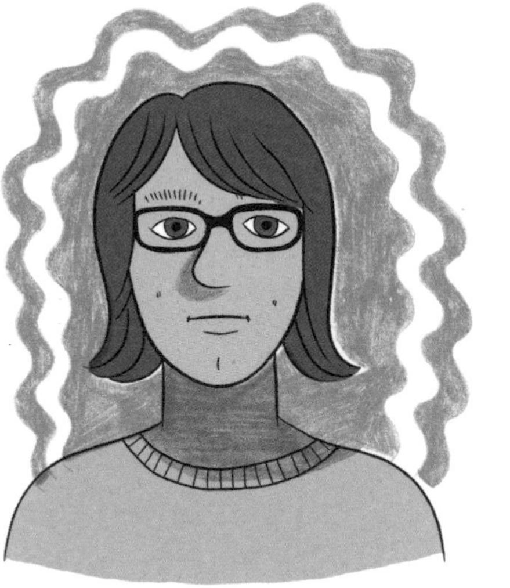

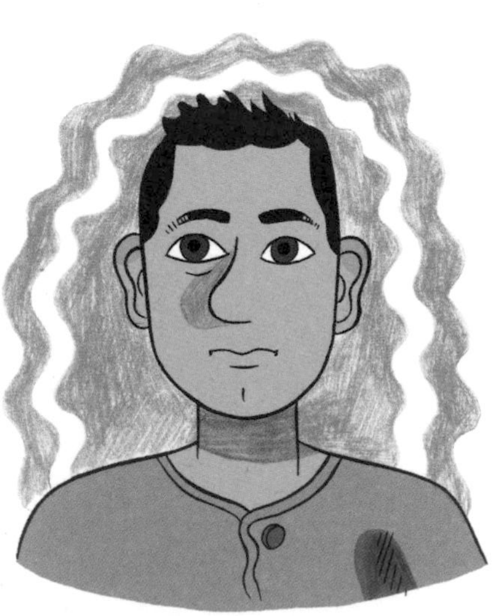

ADAM RUTHERFORD

Dr Adam Rutherford is a British geneticist, author, and broadcaster. He presents the flagship BBC Radio 4 programme *Inside Science*, amongst many other radio documentaries and award-winning television programmes. He has also written best-selling books, including *A Brief History of Everyone Who Ever Lived*, on history retold through DNA, and *Creation*, on synthetic biology and the origin of life. A film-fanatic, he was also science advisor on Bjork's film *Biophilia Live* and worked on *World War Z*, *Ex Machina*, *Life*, and most recently, *Annihilation*.

REFERENCES AND FURTHER READING

What follows is a list of the books, websites and documentaries that informed and inspired the creation of *Out of Nothing*. This graphic novel is not a work of rigorous scholarship but rather an attempt to tell a compelling story of humanity's quest for understanding and knowledge. As such, this is in no way an exhaustive list. Throughout the book we have attributed a number of statements to famous historical characters. Some of these quotes are based on actual interview sources that we picked up during the course of our research, but more often they are bastardised and should not be treated as verbatim.

Begelman, M., *Turn Right At Orion: Travels Through the Cosmos* (Perseus Publishing, 2000)
Butler, O. E., *Parable of the Sower* (Grand Central Publishing, 2000)
Dick, P. K., *Martian Time-Slip* (Gollancz, 2013)
Harari, Y. N., *Sapiens: A Brief History of Humankind* (Vintage, 2015)
Le Guin, U., *The Dispossessed* (Gollancz, 1999)
Stapledon, O., *Star Maker* (Gollancz, 1999)
Steyerl, H., Aranda, J. (Ed), Wood, B. K. (Ed), Vidokle, A. (Ed),
The Wretched of the Screen (Sternberg Press, 2012)
Tezuka, O., *Phoenix, Vol. 1: Dawn* (Viz Media, 2000)

Lane, N., *The Vital Question: Why is Life the Way it is* (Profile Books, 2015)
Rutherford, A., *Creation: The Origin of Life / The Future of Life* (Penguin, 2014)
Tyson, N. D., Strauss, M. A., and Gott, J. R.,
Welcome to the Universe: An Astrophysical Tour (Princeton University Press, 2016)
Watson, J., *The Double Helix* (W&N, 2010)
Cosmos, television documentary by Carl Sagan, first broadcast September 1980
Wonders of the Universe, television documentary by Brian Cox,
first broadcast March 2011

Man, J., *The Gutenberg Revolution* (Bantam, 2009)
McMurtrie, D. C., *The Book, the Story of Printing and Bookmaking* (Braken Books, 1989)
Steinberg, S. H., *Five Hundred Years of Printing* (Pelican Books, 1974)

Birmant, J. and Oubrerie, C., *Pablo* (Self Made Hero, 2015)

Danchev, A., Georges Braque: *A Life* (Arcade Publishing, 2012)

Gantefuhrer-Trier, A., *Cubism* (Taschen Basic Art Series, 2004)

Hilton, T., *Picasso* (World of Art) (Thames and Hudson Ltd, 1976)

Palermo, C., Cooper, H., Poggi, C., Bourneuf, A., Barry, C., Devolder, B.,
Picasso and Braque: The Cubist Experiment, 1910–1912 (Yale University Press, 2011)

Isaacson, W., *Einstein: His Life and Universe* (Simon & Schuster, 2017)

Maier, C. and Simon, A., *Einstein* (Nobrow, 2016)

Neffe, J., Frish, S. (Trans), *Einstein: A Biography* (Polity Press, 2016)

Inside Einstein's Mind: The Enigma of Space and Time, television documentary
by BBC Four, first broadcast December 2015

Berners-Lee, T., *Weaving the Web: The Past, Present and Future of the
World Wide Web by its Inventor* (Orion Business, 1999)

www.w3.org/History/1989/proposal.html (March 1989); the original proprosal
for the software project at CERN that became the World Wide Web.

https://webfoundation.org/about/

www.ted.com/talks/tim_berners_lee_on_the_next_web; Tim Berners-Lee's next project.

webfoundation.org/about/community/transcript-of-tim-berners-lee-video/;
a transcript of Tim Berners-Lee's speech before the Knight Foundation, where he
outlines the origin of the World Wide Web and his hope for its future.

The Virtual Revolution: The Great Levelling?, radio documentary by BBC One,
first broadcast May 2010

Kelly, C. C., Rhodes, R., *The Manhattan Project: The Birth of the Atomic Bomb
in the Words of Its Creators, Eyewitnesses, and Historians* (Black Dog & Leventhal, 2011)

Race for the World's First Atomic Bomb: A Thousand Days of Fear,
television documentary by BBC Four, first broadcast Aug 2015

Chang, J., Can't Stop, Won't Stop: *A History of the Hip-hop Generation* (Picador USA, 2006)

Piskor, E., *Hip Hop Family Tree* (Fantagraphics Books, 2013)

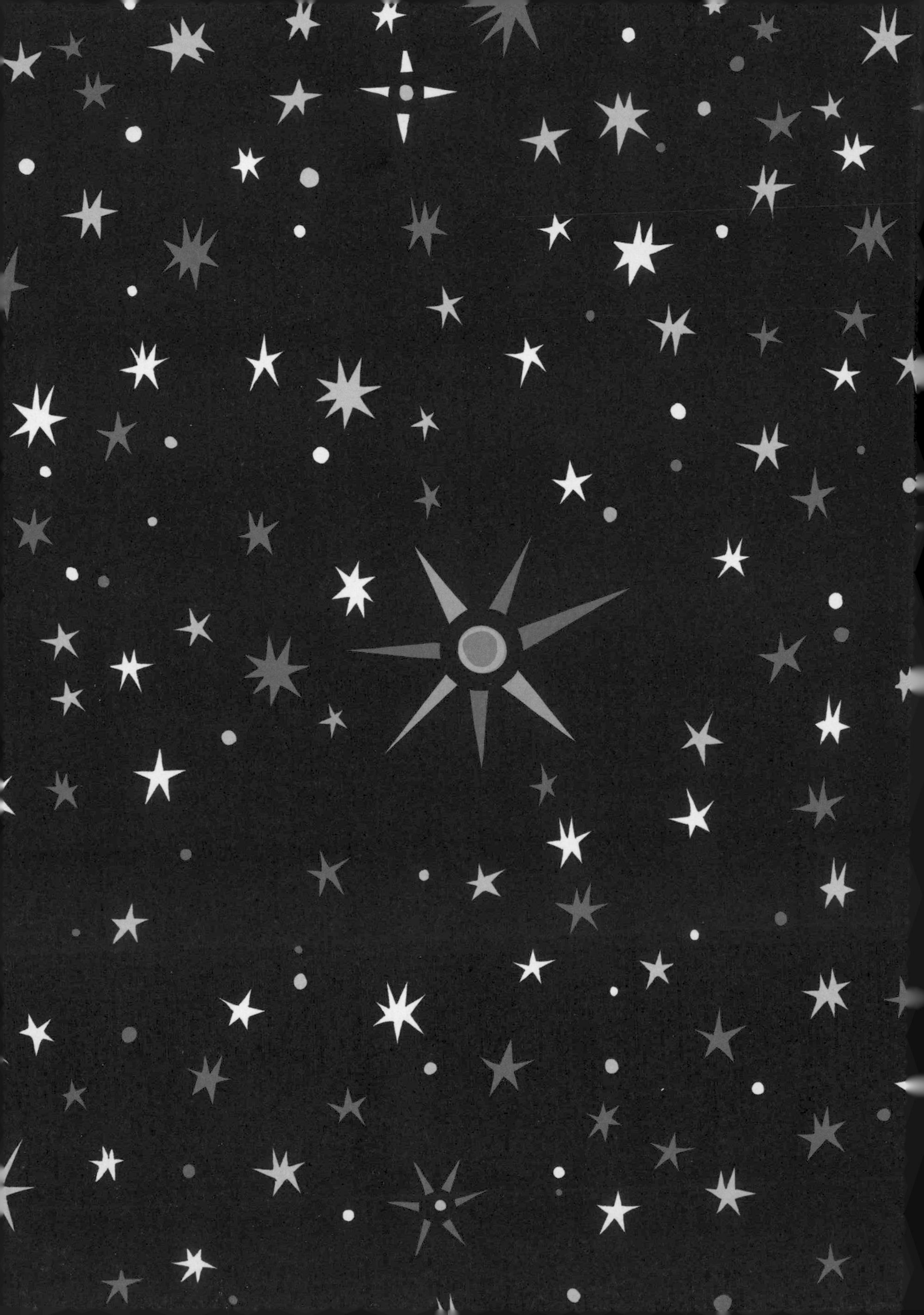